Food Phytochemicals for Cancer Prevention I

ACS SYMPOSIUM SERIES **546**

Food Phytochemicals for Cancer Prevention I

Fruits and Vegetables

Mou-Tuan Huang, EDITOR
Rutgers, The State University of New Jersey

Toshihiko Osawa, EDITOR
Nagoya University

Chi-Tang Ho, EDITOR
Rutgers, The State University of New Jersey

Robert T. Rosen, EDITOR
Rutgers, The State University of New Jersey

Developed from a symposium sponsored
by the Division of Agricultural and Food Chemistry
at the 204th National Meeting
of the American Chemical Society,
Washington, D.C.,
August 23–28, 1992

American Chemical Society, Washington, DC 1994

Library of Congress Cataloging-in-Publication Data

Food phytochemicals for cancer prevention / Mou-Tuan Huang ... [et al.].

p.   cm.—(ACS symposium series, ISSN 0097–6156; 546)

"Developed from a symposium sponsored by the Division of Agricultural and Food Chemistry at the 204th National Meeting of the American Chemical Society, Washington, D.C., August 23–28, 1992."

Includes bibliographical references and index.

ISBN 0–8412–2768–3 (v. 1)

1. Cancer—Chemoprevention—Congresses.   2. Cancer—Nutritional aspects—Congresses. 3. Fruit—Therapeutic use—Congresses. 4. Vegetables—Therapeutic use—Congresses.

I. Huang, Mou-Tuan, 1935–   . II. American Chemical Society. Division of Agricultural and Food Chemistry. III. American Chemical Society. Meeting (204th: 1992: Washington, D.C.) IV. Series.

RC268.15.F655   1994
616.99′4052—dc20

93–33775
CIP

The paper used in this publication meets the minimum requirements of American National Standard for Information Sciences—Permanence of Paper for Printed Library Materials, ANSI Z39.48–1984. ∞

PRINTED IN THE UNITED STATES OF AMERICA

Foreword

THE ACS SYMPOSIUM SERIES was first published in 1974 to provide a mechanism for publishing symposia quickly in book form. The purpose of this series is to publish comprehensive books developed from symposia, which are usually "snapshots in time" of the current research being done on a topic, plus some review material on the topic. For this reason, it is necessary that the papers be published as quickly as possible.

Before a symposium-based book is put under contract, the proposed table of contents is reviewed for appropriateness to the topic and for comprehensiveness of the collection. Some papers are excluded at this point, and others are added to round out the scope of the volume. In addition, a draft of each paper is peer-reviewed prior to final acceptance or rejection. This anonymous review process is supervised by the organizer(s) of the symposium, who become the editor(s) of the book. The authors then revise their papers according to the recommendations of both the reviewers and the editors, prepare camera-ready copy, and submit the final papers to the editors, who check that all necessary revisions have been made.

As a rule, only original research papers and original review papers are included in the volumes. Verbatim reproductions of previously published papers are not accepted.

M. Joan Comstock
Series Editor

Contents

ix

Preface

EFFORTS IN CANCER CHEMOTHERAPY have intensified over the past several decades, but many cancers still remain difficult to cure; cancer prevention could become an increasingly useful strategy in our fight against cancer. Human epidemiology and animal studies have indicated that cancer risk may be modified by changes in dietary habits or dietary components. Humans ingest large numbers of naturally occurring antimutagens and anticarcinogens in food. These antimutagens and anticarcinogens may inhibit one or more stages of the carcinogenic process and prevent or delay the formation of cancer. Recent studies indicate that compounds with antioxidant or antiinflammatory properties, as well as certain phytochemicals, can inhibit tumor initiation, promotion, and progression in experimental animal models. Epidemiological studies indicate that dietary factors play an important role in the development of human cancer. Attempts to identify naturally occurring dietary anticarcinogens may lead to new strategies for cancer prevention.

The two volumes of *Food Phytochemicals for Cancer Prevention* present recent research data and review lectures by numerous prestigious experts. Contributors from academic institutions, government, and industry were carefully chosen to provide different insights and areas of expertise in these fields. Volume I covers many phytochemicals in fruits and vegetables, and their chemical and biological properties as well as their effects on health. Special emphasis is on isolation, purification, and identification of novel phytochemicals from fruits and vegetables. Biological, biochemical, pharmacological, and molecular modulation of tumor development in experimental animal models, and possibly humans, is also included. Volume II explores the chemical, biological and molecular properties of some phytochemicals in teas, spices, oriental herbs, and food coloring agents, as well as their effects on modulation of the carcinogenic process. This book provides valuable information and useful research tools for chemists, biochemists, pharmacologists, oncologists, and molecular biologists, as well as researchers in the field of food science.

Acknowledgments

We are indebted to the contributing authors for their creativity, promptness, and cooperation in the development of this book. We also sincerely appreciate the patience and understanding given to us by our wives, Chiu

xi

Hwa Huang, Mari Osawa, Mary Ho, and Sharon Rosen. Without their support, this work would not have materialized. We thank Thomas Ferraro for his excellent reviews and suggestions and for preparing the manuscripts as camera-ready copy.

We acknowledge the financial support of the following sponsors: Campbell Soup Company; Kalsec, Inc.; Merck Sharp & Dohme Research Laboratories; The Quaker Oats Company; Schering-Plough Research Institute; Takasago USA; Tea Council of the USA; Thomas J. Lipton Company; and the Division of Agricultural and Food Chemistry of the American Chemical Society.

MOU-TUAN HUANG
Laboratory for Cancer Research
Department of Chemical Biology
   and Pharmacognosy
College of Pharmacy
Rutgers, The State University
   of New Jersey
Piscataway, NJ 08855–0789

TOSHIHIKO OSAWA
Department of Food Science
   and Technology
Nagoya University
Chikusa, Nagoya 464–01, Japan

CHI-TANG HO
Department of Food Science
Cook College
Rutgers, The State University
   of New Jersey
New Brunswick, NJ 08903

ROBERT T. ROSEN
Center for Advanced Food
   Technology, Cook College
Rutgers, The State University
   of New Jersey
New Brunswick, NJ 08903

Received August 20, 1993

PERSPECTIVES

Chapter 1

Cancer Chemoprevention by Phytochemicals in Fruits and Vegetables

An Overview

Mou-Tuan Huang[1], Thomas Ferraro[1], and Chi-Tang Ho[2]

[1]Laboratory for Cancer Research, College of Pharmacy, Rutgers, The State University of New Jersey, Piscataway, NJ 08855–0789
[2]Department of Food Science, Cook College, Rutgers, The State University of New Jersey, New Brunswick, NJ 08903

Laboratory animal studies and epidemiological data indicate that dietary factors play an important role in animal and human health and in the development of certain diseases, including cancer. Certain phytochemicals are able to inhibit the development of some cancers in laboratory animals. Epidemiological studies indicate that the frequent and high intake of fresh vegetables and fruits is associated with lower cancer incidence and that high plasma levels of ascorbic acid, α-tocopherol, β-carotene, vitamin A, and certain phytochemicals are inversely related to cancer incidence. This evidence suggests that some constituents of fruits and vegetables may play important roles in inhibiting the carcinogenic process. The phytochemicals that are believed to be cancer preventives generally possess one or more common biological properties — induction of phase I or II detoxification enzymes, modulation of phase I and other enzyme activities, antioxidant activity, electrophile scavenging activity, inhibition of nitrosation, and/or modulation of oncogene or protooncogene expression or function. Change of lifestyle and daily eating habits today may greatly reduce risk of cancer in the future. In this chapter, the influence of intake of certain phytochemicals in fruits and vegetables on carcinogenic processes and risk of cancer is discussed and reviewed.

Cancer, a disease which today remains difficult to cure, is preventable (*1,2*). Since Berenblum (*3*) described the two-stage mouse skin carcinogenesis model in 1944, the secrets of chemically induced carcinogenesis in several animal models have been well studied and the carcinogenic process has become more understandable, especially in the past decade. The first stage, initiation, can result from a single application of a subcarcinogenic dose of a carcinogen. Promotion, the second stage, occurs with repeated application of an irritating agent like the phorbol ester, 12-*O*-tetradecanoylphorbol-13-acetate (TPA).

0097–6156/94/0546–0002$06.00/0
© 1994 American Chemical Society

Initiation can result from exposure to carcinogens and is permanent damage to genetic material and is virtually irreversible. Promotion involves cellular proliferation and selective clonal expansion, and during its early stages is reversible, but becomes irreversible with time.

Data from many laboratory animal studies clearly indicate that many cancers can be prevented by certain chemicals. Several excellent reviews of laboratory animal studies are available (*1,4,5*). Epidemiological data have also been comprehensively reviewed and published elsewhere (*6,7*).

Many phytochemicals in fruits and vegetables have been isolated and identified and have been demonstrated to block different stages of the carcinogenic process in several animal models (*1,4,5*). Chemicals that are able to prevent the formation of carcinogens from precusor substances or to prevent carcinogens from reaching or reacting with critical target DNA sites in the tissues are "blocking agents." Chemicals that act by suppressing the expression of neoplasia in cells previously exposed to doses of a carcinogenic agent, are "suppressing agents" (*1*).

Administration of certain vegetables and/or fruits or their constituents in the diet to animals can reduce chemically-induced tumor incidence (*1*). Table I is a list of some phytochemicals in fruits and vegetables that are able to inhibit carcinogenesis in experimental animal models. Their structures are shown in Figure 1. In this chapter, the influence of intake of some of these phytochemicals in frutis and vegetables on the carcinogenic process is discussed and reviewed. Emphasis is on occurrence, analysis, identification of certain phytochemicals in fruits and vegetables.

Sulfur-containing Chemicals in Garlic and Onions

Plants of the genus *Allium* have been cultivated in the Middle and Far East for at least five thousand years. For many centuries they have been grown for their characteristic pungent flavor and medicinal properties. *Alliums*, especially onions and garlic, were highly prized as food stuffs in ancient China, Egypt, and India. In addition to their application as food flavorants, the medicinal properties of garlic and onions have been recognized for centuries in some parts of the world (*8–10*). The physiological activity of the components of these plants includes: antimicrobial activity, insect and animal attraction/repulsion, effects on lipid metabolism, lipid-lowering effect, hypocholesteremic activity, lipoxygenase and tumor inhibition, antithrombotic effect, platelet-aggregation inhibition activity, hypoglycemic activity, and olfactory/gustatory/lachrymatory effects. The biologically active sulfur compounds responsible for these properties, however, have been isolated from garlic and onions only very recently (*11–13*).

The characteristic aromas of the *Allium* species are also attributed to the sulfur-containing volatiles in these plants. The composition and formation of volatiles in garlic and onions have been extensively studied and reviewed (*9,10,14, 15*). The volatile components of the genus *Allium* are released from their nonvolatile precursors, S-alk(en)ylcysteine sulfoxides, by an enzymatic-mediated degradation which takes place when the plants are disrupted. The alk(en)yl groups are mainly a combination of propyl, allyl, 1-propenyl and methyl groups, depending on the species.

The chemical structures, properties, origin, and formation of the compounds responsible for the flavor properties of *Alliums* have been well studied. The

Table I. Inhibitory Effects of Some Phytochemicals in Fruits and Vegetables on Chemically Induced Carcinogenesis in Animal Models

Group	Phytochemicals	Source
Allylic compounds	Allyl mercaptan	*Allium sp.* vegetable (garlic and onion)
	Allyl methyl disulfide	”
	Allyl methyl trisulfide	”
	Diallyl sulfide	”
	Diallyl disulfide	”
	Diallyl trisulfide	”
Isothiocyanates	Benzyl isothiocyanate	Cruciferous vegetables
	Phenethyl isothiocyanate	”
Indoles	Indole-3-cabinol	Cruciferous vegetables
	Indole-3-acetonitrile	”
Monoterpenes	D-Limonene	Citrus fruit oils
	D-Carvone	Caraway seed oil
Vitamins	Ascorbic acid	Fruits and vegetables
	α-Tocopherol	Vegetable oils
	Vitamin A	Vegetables
Carotenoids	β-Carotene	Orange-yellow vegetables
Chlorophyll	Chlorophyll	Green vegetables
	Chlorophyllin	”
Flavonoids	Quercetin	Vegetables and fruits
	Rutin	”
	Tangeretin	Citrus
	Nobiletin	”
Cinnamic acids	Caffeic acid	Fruits, coffee bean and soybean
	Ferulic acid	Fruits and soybean
	Chlorogenic acid	Fruits, coffee bean and soybean

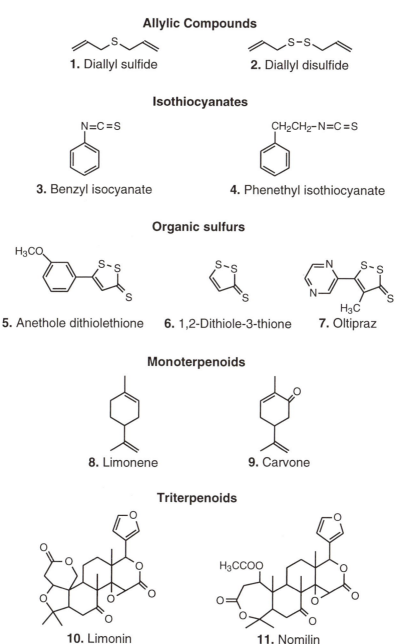

Figure 1. Structures of some phytochemicals. Continued on next page.

Vitamins

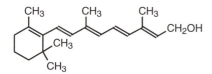

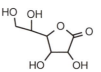

12. Vitamin A (retinol)   **13.** Vitamin C (ascorbic acid)

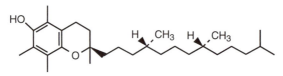

14. Vitamin E (α-tocopherol)

Carotenoids

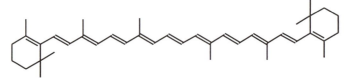

15. β-Carotene

Chlorophyll compounds

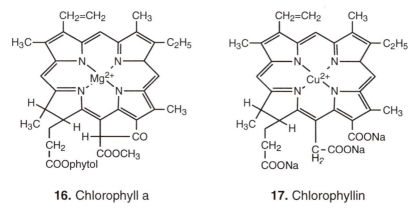

16. Chlorophyll a   **17.** Chlorophyllin

Figure 1 (continued). Structures of some phytochemicals.

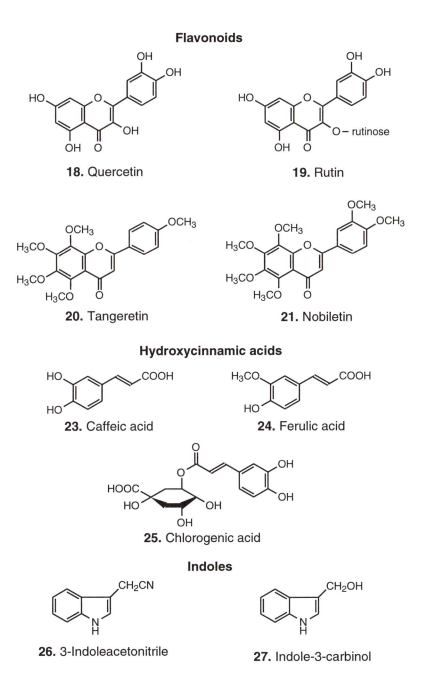

Figure 1 (continued). Structures of some phytochemicals.

chemistry of *Allium* vegetables (garlic and onions) has been previously reviewed (*9,16*).

Garlic. In 1844 and 1845, Wertheim conducted the first important study on the composition of steam distilled garlic oil (*9*). Semmler (*17*) used fractional distillation of a steam volatile oil to identify diallyl disulfide which was characterized as a key aroma compound of garlic. The breakdown of allicin and other thiosulfinates plays a major role in the formation of flavor compounds of garlic (*18*). Table II lists some major or important volatile compounds identified in garlic.

Table II. Major and Important Trace Flavor Compounds Found in Garlic

Major compounds	
Diallyl disulfide	Diallyl trisulfide
Methyl allyl trisulfide	Allyl alcohol
Methyl allyl disulfide	Diallyl sulfide
Allyl 1-propenyl disulide	Methyl 1-propenyl trisulfide
3,5-Diethyl-1,2,4-trithiolane	Allyl 1-propenyl trisulfide
Important trace compounds	
3-Vinyl-4*H*-1,2-dithiin	2-Vinyl-4*H*-1,3-dithiin
Allyl mercaptan	(*E,Z*)-ajoene

Onions. Semmler first investigated the essential oil of onions and concluded that disulfides and polysulfides were present (*17*). Methyl- and propyl-1-propenyl disulfides have been recognized as important constituents of onion oil by Brodnitz *et al.* (*19*) and Boelens *et al.* (*20*). Table III lists some major or important volatile compounds identified in onion.

Table III. Major and Important Trace Flavor Compounds Found in Onion

Major compounds	
Methyl 3,4-dimethyl-2-thienyl disulfide*	Dipropyl disulfide
Methyl propyl disulfide	Dipropyl trisulfide
1-Propenyl propyl disulfide	Methyl propyl trisulfide
1-Propenyl propyl trisulfide	Dipropyl tetrasulfide
1-Propenyl methyl disulfide	Dimethyl trisulfide
Important trace compounds	
1-Propanethiol	Methanethiol
1,3-Propanedithiol	3,4-Dimethylthiophene
3-Ethyl-1,2-dithi-4-ene	3-Ethyl-1,2-dithi-5-ene
2-Methyl-2-butenal	3,5-Diethyl-1,2,4-trithiolane
Propanal	Triethyl-dihydrodithiazine
Zwiebelanes	2,3-Dimethyl-1,4-butanethial S,S'-dioxide

*Isomers; only found in in supercritical CO_2 onion extract.

In search of antiasthmatic agents, Bayer *et al.* discovered isomeric biologically active compounds — zwiebelanes — in onion juice extract (*12*). In addition, a novel biologically active organosulfur compound, (*Z,Z*)-d,l-2,3-dimethyl-1,4-butane-dithial *S,S'*-dioxide was found in onion extract. This *bis*(thial-S-oxide) shows a moderate *in vitro* inhibition of 5-lipoxygenase.

The isolation and identification of several α-sulfinyl disulfides from onions are discussed by Kawakishi and Morimitsu (Vol. I, Chapter 8). These α-sulfinyl disulfides strongly inhibited prostaglandin endoperoxide synthase of the arachidonic acid cascade in platelets. They also had strong inhibitory effect on human 5-lipoxygenase which is concerned with the biosynthesis of leukotriene from arachdonic acid in leukocytes.

Cancer Chemopreventive Effects of *Allium* Chemicals in Animal Studies.
Inhibition of gastrointestinal cancer by organosulfur compounds in garlic has been studied and reviewed by Wargovich (*21*). Onion and garlic oils have been demonstrated to inhibit formation tumors in mouse skin by Belman (*22*). Subsequently, dialllyl sulfide, a flavor component of garlic (*Allium sativum*), and its analogues have been shown to inhibit 1,2-dimethylhydrazine (DMH)-induced colon tumorigenesis in mice and rats (*23,24*), to inhibit benzo[*a*]pyrene (BP)-induced forestomach tumorigenesis in A/J mice (*24*), to inhibit nitrosomethyl-benzylamine-induced formation of esophageal tumors in rats (*24*), to inhibit DMH-induced formation of liver tumors (*25*), to inhibit 3-methylcholanthrene-induced uterine cervix tumors in mice (*26*), to inhibit benzoyl peroxide-induced tumor promotion in Sencar mice previously initiated with 7,12-dimethylbenz[*a*]-anthracene (DMBA) (*27*). Some studies on the inhibitory effects of garlic and onions and their constituents on carcinogen-induced tumorigenesis in animals are shown in Table IV.

Diallyl sulfide has been shown to inhibit DMH-induced colon cancer in C57BL/6J mice (*23*). The study of Hong *et al.* (Vol. I, Chapter 6) indicated that diallyl sulfide inhibits the formation of lung tumors in mice by reducing the metabolic activation of the tobacco-specific nitrosamine NNK.

Other compounds having biological activity in Table II are the *E* and *Z* isomers of ajoene. Ajoene has been shown to be a potent inhibitor of platelet aggregation.

Epidemiological studies. Two separate studies from China and Italy have indicated that frequent and high consumption of *allium* vegetables reduces stomach cancer incidence (*31,32*). More epidemiological studies are needed to establish whether consumption of allium vegetables is related to risk of cancer at other sites.

Isothiocyanates and Glucosinolates in Vegetables

Glucosinolates are naturally occurring constituents of cruciferous vegetables. The term refers to a class of more than 100 sulfur-containing glycosides that yield thiocyanate, nitrile and isothiocyanate derivatives upon hydrolysis (Vol. I, Chapter 20). Indole-3-carbinol (3-indolemethanol), a product derived from indole glucosinolates of cruciferous vegetables, has been shown to inhibit DMBA-induced mammary tumors in rats (*33*). In this volume, Michnovicz and Bradlow (Vol. I, Chapter 23) discuss the results of the first clinical investigation of a cytochrome P450 inducer. They show that indole-3-carbinol is an inducer of estradiol

2-hydroxylase in human subjects. In another chapter, detailed discussions on the analysis and biological activity of glucosinolates are given by Betz and Fox (Vol. I, Chapter 14).

Table IV. Some Studies on the Inhibitory Effects of Garlic and Its Constituents on Tumorigenesis in Animals

Compound	Organ, species	Carcinogen	Investigator
Diallyl sulfide	Skin, mouse	DMBA	Athar *et al.* (*27*)
Allyl methyl disulfide Allyl methyl disulfide Diallyl sulfide Diallyl trisulfide	Forestomach, mouse	BP	Sparnins *et al.* (*28*)
Diallyl disulfide Allyl mercaptan Allyl methyl disulfide	Forestomach, mouse	DEN	Wattenberg *et al.* (*29*)
Diallyl sulfide	Liver foci, rat	DEN	Jang *et al.* (*30*)
Diallyl sulfide	Skin	DMBA/TPA	Belman *et al.* (*22*)
Diallyl sulfide	Colon	DMH	Wargovich (*23*)
Diallyl sulfide	Lung	NNK	Hong *et al.* (Vol. I, Chapter 6)
Diallyl sulfide	Liver	DMH	Hayes *et al.* (*25*)
Garlic	Uterine cervix	3-MC	Hussain *et al.* (*26*)

DMBA, 7,12-dimethylbenz[*a*]anthracene; BP, benzo[*a*]pyrene; DEN, diethylnitrosamine; TPA, 12-*O*-tetradecanoylphorbol-13-acetate; DMH, 1,2-dimethylhydrazine; NNK, 4-(methylnitrosamino)-1-(3-pyridyl)-1-butanone; 3-MC, 3-methylcholanthrene.

The glucosinolates were relatively heat labile when processed in aqueous solution or within food matrix. Decomposed glucosinolates were converted to 3-indolemethanols with the release of free thiocyanate ion or to 3-indoleacetonitriles. The formation of aldehydes and 3,3'-indolylmethanes was proposed to be the oxidation and condensation of 3-indolemethanol (*34*).

Isothiocyanates, also the enzymatic hydrolysis products of glucosinolates, have been shown to block chemical carcinogenesis. Phenylethyl and benzyl isothiocyanates inhibited carcinogenesis when given shortly before diethylnitrosamine (DEN), DMBA, or BP (*35*). Recently, 4-methylsulfinylbutyl isothiocyanate — isolated and identified in broccoli — has been shown to be the major inducer of phase II detoxication enzymes (*36*). Table V shows the principle isothiocyanates found in glucosinolate form in the crucifers that are grown in quantity for food.

Table V. Selected Isothiocyanates of Cruciferous Vegetables

Vegetable	Isothiocyanate	Relative amount
Cabbage	Allyl isothiocyanate	Major
	3-Methylsulfinylpropyl isothiocyanate	Major
	4-Methylsulfinylbutyl isothiocyanate	Major
	3-Methylthiopropyl isothiocyanate	Minor
	4-Methylthiobutyl isothiocyanate	Minor
	2-Phenylethyl isothiocyanate	Minor
	Benzyl isothiocyanate	Minor
Broccoli	3-Methylsulfinylpropyl isothiocyanate	Major
	3-Butenyl isothiocyanate	Minor
	Allyl isothiocyanate	Minor
	4-Methylsulfinylbutyl isothiocyanate	Minor
Turnips	2-Phenylethyl isothiocyanate	Major
Watercress	2-Phenylethyl isothiocyanate	Major
Garden cress	Benzyl isothiocyanate	Major
Radish	4-Methylthio-3-butenyl isothiocyanate	Major

Cancer Chemopreventive Effects of Cruciferous Vegetables (Isothiocyanates) in Animal Studies.   In human study, frequent consumption of cruciferous vegetables is associated with low cancer incidence (*37,38*). Addition of cruciferous vegetables to animal diets inhibits tumorigenesis in experimental animals (*39-42*). Cruciferous vegetables also have been shown to decrease mammary gland tumorigenesis in rodents (*43-45*). Monooxygenases (cytochrome P450 isozymes) may be involved in metabolic transformations of carcinogens and/or other xenobiotics as well as endogenous hormones, thus offering protective effects against tumor development. Green and yellow vegetables, including cabbage, brussels sprouts and other cruciferous vegetables contain several organosulfur compounds including isothiocyanates and dithiolethiones (*46*). Aromatic isothiocyanates such as benzyl isothiocyanate and phenethyl isothiocyanate are constituents of cruciferous vegetables including cabbage, brussels sprouts, cauliflower, and broccoli (*47,48*). Phenethyl isothiocyanate (PEITC) and benzyl isothiocyanate (BITC) are found in cruciferous vegetables as their glucosinolates, gluconasturtiin, and glucotropaeolin, respectively.

Administration of cabbage, cauliflower dehydrated powder or benzyl isothiocyanate in semipurified diet to Sprague-Dawley rats inhibited DMBA-induced mammary gland tumors in rats both the number of tumors per rat and percent of tumor incidence (*5,49*). Addition of freeze-dried ground cabbage to a purified diet and administered to Fischer rats inhibited aflatoxin B_1 (AFB_1)-induced the formation of liver tumor (*39,50*). Aromatic isocyanates have been shown to inhibit mammary gland, forestomach, and lung tumorigenesis induced by polycyclic aromatic hydrocarbons in mice and rats (*35,51,52*).

Some synthetic derivatives of benzyl isothiocyanate and phenethyl isothiocyanate are potent inhibitors of lung tumorigenesis in A/J mice induced by the tobacco-specific nitrosamine 4-(methylnitrosamino)1-(3-pyridyl)-1-butanone (NNK). This area of study has been reviewed by Chung (53).

Among dithiolethiones, oltipraz, a substituted form of of dithiolthione, [5-(2-pyrazinyl)-4-methyl-1,2-dithiole-3-dithiole-3-thione], has been used as an antischistosomal drug (54)(Chapter 11, Vol. 1).

The inhibitory effects of organic sulfur compounds on chemically induced carcinogenesis in several experimental animal models have been extensively studied (Table VI).

Table VI. Cancer Chemopreventive Effects of Some Organic Sulfides on Chemically Induced Carcinogenesis in Several Animal Models

Organic sulfide	Carcinogen	Organ	Reference
Benzyl isothiocyanate Phenethyl isothiocyanate	NNK	Lung	(53)
Phenethyl isothiocyanate	NNK	Esophagus	Chapter 13
Oltipraz, diallyl sulfide	AOM	Colon	Chapter 12
1,2-Dithiole-3-thione Oltipraz	AFB_1	Liver	Chapter 11
Diallyl sulfide	NNK, DEN	Forestomach	Chapter 6

NNK, 4-(methylnitrosamino)1-(3-pyridyl)-1-butanone; AOM, azoxymethane; AFB_1, aflatoxin B_1; DEN, diethylnitrosamine.

Indole-3-carbinol. Some indole-containing compounds are found in cruciferous vegetables (55). Indole-compounds are inducers of phase I and II enzymes (56–59). Cruciferous vegetables are inducers of cytochrome P450 isozymes in animals as well as in humans (60–63). Indole compounds may be partially responsible for the induction of phase I and II enzymes. Indole-3-carbinol can modulate phase I and II enzymes and thus affect tumorigenesis (for details, see Smith and Yang in Chapter 2, Vol. 1). Modulation of estrogen metabolism and protection against mammary gland tumorigenesis by indole-3-carbinol is well studied and reported by Michnovicz and Bradlow (Chapter 23, Vol. 1).

Monoterpenes

Monoterpenes are widely distributed in a variety of fruit oils, such as sweet orange, grapefruit, lemon, lime, bitter orange, and bergamot oils. D-Limonene is the most widely distributed of the monocyclic terpenes. Limonene occurs in citrus, mint, myristica, caraway, thyme, cardamom, coriander, orange flower, and many other oils.

Laboratory Animal Studies of the Chemopreventive Effects of the Monoterpenes D-Limonene and D-Carvone. The effects of citrus fruit, orange and lemon oils, as well as D-limonene and D-carvone, on chemically induced formation of tumors in mice and rats have been extensively studied by Wattenberg (*29*) and Gould (*64*). Some results are summarized in Table VII.

D-Limonene (0.2 mmol) or D-carvone (0.2 mmol) were given to A/J mice by gavage 1 hour before administration of 20 mg/kg *N*-nitrosodiethylamine (NDEA) inhibited NDEA-induced forestomach tumorigenesis by 60% and lung tumorigenesis by 35% (*29*). In an additional study, administration of lemon oil or orange oil one hour before NNK markedly inhibited NNK-induced forestomach and lung tumorigenesis (*65*). Both lemon and orange oils contain greater than 90% D-limonene. The inhibitory effects of these crude citrus fruit oils can be accounted for on the basis of their content of D-limonene. A single administration of D-limonene (25 mg) 1 hour before a single i.p. dose of NNK (2 mg/mouse) inhibited NNK-induced pulmonary tumorigenesis by 78% (*65*).

Gould and coworkers reported that the monoterpenoid D-limonene, the major component of orange peel oil, could prevent chemically induced mammary tumors in the rat in a dose-dependent manner. Dietary limonene inhibited mammary tumor formation induced by the indirect acting carcinogen DMBA in rats when the monoterpene was fed either before or after carcinogen treatment (*66*). On the other hand, limonene inhibited mammary carcinoma induced by the direct acting carcinogen nitrosomethyl urea (NMU) only when the limonene was given after carcinogen treatment. D-Limonene had no effect on NMU-induced mammary gland carcinoma formation when administered in the diet of rats before and during carcinogen treatment (*67*).

Limonene was very effective in causing the regression rat mammary carcinoma that were induced either by DMBA or by NMU (*68*).

Dietary D-limonene can induced both phase I and phase II hepatic detoxification enzymes in rats. Based on structure-activity studies, induction of phase II hepatic glucuronosyl transferase and glutathione transferase may be responsible for inhibition of DMBA-induced initiation of mammary carcinomas in rats by dietary D-limonene (*64,69*). Although dietary D-limonene also induces hepatic cytochrome P450 enzymes, it appears that cytochrome P450 enzymes are not involved in the mechanism of inhibition of DMBA induction of mammary carcinoma initiation by D-limonene (*64*).

More interestingly, D-limonene and its metabolites, sobrerol, perillic acid and dihydroperillic acid, have been shown to affect ras-p21 or G protein prenylation. Thus, D-limonene was shown to be able to post-translationally modify isoprenylation in mammary cell line (184B5) and NIH 3T3 cells (*70*). D-Limonene selectively inhibits isoprenylation of a subset of cellular growth control-associated proteins. This inhibition of isoprenylation may be the mechanism responsible for the chemopreventive and chemotherapeutic activities of limonene against mammary and other cancers.

Summary and Conclusions

Fresh fruits and vegetables are generally rich in vitamins A, C, and E, β-carotene, flavonoids and other constituents that have been studied as cancer chemopreventive agents. Eating more fresh fruits and vegetables has many health benefits, including the reduction of cancer risk. Although we have examined the evidence for cancer chemoprevention by many individual phytochemicals in this review, there is not a magic bullet. The synergistic effects of compounds in fruits and vegetables, the presence of fiber, and the value of fruits and vegetables as low fat, zero cholesterol sources of vitamins, antioxidants, micronutrients and chemopreventive functions cannot be overlooked.

Table VII. Some Studies on the Inhibitory Effects of Fruit Oils and Their Constituents on Tumorigenesis in Experimental Animal Models

Organ	Species	Chemopreventive agent	Carcinogen	Observed effect	Reference
Forestomach	Mouse	Lemon oil Orange oil D-Limonene	NNK NNK NNK	Initiation	(65)
		Caraway seed oil D-Limonene D-Carvone	NDEA NDEA NDEA	Initiation	(29)
Lung	Mouse	Lemon oil Orange oil D-Limonene	NNK NNK NNK	Initiation	(65)
		Caraway seed oil D-Limonene D-Carvone	NDEA NDEA NDEA	Initiation	(29)
Mammary gland	Rat	D-Limonene	DMBA	Initiation Post-initiation Regression	(66)
	Rat	D-Limonene Orange oil	NMU	Promotion Progression	(67)
	Rat	D-Limonene	DMBA NMU	Regression	(68)

NNK, 4-(methylnitrosamino)1-(3-pyridyl)-1-butanone; NDEA, N-nitrosodiethyl-amine; DMBA, 7,12-dimethylbenz[a]anthracene; NMU, nitrosomethylurea.

Literature Cited

1. Wattenberg, L. W. *Cancer Res.* **1992**, *52*, 2085s–2091s.
2. Weinstein, I. B. *Cancer Res.* **1991**, *51*, 5080s–5085s.
3. Berenblum, I. *Arch. Pathol.* **1944**, *38*, 233–271.
4. Boone, C. W.; Kelloff, G. J.; Malone, W. E. *Cancer Res.* **1990**, *50*, 2–9.
5. Wattenberg, L. W. *Cancer Res. (Suppl.)* **1983**, *43*, 2448s–2453s.
6. Block, G. *Am. J. Clin. Nutr.* **1991**, *53*, 270s–282s.

7. Block, G.; Patterson, B.; Subar, A. *Nutr. Cancer* **1992**, *18*, 1–29.
8. Block, E. *Angew. Chem. Int. Ed.* **1992**, *31*, 1135–1178.
9. Fenwick, G. F.; Hanley, A. B. *CRC Crit. Rev. Food Sci. Nutr.* **1985**, *22*, 273.
10. Whitaker, J. R. *Adv. Food Res.* **1976**, *22*, 73–133.
11. Block, E.; Admad, S.; Catalfamo, J. L.; Jain, M. K.; Apitz-Castro, R. *J. Am. Chem. Soc.* **1986**, *108*, 7045–7055.
12. Bayer, T.; Wanger, H.; Block, E.; Grisoni, S.; Zhao, S. H.; Neszmelyi, A. *J. Am. Chem. Soc.* **1989**, *111*, 3085–3086.
13. Bayer, T.; Breu, W.; Seligman, O.; Wray, V.; Wanger, H. *Phytochem.* **1989**, *28*, 2373–2377.
14. Freeman, G. G.; Whenham, R. J. *J. Sci. Food Agri.* **1975**, *26*, 1333–1345.
15. Carson, J. F. *Food Rev. Internat.* **1987**, *3*, 71–103.
16. Block, E. *Sci. Am.* **1985**, *252*, 114.
17. Semmler, F. W. *Arch. Pharmaz.* **1982**, *230*, 434–443.
18. Brodnitz, M. H.; Pascale, J. V.; Dersilce, L. V. *J. Agri. Food Chem.* **1971**, *19*, 273–275.
19. Brodnitz, M. H.; Pollock, C. L.; Vallon, P. P. *J. Agr. Food Chem.* **1969**, *17*, 760–763.
20. Boelens, H.; de Volois, P. J.; Wobben, H.; van der Gen, A. *J. Agri. Food Chem.* **1971**, *19*, 984–991.
21. Wargovich, M. J. In *Cancer Chemoprevention*; Wattenberg, L., Lipkin, M., Boone, C. W.; Kelloff, G. J., Ed.; CRC Press: Boca Raton, FL, 1992; pp 195–203.
22. Belman, S. *Carcinogenesis* **1983**, *4*, 1063.
23. Wargovich, M. J. *Carcinogenesis* **1987**, *8*, 487–489.
24. Wargovich, J. F.; Woods, C.; Eng, V. W. S.; Stephens, L. C.; Gray, K. N. *Cancer Res.* **1988**, *48*, 5937.
25. Hayes, M. A.; Rushmore, T. H.; Goldberg, M. T. *Carcinogenesis* **1987**, *8*, 1155.
26. Hussain, S. P.; Jannu, L. N.; Rao, A. R. *Cancer Lett.* **1990**, *49*, 175.
27. Athar, M. A.; Raza, H.; Bickers, D. M.; Muktar, H. *J. Invest. Dermatol.* **1990**, *94*, 162.
28. Sparnins, V. L.; Barany, G.; Wattenberg, L. W. *Carcinogenesis* **1988**, *9*, 131.
29. Wattenberg, L. W.; Sparnins, V. L.; Barany, G. *Cancer Res.* **1989**, *49*, 2689.
30. Jang, J. J.; Cho, K. J.; Kim, S. H. *Anticancer Res.* **1989**, *9*, 273.
31. Buiatti, E.; Palli, D.; Decarli, A.; Amadori, D.; Avellini, C.; Bianchi, S.; Biserni, R.; Cipriani, F.; Cocco, P.; Giacosa, A., et al. *Int. J. Cancer* **1989**, *44*, 611.
32. You, W. C.; Blot, W. J.; Chang, Y. S.; Ershow, A.; Yang, Z. T.; An, Q.; Henderson, B. E.; Fraumeni, J.; Wang, T. G. *J. Natl. Cancer Inst.* **1989**, *81*, 162.
33. Wattenberg, L. W.; Loub, W. D. *Cancer Res.* **1978**, *38*, 1410–1413.
34. Slominski, B. A.; Campbell, L. D. *J. Agri. Food Chem.* **1989**, *37*, 1297–1302.
35. Wattenberg, L. W. *Carcinogenesis* **1987**, *12*, 1971–1973.
36. Zhang, Y.; Talalay, P.; Cho, C.-G.; Posner, G. H. *Proc. Natl. Acad. Sci. USA* **1992**, *89*, 2399–2403.
37. Steinmetz, K. A.; Potter, J. D. *Cancer Causes Const.* **1991**, *2*, 325–357.
38. Steinmetz, K. A.; Potter, J. D. *Cancer Causes Cont.* **1991**, *2*, 427–442.
39. Boyd, J. N.; Babish, J. D.; Stoewsand, G. S. *Fd. Chem. Toxic.* **1982**, *20*, 47–52.
40. Fong, A. T.; Hendricks, J. D.; Dashwood, R. H.; Van Winkle, S.; Lee, B. C.; Bailey, G. S. *Toxicol. Appl. Pharmacol.* **1988**, *96*, 93–100.

41. Dashwood, R. H.; N., A. D.; Fong, A. T.; Pereira, C.; Hendricks, J. D.; Bailey, G. S. *Carcinogenesis* **1989**, *10*, 175–181.
42. Bresnick, E.; Birt, D. F.; Wolterman, K.; Wheeler, M.; Markin, R. *Carcinogenesis* **1990**, *11*, 1159–1163.
43. Stoewsand, G.; Anderson, J. L.; Munson, L. *Cancer Lett.* **1988**, *39*, 199–207.
44. Scholar, E. M.; Wolterman, K.; Birt, D. F.; Bresnick, E. *Nutr. Cancer* **1989**, *12*, 121–126.
45. Stoewsand, G. S.; Anderson, J. L.; Munson, L.; Lisk, D. J. *Cancer Lett.* **1989**, *45*, 43–48.
46. Jirousek, L.; Starka, L. *Nature* **1980**, *45*, 386–387.
47. Kiaer, A. In *Chemistry of Organic Sulfur Compounds*; Kharasch, N., Ed.; Pergamon Press: Elmsford, New York, 1961, Vol. 1; pp 409–420.
48. Virtanen, A. *Angew. Chem.* **1962**, *1*, 299–306.
49. Wattenberg, L. W. *Cancer Res.* **1981**, *41*, 2991–2994.
50. Stoewsand, G. S.; Baboish, J. B.; Wimberly, H. C. *J. Envir. Path. Toxicol.* **1978**, *2*, 399.
51. Wattenberg, L. W. *Cancer Res.* **1985**, *45*(1), 1–8.
52. Morse, M. A.; Wang, C.-X.; Stoner, G. D.; Mandal, S.; Conran, P. B.; Shantu, G. A.; Hecht, S.; Chung, F.-L. *Cancer Res.* **1989**, *49*, 549–553.
53. Chung, F.-L.; Morse, M. A.; Eklind, K. I. *Cancer Res.* **1992**, , 2719s–2722s.
54. Bueding, E.; Dolan, P.; Leroy, J. P. *Res. Commun. Chem. Pathol. Pharmacol.* **1982**, *37*, 297–303.
55. Wattenberg, L. W. *J. Natl. Cancer Inst.* **1974**, *52*, 1583–1586.
56. Loub, W. D.; Wattenberg, L. W.; Davis, D. W. *J. Natl. Cancer Inst.* **1975**, *54*, 985–988.
57. Sparnins, V. L.; Venegas, P. L.; Wattenberg, L. W. *J. Natl. Cancer Inst,* **1982**, *68*, 493–496.
58. Godlewski, C. E.; Boyd, J. N.; Sherman, W. K.; Anderson, J. L.; Stoewsand, G., S. *Cancer Lett.* **1985**, *28*, 151–157.
59. Ramsdell, H. S.; Eaton, D. L. *J. Toxicol. Env. Health* **1988**, *25*, 269–284.
60. Pantuck, E. J.; Pantuck, C. B.; Garland, W. A.; Min, B. H.; Wattenberg, L. W.; Anderson, K. E.; Kappas, A.; Conney, A. H. *Clin. Pharmacol. Ther.* **1979**, *25*, 88–95.
61. Hendrich, S.; Bjeldanes, L. F. *Food Chem. Toxicol.* **1983**, *21*, 479–486.
62. Pantuck, E. J.; Pantuck, C. B.; Anderson, K. E.; Wattenberg, L. W.; Conney, A. H.; Kappas, A. *Clin. Pharmacol. Ther.* **1984**, *35*, 161–169.
63. McDanell, R.; McLean, A. E. M.; Hanley, A. B.; Heaney, R. K.; Fenwick, G. R. *Food Chem. Toxicol.* **1987**, *25*, 363–368.
64. Gould, M. N. *Pro. Am. Assoc. Cancer Res.* **1991**, *32*, 474–475.
65. Wattenberg, L. W.; Coccia, J. B. *Carcinogenesis* **1991**, *12*, 115–117.
66. Elson, C. E.; Maltzman, T. H.; Boston, J. L.; Tanner, M. A.; Gould, M. N. *Carcinogenesis* **1988**, *9*(2), 331–332.
67. Maltzman, T. H.; Hurt, L. M.; Elson, C. E.; Tanner, M. A.; Gould, M. N. *Carcinogenesis* **1989**, *10*, 781–783.
68. Elegbede, J. A.; Elson, C. E.; Qureshi, A.; Tanner, M. A.; Gould, M. N. *Carcinogenesis* **1984**, *5*(5), 661–664.
69. Maltzman, T. H.; Christou, M.; Gould, M. N.; Jefecoate, C. R. *Carcinogenesis* **1991**, *12*(11), 2081–2087.
70. Crowell, P. L.; Chang, R. R.; Ren, Z.; Elson, C. E.; Gould, M. N. *J. Biol. Chem.* **1991**, *266*(26), 17679–17685.

RECEIVED August 20, 1993

Chapter 2

Effects of Food Phytochemicals on Xenobiotic Metabolism and Tumorigenesis

Theresa J. Smith and Chung S. Yang

Laboratory for Cancer Research, Department of Chemical Biology and Pharmacognosy, College of Pharmacy, Rutgers, The State University of New Jersey, Piscataway, NJ 08855–0789

Many food phytochemicals are known to affect the biotransformation of xenobiotics, and may influence the toxicity and carcinogenicity of environmental chemicals. In this article, some of the basic mechanisms of these actions are reviewed. Special attention is placed on studies with indoles, isothiocyanates, allium organosulfur compounds, flavonoids, phenolic acids, terpenoids, and psoralens. These compounds may alter the levels of phase I and phase II drug-metabolizing enzymes by affecting the transcriptional rates of their genes, the turnover rates of specific mRNAs or enzymes, or the enzyme activity by inhibitory or stimulatory actions. In many cases, the actions can be rather selective via their actions on specific enzymes, especially on the different forms of cytochrome P450 enzymes. The inhibitory actions of these phytochemicals against tumorigenesis have been studied extensively in animal models. The results help us to understand the possible beneficial or harmful effects of these compounds. Caution has to be applied when extrapolating the results to humans, however, because of species differences and the large doses used in animal studies.

The close relationship between food phytochemicals and xenobiotic metabolizing enzymes may be traced back to prehistoric days in "animal-plant warfare" during evolution (*1*). Plants synthesized chemicals for self-protection and animals had to develop xenobiotic-metabolizing enzymes such as cytochrome P450 (P450[1]) for

---

[1] Abbreviations used are: cytochrome P450, P450; indole-3-carbinol, I3C; aryl hydrocarbon hydroxylase, AHH; ethoxyresorufin *O*-deethylase, EROD; aromatic hydrocarbon receptor, *Ah* receptor; *N*-nitrosodimethylamine, NDMA; pentoxyresorufin *O*-dealkylase, PROD; 4-(methylnitrosamino)-1-(3-pyridyl)-1-butanone, NNK; benzo[*a*]pyrene, B[*a*]P; phenethyl isothiocyanate, PEITC; diallyl sulfide, DAS; 7,12-dimethylbenz[*a*]anthracene, DMBA; (-)-epigallocatechin-3-gallate, EGCG.

0097–6156/94/0546–0017$09.00/0

the detoxification of these chemicals. The evolution of the large number of P450 genes 400 million years ago may correspond to the advance of animals on to land where they encountered new terrestrial plants and phytochemicals. The work of many investigators in the past 30 years has clearly established that various dietary chemicals have marked effects on the metabolism of drugs, environmental chemicals, and certain endogenous substrates. In this review, the effects of food phytochemicals on phase I and phase II metabolism of xenobiotics and tumorigenesis will be discussed.

Xenobiotic-metabolizing Enzymes

The metabolism of xenobiotics are catalyzed by a number of enzymes. These xenobiotic-metabolizing enzymes are involved in phase I and phase II reactions (Figure 1). Phase I reactions include oxidation, hydroxylation, reduction, and hydrolysis, resulting in more water soluble metabolites to facilitate subsequent conjugation reactions and their excretion. The cytochrome P450-dependent monooxygenase, also known as mixed-function oxidase, is the most extensively studied phase I enzyme system responsible for the oxidative metabolism of a large number of xenobiotics. P450s are a large group of enzymes encoded by the superfamily of CYP genes (2). In the monooxygenase system NADPH:P450 oxidoreductase transfers electrons from NADPH to P450 forming ferro-cytochrome P450 which catalyzes the activation of molecular oxygen, and one of the oxygen atoms is added to the substrate (3) (Figure 1). Other phase I enzymes include: microsomal flavin-containing monooxygenase, cyclooxygenase, lipoxygenase, hydrolases, monoamine oxidases, dehydrogenases, aromatases, and reductases (3,4). The products of the phase I reactions are usually substrates for phase II enzymes (Figure 1), but some xenobiotics can be directly conjugated, bypassing phase I metabolism.

Phase II enzymes are involved primarily in conjugating reactions such as glucuronidation, sulfation and glutathione conjugation (Figure 1). The conjugated drug can then be excreted. UDP-glucuronosyltransferase (glucuronyl transferase) catalyzes the transfer of glucuronic acid from UDP-glucuronic acid to the compound, forming a glucuronide conjugate. Sulfotransferase catalyzes the sulfation of xenobiotics containing a hydroxyl or amino group using 3′-phosphoadenosine-5′-phosphosulfate (PAPS) as the sulfate donor (3,4). Glutathione S-transferase catalyzes the conjugation of epoxides, alkyl and aryl halides, sulfates, and 1,4-unsaturated carbonyl compounds with glutathione (3). Glutathione S-transferases have been isolated from many sources and exists in multiple forms (isoenzymes). Mammalian glutathione S-transferase isoenzymes are grouped into three classes (α, μ, and π) based upon their substrate specificities, structural homologies and immunological cross-reactivities (5) and there are at least 7 subunits (6). Transmethylases catalyzes the methylation of compounds containing O-, S-, and N-groups using S-adenosine-L-methionine as the methyl donor (3,4). NAD(P)H:quinone oxidoreductase, also known as DT-diaphorase, is a phase I enzyme by definition, but is considered a phase II enzyme by some authors (7,8). NAD(P)H:quinone oxidoreductase is involved in the detoxification of quinones through a two-electron reduction.

Although the phase I and phase II enzymes are believed to be evolved for the detoxification of xenobiotics, they are also known to be involved in the generation of reactive intermediates, which attack cellular macromolecules, leading to toxicity and carcinogenesis. The roles of P450 enzymes in the activation of a

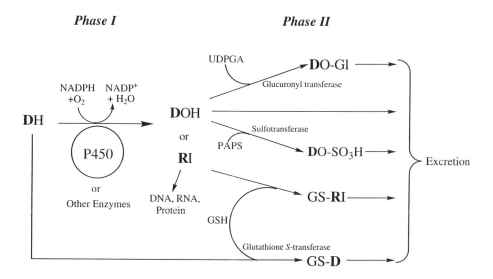

Figure 1. Phase I and phase II reactions involved in xenobiotic metabolism.

variety of toxicants and carcinogens are well recognized (*9,10*). The roles for phase II enzymes in the detoxification of many xenobiotics have been illustrated. In certain cases, however, they may be involved in the activation of carcinogens or toxicants; for example, the activation of certain arylamines by sulfotransferase (*11*). In addition, conjugation may also be a means of transporting activated metabolites to different tissues where it could be reactivated into reactive metabolites (*12*). Glutathione, a cofactor required for the glutathione *S*-transferase reaction is known to be involved in the activation of certain halogenated compounds (*6,13–15*).

Mechanisms by Which Food Phytochemicals Affect Xenobiotic Metabolism

Many food phytochemicals can alter the levels of enzymes involved in the phase I and phase II reactions. These naturally occurring constituents or their metabolites can regulate certain specific P450s but not affect others. Dietary compounds may affect rates of P450 gene transcription and translation, as well as the degradation of P450 mRNA and protein (*16*). These compounds can also inactivate P450s by covalently binding to the P450 apoprotein or heme moiety leading to the inactivation of these enzymes. Some dietary constituents can bind directly to the P450, whereas others require metabolic activation by specific P450s to form reactive intermediates which can then attack the P450 molecules. Food phytochemicals can also bind reversibly to the active sites of P450s, serving as competitive inhibitors.

Food phytochemicals can interact with NADPH:P450 oxidoreductase and alter the levels of the reductase. Since NADPH:P450 oxidoreductase is required to transfer electrons from NADPH to P450, a decrease or increase in the reductase level can either impair or stimulate the flow of electrons to P450, thus leading to an alteration in monooxygenase activities. With xenobiotics that are metabolized by different competing pathways, food phytochemicals may selectively affect certain pathways and alter the physiological effects, such as the toxicity or carcinogenicity of these compounds.

The phase II enzymes can also be induced by various food phytochemicals. In many cases, the dietary inducers contain electrophilic centers (or acquire them by cellular metabolism) which elicit an electrophilic chemical signal that activates the transcription of genes coding for phase II enzymes (*17,18*). An induction in phase II enzymes can lead to increased conjugation reaction and faster excretion of the drugs or environmental chemicals. The rates of the phase II reactions are also affected by the availability of cellular glutathione, UDP-glucuronic acid and PAPS.

A large number of food phytochemicals are known to affect xenobiotic metabolism. The effects of some of the most extensively studied chemicals, most of them dietary constituents, are summarized in Table I. The specific effects of these chemicals on xenobiotic metabolism and carcinogenesis are discussed in subsequent sections.

Indoles

Cruciferous vegetables, such as cabbage, broccoli, cauliflower and Brussels sprouts contain glucosinolates. It has been estimated that approximately 30 mg of glucosinolates are consumed daily per person from cruciferous vegetables in the United Kingdom (*19*). Indole-3-carbinol (I3C) is present in cruciferous vegetables in the form of 3-indolylmethyl glucosinolate (glucobrassicin) (*20*). The glucobrassicin intake is estimated to be 12.5 mg/person from fresh sources and 7

Table I. Effects of Food Phytochemicals on Xenobiotic Metabolism

Compound		Xenobiotic Metabolism and Enzymes[a]	References
Indole-3-carbinol	↑	P450s 1A1, 2B1, aryl hydrocarbon hydroxylase, ethoxycoumarin O-deethylase, benzo[a]pyrene oxidase, p-nitroanisole O-demethylase, aniline C-hydroxylase, pentoxyresorufin O-dealkylase, testosterone 6α-, 16α- and 16β-hydroxylase, estradiol 2-hydroxylase activities and demethylation of NNK and NDMA	24–34,37,39, 40,46
	↑↑↑	Ethoxyresorufin O-deethylase, NAD(P)H:quinone oxidoreductase and UDP-glucuronyl transferase activities	
Isothiocyanates	↓	P450 2E1, ethoxyresorufin O-deethylase, erythromycin N-demethylase activities and metabolism of N-nitrosonornicotine and N-nitrosopyrrolidine	8,33,51,52, 56–64,68–71
	↓↓↓	NDMA demethylase, and metabolism of NNK, N-nitrosobenzylmethylamine and N-nitrosomethylamylamine	
	↑	P450 2B1, pentoxyresorufin O-deethylase, glutathione S-transferase, NAD(P)H:quinone oxidoreductase and UDP-glucuronyl transferase activities	
Diallyl sulfide	↓	P450 2E1, 6-testosterone hydroxylase activity	74,78,80–82, 84–89,91–92
	↓↓↓	NDMA demethylase activity and metabolism of N-nitrosobenzylmethylamine, N-nitrosodiethylamine and NNK	
	↓↓	p-nitrophenol hydroxylase activity and metabolism of aflatoxin B1	
	↑	P450 2B1, pentoxyresorufin O-dealkylase, ethoxyresorufin O-deethylase, 16α- and 16β-testosterone hydroxylase, glutathione S-transferase, glutathione peroxidase and glutathione reductase activities	
Naringenin	↓↓	Ethoxyresorufin O-deethylase, aminopyrine N-demethylase and biphenyl 4-hydroxylase activities	98,100–102
	↓↓	Metabolism of nifedipine, aflatoxin B1, and benzo[a]pyrene	

[a]A single arrow indicates a pretreatment effect; double arrows indicates an *in vitro* effect; triple arrows indicate both effects.
Continued on next page

Table I. Continued

Compound		Xenobiotic Metabolism and Enzymes[a]	References
Quercetin	↓↓↓	Aryl hydrocarbon hydroxylase, ethoxycoumarin O-deethylase and ethoxyresorufin O-deethylase activities	95,99,101, 102,107–110
	↓↓	7-pentoxyresorufin dealkylase, lipoxygenase, cyclooxygenase, cytochrome c reductase, p-nitroanisole O-demethylase activities and metabolism of benzo[a]pyrene, nifedipine and aflatoxin B$_1$	
Flavone	↑	P450, pentoxyresorufin O-dealkylase, glutathione S-transferase and UDP-glucuronyl transferase activities	97,101,105, 112,114,115, 117
	↑↑↑	NADPH cytochrome c reductase,ethoxycoumarin O-deethylase, ethoxyresorufin O-deethylase activities and zoxazolamine metabolism	
	↑↑	Metabolism of benzo[a]pyrene and aflatoxin B$_1$	
	↓↓	Ethoxyresorufin O-deethylase, estrogen synthetase activities and metabolism of benzo[a]pyrene	
Catechin	↓↓	NADPH cytochrome c reductase, ethoxyresorufin-O-deethylase, benzphetamine N-demethylase activities and NNK α-hydroxylation	119–122
	↑↑	N-hydroxylation and deacetylation of 2-acetylaminofluorene	
Epicatechins	↓↓	NADPH cytochrome c reductase, pentoxyresorufin O-dealkylase, ethoxyresorufin O-deethylase, p-nitrophenol hydroxylase, aryl hydrocarbon hydroxylase, ethoxycoumarin O-deethylase activities and NNK metabolism	123–125
Isoflavone	↑↑	P450 and prostaglandin synthase	117,127–130
	↓↓	Benzo[a]pyrene metabolism	
Tannic acid	↓↓↓	Aryl hydrocarbon hydroxylase, ethoxycoumarin O-deethylase and ethoxyresorufin O-deethylase activities	107,138,147
	↓↓	Epoxide hydrolase activity and metabolism of benzo[a]pyrene	
	↑	Glutathione S-transferase and NAD(P)H:quinone oxidoreductase activities	

Compound	Effect[a]	Activity/Metabolism	Refs.
Ellagic acid	↓	P450	133,141,143, 146-148
	↓↓↓	Aryl hydrocarbon hydroxylase, ethoxycoumarin O-deethylase activities and metabolism of benzo[a]pyrene	
	↓↑	NNK hydroxylation and N-nitrosobenzylmethylamine metabolism	
	↑	Glutathione S-transferase activity	
Caffeic acid	↓↓	Lipoxygenase activity and aflatoxin B$_1$ metabolism	153,156,157
	↑↑	Prostaglandin synthase activity	
Curcumin	↑	Aryl hydroxylase activity	159,163,164
	↓↓	Lipoxygenase and cyclooxygenase activities	
Piperine	↓	Cytochrome b$_5$, N,N-dimethylaniline N-demethylase and UDP-glucuronyl transferase activities	163, 166
	↓↓↓	Ethoxycoumarin O-deethylase and ethylmorphine N-demethylase activities	
	↓↓↓	Aryl hydrocarbon hydoxylase activity	
Limonene	↑↑	P450, P450s 2B, 2C and epoxide hydratase activity	173, 174
	↑↑	DMBA metabolism	
Psoralens	↓↓↓	P450, aminopyrine N-demethylase, benzo[a]pyrene hydroxylase, hexobarbital hydroxylase and ethoxycoumarin O-deethylase activities	182,184–186, 188
	→	Acetominophen metabolism	
	↑	Aryl hydrocarbon hydroxylase and ethylmorphine N-demethylase activities	

[a] A single arrow indicates a pretreatment effect; double arrows indicate an *in vitro* effect; triple arrows indicate both effects.

mg/person from cooked sources in the United Kingdom (20). When a vegetable that contains 3-indolylmethyl glucosinolate is cut or chewed, the compound is hydrolyzed by myrosinase to form I3C. I3C has been shown to modulate chemically-induced carcinogenesis (21–23). It exerts its chemopreventive effects partially by altering the P450-mediated oxidative metabolism of carcinogens. I3C can induce several P450s and other drug metabolizing enzymes.

The induction of P450 1A by I3C and I3C-containing food items has been extensively studied (24–28). Rats fed diets containing I3C, 25% Brussels sprouts or 25% cabbage (freeze-dried) had a marked increase in intestinal and hepatic aryl hydrocarbon hydroxylase activity (AHH; due to P450 1A) and ethoxycoumarin-O-deethylase activity (due to P450s 1A and 2B), with the induction of the intestinal enzyme activities being more prominent (22,24,29–31). The hepatic and intestinal P450 1A1 protein levels and ethoxyresorufin-O-deethylase (EROD; due to P450 1A) activity was increased in rats and mice fed I3C (20,25,26). Furthermore, I3C has been demonstrated to increase the 2-hydroxylation of estradiol (P450 1A2 is the major enzyme involved in this activity) in humans (27) and animals (25,28). Since this activity is believed to be associated with lowering of mammary cancer incidence, it has been suggested that doses of 6–7 mg I3C/kg/day may be effective in the chemoprevention of breast cancer in women (27).

Other P450s which are altered by I3C include P450s 2E1, 2A1 and 2B1 (25, 26,32). Acute and chronic treatment of rats with I3C resulted in an induction of hepatic N-nitrosodimethylamine (NDMA) demethylation (33), an activity indicative of P450 2E1. Furthermore, administration of diets containing 10% broccoli to rats induced hepatic and colonic P450 2E1 levels (32). Dietary I3C elevated testosterone 6α-hydroxylase (due to P450 2A1) activity, whereas, testosterone 6β-hydroxylase (due to P450 3A) activity was not affected (25). The P450 2B1 level and its activity as 7-pentoxyresorufin O-dealkylase (PROD), and 16α- and 16β-hydroxylation of testosterone (preferentially catalyzed by P450 2B) activities were increased in the liver and intestine of rats fed I3C for 2–28 days (26). When rat hepatocytes were treated with I3C, however, there was no effect on P450-mediated activities (26,34–36). Apparently I3C is not the active component for the *in vivo* effects of I3C; rather acid condensation products are believed to be the inductive agents.

Acid condensation products may be formed from I3C during passage through the stomach. The acid condensation products, 3,3'-diindolylmethane and 2,3-bis[3-indolylmethyl]indole, were detected in gastric contents, stomach tissue, small intestine and liver an hour after rats received an oral dose of I3C (35). The presence of these acid reaction products in the liver suggests they are absorbed from the gastrointestinal tract. Furthermore, addition of I3C to an acidic solution (pH 4.5–5) produced a similar acid condensation product profile as was found in the gastric contents of rats treated with I3C (35). Moreover, the importance of the acid condensation products is apparent when the route of I3C administration becomes a factor. When rats were given I3C intraperitoneally (bypassing the acidic environment of the stomach) there was no effect on hepatic EROD activity. If acid condensation products were given intraperitoneally or intragastrically, however, EROD activity was increased (36). Therefore, the potency of I3C in altering monooxygenase activities could be attributed to the action of the acid condensation products formed upon introduction of I3C into the stomach.

The induction of P450 1A is probably due to the activation of the transcription of the CYP1A genes by the acid-condensation products of I3C (37). Although

I3C binds weakly to the *Ah* receptor, indolo[3,2-*b*]carbazol and other acid products of I3C bind with high affinity to the *Ah* receptor (*37,38*) and activate P450 1A genes. Acute administration of I3C induced colonic P450 1A1 mRNA, whereas, both P450s 1A1 and 1A2 mRNAs were induced in the liver (*37*). A similar P450 1A mRNA induction profile was observed in the colon and liver of rats fed 10% broccoli for 1 week (*32*).

Treatment of rat hepatocytes with I3C condensation products increased P450 1A, EROD activity and 7α-hydroxylation of testosterone (due to P450 2A), whereas, 16α- and 2α-hydroxylation of testosterone (attributed to P450s 2B1 and 2C11) activities were decreased by the I3C condensation products (*26,34*). Testosterone 6β-hydroxylase (due to P450 3A) activity was also inhibited *in vitro* by acid condensation products of I3C (*25,34*). The decrease in the testosterone hydroxylation activities may be due to the down-regulation of constitutive P450 forms (*34*). The acid condensation products can also serve as inhibitors of testosterone hydroxylation. The mechanism by which I3C modulates mono-oxygenase activities appears to be mediated via acid condensation products of I3C.

Indole-3-acetonitrile is another major indole present in cruciferous vegetables. The demethylation of NDMA and 4-(methylnitrosamino)-1-(3-pyridyl)-1-butanone (NNK) were induced in rats chronically fed indole-3-acetonitrile and I3C, but the extent of induction by indole-3-acetonitrile was only half that of the induction by I3C (*33*). Oral administration of indole-3-acetonitrile to rats and mice had no effect on P450, EROD or benzo[*a*]pyrene (B[*a*]P) oxidase activity, whereas, I3C increased P450, B[*a*]P oxidase, EROD, *p*-nitroanisole *O*-demethylase and aniline C-hydroxylase activities (*39,40*). Furthermore, indole-3-acetonitrile binds very weakly, if at all, to the *Ah* receptor (*38*).

When cruciferous vegetables are used in animal diets, they are usually in the freeze-dried form. Humans, on the other hand, are more likely to consume cooked cruciferous vegetables. Cooking can reduce the levels of hydrolyzed glucosinolates due to the inactivation of myrosinase. It has been demonstrated that boiling cabbage reduced the indole glucosinolate content by 50%; fermentation had no effect on the indole glucosinolate level (*41*). Administration of diets supplemented with cooked Brussels sprouts to rats for 2–28 days induced hepatic P450s 1A1 and 1A2 protein levels, EROD and PROD activities, and the formation of 6β-hydroxytestosterone, but decreased the formation of 2α-hydroxytestosterone. Cooked Brussels sprouts markedly increased the P450 2B protein level, PROD and EROD activities, and the 16α- and 16β-hydroxylation of testosterone in intestinal microsomes (*42*). Feeding cooked Brussels sprouts for as little as 2 days alters the metabolic activities of P450 enzymes similar to raw cruciferous vegetables and I3C. Overall, I3C has both inductive and suppressive effects on constitutive P450s which are involved in carcinogen and steroid hormone metabolism.

Cabbage and Brussels sprouts (cooked or freeze-dried) induce hepatic and intestinal glutathione *S*-transferase activity (*24,29–31,42*). In contrast, I3C and I3C-condensation products were shown to have no effect on glutathione *S*-transferase activity, suggesting that compounds other than I3C in cruciferous vegetables are responsible for the induction of glutathione *S*-transferase activity (*24,34*). Goitrin, a major constituent of cruciferous vegetables, may be responsible for the induction of glutathione *S*-transferase (*43*). Dietary *R*-goitrin was shown to increase hepatic and intestinal mucosa glutathione *S*-transferase and epoxide hydratase activities in rats (*44*). Benzyl isothiocyanate, present in cruciferous vegetables as glucotropaeolin (a glucosinolate), has also been shown to induce glutathione *S*-transferase activity

(45). On the other hand, NAD(P)H:quinone oxidoreductase activity was induced by I3C, cooked Brussels sprouts, and acid reaction products of I3C in rat hepatic and intestinal microsomes, and rat and monkey hepatocytes (26,34,35,42,46). An induction of NAD(P)H:quinone oxidoreductase by I3C may offer protection against quinone toxicity and possible carcinogenicity. It has been suggested that induction of both NAD(P)H:quinone oxidoreductase and EROD activities by acid condensation products of I3C is due to these compounds acting by means of *Ah* receptor binding because inducers which induce P450 1A and NAD(P)H:quinone oxidoreductase requires *Ah* receptor binding for their action (35). UDP-glucuronyl transferase has also been shown to be induced by I3C, cooked Brussels sprouts and I3C-condensation products (26,34,42). The effect of I3C — or more likely the acid condensation products of I3C — on both phase I and phase II enzymes may be partially responsible for the modulation of carcinogenesis by cruciferous vegetables.

Isothiocyanates

In addition to indoles, isothiocyanates are also present in cruciferous vegetables in the form of glucosinolates. The naturally occurring aromatic isothiocyanates benzyl isothiocyanate and phenethyl isothiocyanate (PEITC) are myrosinase-catalyzed products of their respective glucosinolates glucotropaeolin and gluconasturtiin. Administration of diets containing gluconasturtiin and myrosinase to mice resulted in approximately 21% of the gluconasturtiin being converted to PEITC; whereas, in the absence of myrosinase, less than 1% of the gluconasturtiin was converted to PEITC (47). PEITC is readily distributed to all major organs, with the highest concentrations of the isothiocyanate in the tissues found at 4–8 h after treatment (48). The major route for excretion is the urine (48). Analyses of urine samples from individuals consuming watercress, a vegetable rich in gluconasturtiin, indicated the presence of an *N*-acetylcysteine conjugate of PEITC (47).

Isothiocyanates and glucotropaeolin have been demonstrated to inhibit chemically-induced carcinogenesis and DNA adduct formation in animals (45,47, 49–54), but the isothiocyanates and glucosinolate must be given prior to administration of the carcinogen in order to inhibit tumor formation (45,55). The mechanism involved in the inhibition of carcinogenesis by isothiocyanates is most likely to be due to the blocking of the activation of the carcinogens. Isothiocyanates have been shown to inhibit the metabolism of NNK (33,52,56–61), NDMA (33,56, 57,62), *N*-nitrosobenzylmethylamine (51 and unpublished results), *N*-nitrosornicotine (63), *N*-nitrosopyrrolidine (63) and *N*-nitrosomethylamylamine (64). The inhibition of the metabolism of nitrosamines by isothiocyanates appears to be due to its inhibitory action on P450 enzymes. Isothiocyanates are highly reactive compounds which can inactivate P450s. It has been demonstrated that PEITC decreased the oxidation of NNK by inactivation of P450 and by a competitive mechanism (56,58,61).

PEITC has selective effects on P450 enzymes. Acute administration of PEITC decreased the liver microsomal P450 2E1 level, NDMA demethylase (due to P450 2E1), EROD and erythromycin *N*-demethylase (due to P450 3A) activities in rats and mice (56,57,62). PEITC was metabolized by P450 to a reactive intermediate which then inactivated P450 2E1 (suicide inhibition) (62). In contrast, the hepatic P450 2B1 level and PROD activity were markedly increased (56,57,61, 62), while there was no appreciable effect on benzphetamine *N*-demethylase (due to

P450 2C11 and other P450s) activity (*62*) in animals treated with PEITC. The induction of hepatic P450 2B1 by PEITC correlated with an increase in NNK oxidation (*56*). P450 2B1 is known to be one of the enzymes which catalyzes the oxidation of NNK (*61,65,66*). It is possible that PEITC is activating the transcription of the P450 2B gene. More work is needed, however, to clearly establish the mechanism for the induction of P450 2B1 by PEITC. The effects of PEITC on P450 2B1 appears to be tissue specific. Although the P450 2B1 level and PROD activity was increased in liver microsomes, the PROD activity was decreased in lung and nasal mucosa microsomes of rats treated with a single dose of PEITC (*56*) and the P450 2B level was decreased in lung microsomes of mice chronically fed PEITC (*61*), resulting in a decrease in the oxidation of NNK.

A structure-activity relationship in the inhibitory potency of aromatic isothiocyanates has been observed. As the alkyl chain length of the isothiocyanate increases, the extent of the inhibition of NNK-induced tumor formation, DNA adduct formation, oxidation of NNK and erythromycin *N*-demethylase activity is increased (*47,52–54,57,58*). The order of potency is 6-phenylhexyl isothiocyanate > 4-phenylbutyl isothiocyanate > 3-phenylpropyl isothiocyanate > PEITC > benzyl isothiocyanate. The increased inhibitory potency is probably due to the increased lipophilicity and stability associated with the increase in alkyl chain length (*47,54*). An increased alkyl chain length may favor binding of the isothiocyanates to the active sites of the P450 enzymes. PEITC and 6-phenylhexyl isothiocyanate (a synthetic isothiocyanate) both decreased the oxidation of NNK in a competitive manner, but 6-phenylhexyl isothiocyanate exhibited much lower K_i values than PEITC (11–16 nM vs 51–93 nM), suggesting 6-phenylhexyl isothiocyanate has a higher affinity for binding at the active site of the P450 involved in the bioactivation of NNK (*57,61*).

Isothiocyanates can increase phase II enzyme activities, alter the microsomal metabolism pattern and increase tissue glutathione levels (*45,67*). Benzyl isothiocyanate and PEITC have been shown to increase the activities of glutathione *S*-transferase, NAD(P)H:quinone oxidoreductase and UDP-glucuronyl transferase (*56,57,68–71*). Furthermore, dietary benzyl isothiocyanate increased sulfhydryl levels and the glutathione *S*-transferase α-class subunit 2 level (*68,70,71*). Allyl isothiocyanate, a hydrolysis product of the glucosinolate sinigrin, has also been shown to strongly induce glutathione *S*-transferase activity and the glutathione *S*-transferase subunit 2 level (*72*). The induction of glutathione *S*-transferase subunit 2 by allyl isothiocyanate was similar to the induction pattern in rats receiving a diet containing 30% Brussels sprouts (*72*). Glutathione *S*-transferase subunit 2 appears to be sensitive to induction by isothiocyanates. The presence of an α-hydrogen is required for the inductive activity of isothiocyanates (*17*).

Sulforaphane [(-)-1-isothiocyanato-(4*R*)-methylsulfinyl)butane] is an isothiocyanate which has been isolated from SAGA broccoli (*Brassica oleracea italica*) and identified as a very potent and major monofunctional inducer of NAD(P)H: quinone oxidoreductase (*8*). Monofunctional inducers are compounds, such as benzyl isothiocyanate, which induce phase II enzymes without induction of phase I enzymes (*8,18,73*). Sulforaphane and its sulfide and sulfone analogues induced NAD(P)H:quinone oxidoreductase and glutathione *S*-transferase activities, with sulforaphane and the sulfone being the most potent (*8*). The presence of an oxygen on the sulfur was shown to be an important structural feature in the potency of induction (*8*). The induction of the phase II enzymes by this monofunctional inducer is independent of an *Ah* receptor, instead an electrophilic chemical signal is involved in the regulation of the synthesis of these enzymes (*17,18*).

Allium Organosulfur Compounds

Allium vegetables, including garlic, onions, leeks, chives and shallots, are rich in organosulfur compounds (*74*). These plants have been grown for centuries and used as foodstuff and medicines. Epidemiological studies in certain areas of China have shown an association between frequent dietary intake of allium vegetables with decreased risk for gastric cancer (*75,76*). Garlic consumption appeared to have a stronger protective effect than onions (*76*). Garlic contains organosulfur compounds with allyl groups, whereas onion contains organosulfur compounds with propyl groups (*77*).

Diallyl sulfide (DAS), a flavor component of garlic, is derived from oxidized allicin after a clove of garlic is crushed (*77*). DAS, diallyl disulfide, and related organosulfur compounds have been shown to inhibit chemically induced tumorigenesis in the forestomach (*78,79*), lung (*78*), esophagus (*80,81*) and colon (*82*) of animals. Recently, DAS (a blocking agent) given to animals in combination with Se-methylselenocysteine or quercetin (suppressing agents) was shown to have a greater inhibitory effect on 7,12-dimethylbenz[*a*]anthracene (DMBA)-induced mammary tumors than DAS treatment alone (*83*). Since blocking agents and suppressing agents are targeted at different stages of carcinogenesis, a greater chemopreventive effect is achieved. In addition, the structure of organosulfur compounds can affect the inhibitory potency. An allyl group coupled to a sulfur atom may play an important structural role in the inhibition of carcinogenesis (*82*).

A possible mechanism for the inhibition of tumor formation by DAS is altering the metabolism of the carcinogen. DAS has been shown to inhibit the metabolism of *N*-nitrosobenzylmethylamine (*81,84*), NDMA (*85–87*), *N*-nitroso-diethylamine (*85*), aflatoxin B_1 (*88*) and NNK (*85*). Inactivation of selected P450s is implicated in the inhibition against the metabolic activation of nitrosamines. Acute administration of DAS to rats decreased hepatic P450 2E1, the activities of NDMA demethylase and *p*-nitrophenol hydroxylase (indicative of P450 2E1), and 6β-testosterone hydroxylase activity (due to P450 3A) (*86,87,89*). Furthermore, treatment of rats with metabolites of DAS (diallyl sulfoxide and diallyl sulfone) displayed similar effects on monoxygenase activities (*87,89*). The decrease in hepatic P450 2E1 was not mediated by a decrease in P450 2E1 mRNA. Recent *in vitro* studies demonstrated that the inactivation of P450 2E1 by diallyl sulfone was NADPH-, time- and dose-dependent, having the characteristics of suicide inhibition (*87,89*). Apparently, diallyl sulfone is metabolized by P450 2E1 to a reactive metabolite which modifies the heme moiety of P450 2E1 and thus inactivates this enzyme.

In contrast to the inactivation of P450 2E1, oral administration of DAS and its metabolites were shown to induce hepatic P450 2B1, PROD, EROD and 16α- and 16β-testosterone hydroxylase (due to P450 2B) activities (*74,87*). The increase in P450 2B1 by DAS was mediated by an increase in P450 2B1 mRNA. Furthermore, the transcriptional rate of P450 2B1/2 genes was increased 13-fold 6h after DAS treatment (*90*). Apparently, the induction of P450 2B1/2 in rat liver by DAS is due to transcriptional activation. The induction of P450 2B1/2 is tissue specific. P450 2B1/2 mRNA is induced by DAS in rat liver, stomach and duodenum, but not in nasal mucosa and lung (*90*). The mechanism for the tissue-selective induction of P450 2B1/2 genes by DAS is not known.

Another mechanism by which DAS may exert its anticarcinogenic action is via an induction of detoxification enzymes. Pretreatment of rats and mice with DAS has been shown to increase (1.3- to 2.5-fold) glutathione *S*-transferase, glutathione peroxidase and glutathione reductase activities, as well as the protein level of glutathione *S*-transferase (*78,80,82,91,92*). Furthermore, DAS increased the levels of α, μ and π class glutathione *S*-transferases (*92*). Organosulfur compounds containing an allyl group and sulfur atom induced glutathione *S*-transferase activity more than organosulfur compounds containing a propyl group and/or disulfide linkage (*78,82*).

The induction of detoxification enzymes by organosulfur compounds is a rather slow process. Administration of DAS 3 or 18 h prior to *N*-nitrosobenzyl-methylamine had no effect on the formation of urinary *N*-nitrosobenzylmethyl-amine metabolites in rats (*84*). Hayes *et al.* (*93*) showed that after 18 h of treatment, DAS had no effect on glutathione *S*-transferase, glutathione reductase or gluta-thione peroxidase activities in primary cultured rat hepatocytes. Furthermore, significant induction of hepatic, lung and forestomach glutathione *S*-transferase, glutathione peroxidase and glutathione reductase activities was only observed at 48 or 96 h after administration of allylic garlic sulfides to mice (*78,82,92,94*). The mechanism for the induction of glutathione *S*-transferase by DAS is not known. It appears that the chemopreventive action of organosulfur compounds is mainly due to inhibiting metabolic activation enzymes (*86,87,89*), but induction of detoxifi-cation enzymes may become a factor in other model systems.

Flavonoids

Flavonoids and their glycosides are polyphenolic compounds which are widely distributed in fruits, vegetables and nuts. Naturally occurring flavonoids are classified as flavones, flavonols, flavanones, isoflavones and catechins and all are structurally-related to the parent compound, flavone (2-phenylbenzopyrone) (*95*) (Figures 2 and 3). It has been estimated that humans consume approximately 1 g of mixed flavonoids per day (*95*).

Flavonoids have been shown to modulate P450-mediated monooxygenase activities by exerting different effects on different P450 enzymes. In studies with purified P450s, flavone and α-naphthoflavone (a synthetic flavonoid) activated benzo[*a*]pyrene (B[*a*]P) hydroxylation with rabbit P450s 3A6 and 1A2, but inhibited with P450s 2B4, 2C3 and 1A1 in a reconstituted system (*96*). Secondly, several studies have demonstrated the number and position of hydroxyl groups on the A and B rings (Figure 2) are important in determining the effect of flavonoids on enzyme activity (*95,97–101*). The presence of hydroxyl groups only on the A ring was shown to be the most potent inhibitors (*99*). Increasing the number of hydroxyl groups on the B ring leads to a decrease in the inhibitory effect of flavo-noids. Furthermore, an *ortho*-substituition pattern on the B ring results in an increase in the inhibitory effect (*97,99*). Flavonoids protect P450 against NADPH-initiated lipoperoxidation. When flavonoids interact with P450s in uninduced rat liver microsomes, a modified type II (reverse type I) spectrum is observed, sug-gesting that flavonoids act as a Fe(III) ligand by favoring the low-spin form of the enzyme and preventing its reduction (*100*). This may also partially be responsible

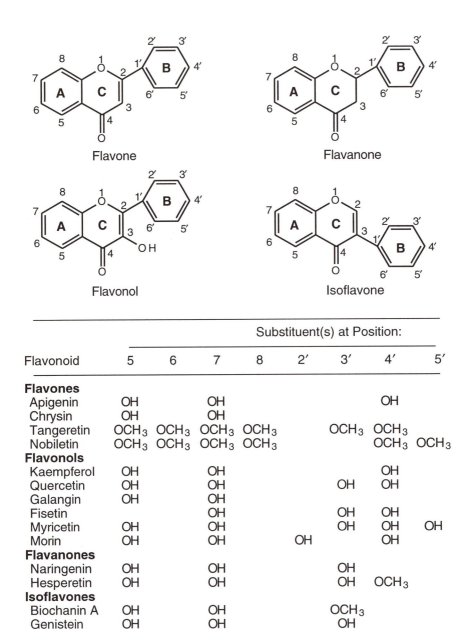

| | Substituent(s) at Position: | | | | | | | |
Flavonoid	5	6	7	8	2'	3'	4'	5'
Flavones								
Apigenin	OH		OH				OH	
Chrysin	OH		OH					
Tangeretin	OCH$_3$	OCH$_3$	OCH$_3$	OCH$_3$		OCH$_3$	OCH$_3$	
Nobiletin	OCH$_3$	OCH$_3$	OCH$_3$	OCH$_3$			OCH$_3$	OCH$_3$
Flavonols								
Kaempferol	OH		OH				OH	
Quercetin	OH		OH			OH	OH	
Galangin	OH		OH					
Fisetin			OH			OH	OH	
Myricetin	OH		OH			OH	OH	OH
Morin	OH		OH		OH		OH	
Flavanones								
Naringenin	OH		OH				OH	
Hesperetin	OH		OH			OH	OCH$_3$	
Isoflavones								
Biochanin A	OH		OH			OCH$_3$		
Genistein	OH		OH			OH		
Daidzein			OH			OH		

Figure 2. Chemical structures of some naturally occurring flavonoids.

for the inhibitory effect of flavonoids on monooxygenase activities. Some of the most extensively studied flavonoids are discussed below.

Flavanones. Naringin is the most abundant flavonoid in grapefruit. Naringin was shown not to be effective in inhibiting the P450 3A4-catalyzed oxidation of nifedipine and felodipine or aflatoxin B_1 activation in human liver microsomes (*102*). Naringenin, a flavanone aglycone present in citrus fruit, however, as well as three other flavonoids found in grapefruit juice (apigenin, hesperetin and kaempferol) were effective in inhibiting the oxidation of nifedipine and the activation of aflatoxin B_1, with naringenin and apigenin being the most effective (*102*). Naringenin has also been shown to inhibit the hydroxylation of B[*a*]P, EROD, aminopyrine *N*-demethylase, and to a lesser extent biphenyl 4-hydroxylase activities (*98,100,101*). Naringenin inhibited aminopyrine *N*-demethylase activity in a noncompetitive manner, whereas the inhibition of biphenyl 4-hydroxylase activity was of a competitive nature (*100*). The differences in the mechanism of inhibition may be due to a combination of electrostatic and lipophilic interactions. The inhibitor simultaneously binds as Fe(III) ligands and to a nearby lipophilic site in the enzyme (substrate binding site) (*100*). The inhibitory potency of naringenin appears to be due to the presence of hydroxyl groups in its structure. It has been suggested that the 7-hydroxyl group is primarily responsible for the inhibitory effect of flavanones and this hydroxyl group preferentially interacts with the Fe(III) of cytochrome P450 due to its steric availability and adequate acidity (*100*). Flavanones are generally considered as not being inducers of xenobiotic metabolism, but detoxification enzymes have been shown to be induced. At low doses, flavanone induced glutathione *S*-transferase and glucuronidation activities (*103*).

Flavonols. Quercetin is a polyhydroxylated flavonol (Figure 2) which is prevalent in fruits, vegetables and cereal grains. Of the estimated 1 gm of dietary flavonoids humans consume daily, approximately 5% is derived from quercetin (*104*). It is extensively metabolized by intestinal bacteria (*95,105*). In addition, the flavonoid glycosides quercitrin and rutin are hydrolyzed by intestinal bacteria to quercetin in humans (*106*).

Quercetin is a potent inhibitor of P450 reactions. It has been shown to be an effective inhibitor of the P450 3A4-catalyzed oxidation of nifedipine and activation of aflatoxin B_1 (*102*). Furthermore, quercetin inhibited AHH, ethoxycoumarin *O*-deethylase, EROD, PROD, lipoxygenase, cyclooxygenase, and *p*-nitroanisole *O*-demethylase activities, as well as hydroxylation of B[*a*]P *in vitro* and *in vivo* (*95, 101,107–110*). Quercetin is a potent inhibitor of P450 1A. Quercetin was shown to be a competitive inhibitor of EROD activity with a K_i of 431 nM, whereas, for PROD activity, a mixed type of inhibition was observed with a K_i of 40 μM (*99*). Quercetin has a greater affinity for binding to P450 1A than to P450 2B1. In human liver microsomes, quercetin was a potent inhibitor of cytochrome c reductase, suggesting that quercetin may be inhibiting the reduction of P450 (*101*).

Although quercetin exhibits an inhibitory effect on monooxygenase activities, this mechanism may only be partially responsible for the anti-carcinogenic action of quercetin. Since quercetin is extensively metabolized by intestinal bacteria, only a limited amount of quercetin may be present in the liver and other extrahepatic tissues. Less than 1% of ingested quercetin was absorbed from the gastrointestinal tract, while more than 50% was metabolized by intestinal bacteria and the remainder was eliminated in the feces (*111*). Therefore, forming adducts with carcinogens and thereby decreasing their bioavailability from the gastrointestinal tract may be another mechanism involved in the anticarcinogenic action of quercetin.

Flavones. Flavone is an unhydroxylated compound which has stimulatory and inhibitory effects on P450. Flavone, nobiletin and tangeretin were shown to increase the metabolism of B[a]P, aflatoxin B$_1$ and zoxazolamine in human and rat liver microsomes ($101,112$). Furthermore, dietary flavone increased hepatic P450, NADPH-cytochrome c reductase, EROD, PROD and ethoxycoumarin deethylase activities in rats, suggesting flavone induces P450s 1A and 2B1 (105). The induction pattern of flavone, however, appears to be tissue specific. Although flavone increased ethoxycoumarin deethylase activity in rat liver microsomes, the enzyme activity was decreased in intestinal microsomes, suggesting that different P450 forms are important in the metabolism of ethoxycoumarin in the liver and intestinal mucosa (98). The effect of flavone on conjugating enzymes is also tissue specific. Hepatic glutathione S-transferase and UDP-glucuronyl transferase activities were markedly induced by flavone, whereas there was no effect on these phase II enzyme activities in the intestines (105). Studies with rabbit and human liver microsomes suggest that flavone exerts its stimulatory effect on monooxygenase activities partially by enhancing the interactions between P450 and NADPH cytochrome P450 reductase, thereby facilitating the flow of electrons to P450 (113).

In contrast to the induction of monooxygenase activities, flavone, α-naphthoflavone and chrysin have been shown to be potent inhibitors of EROD activity in human liver microsomes (97). In addition, estrogen synthetase (aromatase), a P450 enzyme that catalyzes the conversion of androgens to estrogens, was inhibited by flavone, α-naphthoflavone, chrysin and apigenin (114). Structural features of flavones which appear to have an important role in the inhibition of aromatase are the keto group and a 7-hydroxyl group on the A ring ($115,116$). Naturally occurring 5,7-dihydroxyflavones, such as apigenin and chrysin, have been shown to inhibit aromatase by competitive inhibition and metabolism of the flavones is not required for the inhibition (116). Therefore, the flavones directly bind to human cytochrome P450 aromatase. The flavone compounds act as type II binders by serving as the sixth ligand to the heme iron ($115,116$). Naturally occurring and synthetic flavones have been shown to effectively inhibit the metabolism of B[a]P in hamster embryo cells and the presence of the 2,3-double bond and two hydroxyl and/or methoxyl groups at the 5- and 7-positions on ring A were important structural features for the inhibition (117). The stimulatory and inhibitory action of flavones on different P450-mediated monooxygenase activities appears to depend on the species and the monooxygenase activity tested.

Catechins. (+)-Catechin [(+)-cyanidanol] (Figure 3) is a flavanol which is widely distributed in plants and is a major component in tea leaves. (+)-Catechin is also the principal component of catechu, which is used to make betel quid (118). (+)-Catechin is nonmutagenic ($118–121$) and has been shown to decrease the mutagenicity of B[a]P, DMBA and 2-acetylaminofluorene ($118,120$). NNK- and NDMA-induced DNA fragmentation, DNA methylation and DNA binding are also inhibited by (+)-catechin ($119,122$). The action of (+)-catechin may be due to inhibition of activation enzymes rather than acting as a trapping agent since

(+)-catechin was not an effective inhibitor of DNA single strand breaks induced by the direct-acting carcinogen *N*-methyl-*N*-nitrosourea (*122*).

In rat hepatocytes, (+)-catechin (40 μM) markedly inhibited the α-hydroxylation of NNK (activation pathway) but had little or no effect on the pyridine *N*-oxidation (detoxification pathway) or carbonyl reduction of NNK, suggesting that catechin is selectively inhibiting the enzymes involved in NNK activation (*121*). Further studies are needed to substantiate this point. (+)-Catechin has been shown to inhibit NADPH:cytochrome c reductase, benzphetamine *N*-demethylation and EROD activities in a dose-dependent manner (*119,122*). In contrast to the inhibitory action of (+)-catechin, the *N*-hydroxylation and deacetylation pathways of 2-acetylaminofluorene were increased while aryl hydroxylation of 2-acetylaminofluorene (detoxification pathway) was unaffected by (+)-catechin in rat hepatocytes (*120*). The increase in the metabolism resulted in an increased binding of 2-acetylaminofluorene metabolites to hepatocellular DNA (*120*).

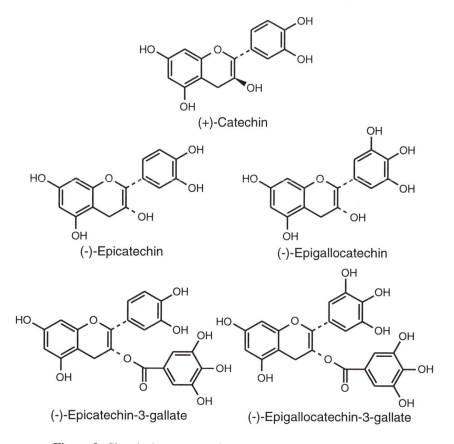

Figure 3. Chemical structures of (+)-catechin and (-)-epicatechins.

Although (+)-catechin modulated the monooxygenase activities *in vitro*, it did not seem to alter monooxygenase enzyme levels *in vivo*. Administration of high doses of (+)-catechin (200 mg/kg) to rats had no effect on hepatic P450 level and NADPH cytochrome c reductase, aniline hydroxylase, aminopyrine *N*-demethylase, benzphetamine *N*-demethylation and EROD activities when assayed *in vitro* (*119, 122*). It has been suggested that the concentration of (+)-catechin (10^{-4} M) used to inhibit these monooxygenase activities *in vitro* cannot be obtained *in vivo* (*119, 122*), therefore, the anticarcinogenic effect of (+)-catechin may be due to direct interaction of (+)-catechin with the proximate/ultimate carcinogen (*119*) thereby decreasing its bioavailability.

The epicatechin derivatives, (-)-epicatechin, (-)-epigallocatechin, (-)-epicatechin-3-gallate and (-)-epigallocatechin-3-gallate (EGCG) (Figure 3) are major polyphenolic compounds present in green tea. Of the epicatechin derivatives, EGCG constitutes the major ingredient in green tea. Theaflavins and thearubigins are the major components of black tea. *In vitro*, green tea and black tea polyphenol fractions inhibited the pulmonary oxidation of NNK, with the black tea polyphenol fraction being the most potent (*123*). Furthermore, EGCG decreased the oxidation of NNK with an IC_{50} of 0.1 mM. The green tea polyphenol fraction and the four epicatechin derivatives also decreased hepatic AHH, EROD, PROD, *p*-nitrophenol hydroxylase (due to P450 2E1), ethoxycoumarin *O*-deethylase, epoxide hydrolase and NADPH cytochrome c reductase activities, with EGCG being the most potent (*124,125* and unpublished results). Of the monooxygenase activities, EGCG was more inhibitory against EROD than PROD or *p*-nitrophenol hydroxylase activities, suggesting EGCG binds with greater affinity to P450 1A than to P450s 2B1 or 2E1 (unpublished results). Therefore, the decrease in NNK oxidation by the green tea polyphenol fraction is most likely due to EGCG binding to P450s 1A and 2B1 since these P450s are involved in the oxidation of NNK (*61,65,66*). The presence of galloyl and hydroxyl groups on the structure of the parent compound (+)-catechin has been suggested to be an important structural feature for the inhibitory potency of the epicatechin derivatives (*124,125*).

Isoflavones. Isoflavones and their glucosides are present in high concentration in soy products. The reduced derivatives of isoflavones are isoflavanones and isoflavans. Large amounts of isoflavonoids are excreted in the urine by humans and animals (*126*). Isoflavonoids have been shown to modulate the metabolism of estradiol by competing with estradiol for the type II estrogen binding sites. Women with a low rate of the 2-hydroxylation of estrogen and a high rate of the 16α-hydroxylation of estrogen has been proposed to be at a greater risk for breast and endometrial cancers (*126*).

Biochanin A, an isoflavone in red clover (*Trifolium pratense* L. Leguminosae), has been shown to inhibit the metabolism of B[*a*]P and the binding of B[*a*]P to DNA in hamster embryo cells and rat liver S-9 homogenate by decreasing the oxidation of B[*a*]P and glucuronidation of B[*a*]P metabolites (*117, 127,128*). Structure-activity studies demonstrated that isoflavones with hydroxyl groups at the 5- and 7-positions were the most potent inhibitors of B[*a*]P metabolism (*117*).

Genistin, the glucoside conjugate of genistein, is the major isoflavonoid in soybean flour and may be the active component responsible for the induction of P450 (*129*). Genistein and soybean flour were shown to induce the P450 level in *Streptomyces griseus*, whereas daidzein and coumestrol (isoflavones in soybeans) had no effect on the P450 level (*129*). In addition, genistein, daidzein, coumestrol and equol (product of intestinal metabolism) enhances microsomal prostaglandin synthase activity *in vitro*, whereas biochanin A and formononetin were slightly inhibitory to prostaglandin synthase (*130*). Prostaglandin synthase preferentially oxidizes the hydroxyl group on the B ring of isoflavones, therefore, these iso-flavones may be co-substrates for the peroxidase component of prostaglandin synthase (*130*). The peroxidase component of prostaglandin synthase is involved in the cooxidation of xenobiotics.

Phenolic Acids

Most plants contain antioxidants which are phenolic or polyphenolic compounds. The most commonly occurring phenolic compounds in foods are flavonoids, hydroxycinnamic acid derivatives, and simple phenols and phenolic acids (*131, 132*). Tannic acid is a naturally occurring polyhydroxyl phenol esterified to gallic acid. Ellagic acid is present in the form of ellagitannins in woody dicotyledonous plants; it is also found in fruits such as grapes, strawberries and raspberries, as well as certain nuts (*133,134*). Tannic acid and ellagic acid are antioxidants capable of inhibiting skin tumor initiation and carcinogenesis (*135,136*). Tannic acid and ellagic acid have been shown to inhibit B[*a*]P-induced lung and forestomach tumor formation in mice (*137,138*), *N*-nitrosobenzylmethylamine-induced esophageal tumors in rats (*139*), 3-methylcholanthrene-induced skin tumors in mice (*140*) and the mutagenicity of aflatoxin B$_1$ (134), as well as DNA binding (*134,138,140–143*). Chang *et al*. (*144*), however, reported that topical application of ellagic acid to the backs of mice did not inhibit 3-methylcholanthrene-induced formation of skin tumors in CD-1 and Sencar mice. It is possible that ellagic acid is poorly absorbed and has a very short half life in blood (*145*).

The inhibition of carcinogenesis by tannic acid and ellagic acid may be due to decreased metabolic activation and increased conjugation reactions. Tannic acid and ellagic acid have been found to inhibit the metabolism of B[*a*]P (*107,133,141, 143*), NNK (*146*) and *N*-nitrosobenzylmethylamine (*147,148*). Tannic acid *in vitro* and pretreatment of mice with tannic acid inhibited AHH, 7-ethoxycoumarin *O*-deethylase and EROD activities (activities reflective of P450 1A) in mouse epidermal, lung and forestomach microsomes (*138,147*). Similarly, ellagic acid has been demonstrated to decrease hepatic, pulmonary, esophageal and epidermal cyto-chrome P450, AHH and 7-ethoxycoumarin *O*-deethylase activities (*133,141,148*). Aminopyrine *N*-demethylase and epoxide hydrolase activities, however, were not affected by ellagic acid (*133*). Alternatively, tannic acid and ellagic acid may be exerting their effect by enhancing the detoxification and/or elimination of the car-cinogenic intermediates. Glutathione *S*-transferase activity was increased (1.5- to 1.8-fold) in the liver and forestomach of mice by tannic acid and ellagic acid (*133, 138*). An increase in glutathione *S*-transferase could result in increased conjugation leading to a faster excretion of the reactive carcinogenic metabolites. Furthermore, NAD(P)H:quinone oxidoreductase activity was increased by tannic acid in mouse forestomach (*138*). A reduction in the levels of toxic quinone derivatives may be possible by inducing NAD(P)H:quinone oxidoreductase.

Another mechanism for the action of ellagic acid is as a scavenger, in which ellagic acid may be forming adducts with the reactive metabolites of the carcino-gen. In aqueous solution, ellagic acid has been shown to form *cis* and *trans* adducts with the ultimate carcinogenic metabolite of B[*a*]P, B[*a*]P-7,8-diol-9,10-epoxide-2.

These adducts undergo hydrolysis to inactive tetrols (*149*). For *N*-nitrosamines, however, ellagic acid may be less effective as a scavenger since the alkyldiazonium ion is very unstable in solution. Therefore, there are several possible mechanisms for inhibition of *N*-nitrosamine-induced DNA damage by ellagic acid: (a) ellagic acid occupies site(s) in the DNA which would react with the reactive species; (b) binding of the reactive metabolites to sites other than the N^7 and O^6 positions of guanine in DNA; (c) scavenging reactive intermediates which approach the N^7 and O^6 positions of guanine; (d) catalyzing the removal of the DNA alkylated bases (*147*). Josephy *et al.* (*150*) showed that ellagic acid and its synthetic analogues had no effect on NDMA- or *N*-methyl-*N*-nitrosourea-induced mutagenesis and had not significantly changed the ratio of O^6- to N^7-methylguanine formation.

Caffeic acid and ferulic acid (Figure 4) are hydroxycinnamic acids which are present in a wide variety of fruits and vegetables. Chlorogenic acid, an ester of caffeic acid and quinic acid, is a major component of green coffee beans and fruits (*151,152*). Upon hydrolysis, chlorogenic acid forms caffeic acid. The phenolics are readily oxidized in the free state, therefore, they can be destroyed by storage or processing conditions (*152*). Caffeic acid and chlorogenic acid can have detrimental effects on food such as nutrient loss, and enzymatic browning which can result in undesirable color and flavor (*131*). Caffeic acid and ferulic acid have been shown to be potent blocking agents that prevent nitrosamine formation *in vitro* and *in vivo* (*151*). Therefore, these plant phenolics may play an important role in preventing *in vivo* nitrosation in the stomach when high levels of nitrate are consumed.

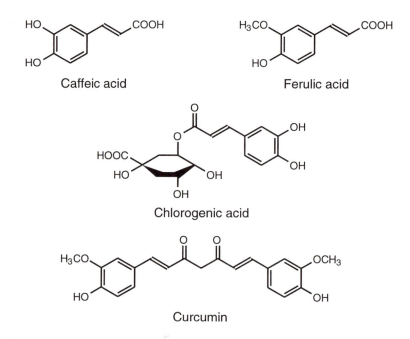

Caffeic acid

Ferulic acid

Chlorogenic acid

Curcumin

Figure 4. Chemical structures of some naturally occurring phenolic compounds.

Aflatoxin B_1 is a procarcinogen which must be metabolically activated in order to exert its carcinogenic effects. Chlorogenic acid and caffeic acid inhibited aflatoxin B_1 metabolism in rat liver microsomes (153). A reduction in the metabolism of aflatoxin B_1 by chlorogenic acid and caffeic acid may be due to interaction with aflatoxin B_1, inhibition of the activation enzymes, or to the scavenging of the activated aflatoxin B_1 metabolite(s). It has been demonstrated that the phenolic compounds do not react covalently with aflatoxin B_1, rather the inhibitory effect is due to the inhibition of the activation enzymes (153).

In addition to the P450 enzyme system, xenobiotics can undergo cooxidation to reactive intermediates during prostaglandin biosynthesis (154,155). Lipoxygenase and prostaglandin synthase are the two enzymes involved in the synthesis of prostaglandins and leukotrienes from arachidonic acid. Lipoxygenase, involved in the formation of leukotrienes, has been shown to be inhibited by caffeic acid (156,157) in a noncompetitive manner (154). Furthermore, a methyl ester of caffeic acid and a caffeic acid enone-type compound were more potent inhibitors of lipoxygenase activity than caffeic acid (156,158) and the enone-type compound had a high specificity for lipoxygenase in polymorphonuclear leukocytes from the peritoneal cavity of guinea pigs. Apparently, the active site of lipoxygenase may be inactivated by the hydroxyl groups (156). It has been suggested that at least three factors are necessary for the inhibition of lipoxygenase by caffeic acid derivatives: an appropriate alkyl chain length, a hydrophobic binding site for the straight alkyl chain, and hydroxyl groups attached to the benzene ring (158). In contrast to these studies, Huang *et al.* (159) demonstrated that caffeic acid had little effect on lipoxygenase activity in the mouse epidermis. On the other hand, high concentrations of caffeic acid and a methyl ester of caffeic acid (>1 mM) enhanced prostaglandin synthase activity (156). The enhancement of prostaglandin synthase by caffeic acid has been suggested to be due to the stimulation of the cyclooxygenase activity of prostaglandin synthase (156). Depending on the concentration, caffeic acid can be both inhibitory and stimulatory on enzyme activity. Phenolic acids which modulate arachidonic acid metabolism may affect the promotion process rather than the initiation stage of carcinogenesis (160).

Curcumin (diferuloylmethane), the principal phenolic constituent in turmeric, curry and mustard, is found in the rhizomes of the plant *Curcuma longa* Linn (159,161,162). Curcumin contains two molecules of ferulic acid linked via a methylene bridge to form a β-diketone (Figure 4) (159). Administration of a 0.5% curcumin diet to rats markedly increased hepatic aryl hydroxylase activity, but had no effect on total P450 content, cytochrome b_5, NADPH:cytochrome c reductase, N,N-dimethylaniline *N*-demethylase and UDP-glucuronyl transferase activities (163). On the other hand, a 5% turmeric diet decreased hepatic P450 and cytochrome b_5 levels and increased glutathione *S*-transferase activity in mice (161). There may be other active constituents present in turmeric. *In vitro*, curcumin has been shown to be a potent inhibitor of lipoxygenase and cyclooxygenase activities in mouse epidermis (159,164). As an epidermal cyclooxygenase inhibitor, curcumin is more potent than quercetin, but as an epidermal lipoxygenase inhibitor, curcumin is not as effective as quercetin (159). It has been suggested that the presence of the hydroxyl groups, double bonds in the alkene portion, and/or the β-diketone moiety may be important for the activity of curcumin (159).

Piperine is the active phenolic component of black pepper (*Piper nigrum* Linn) and long pepper (*Piper longum* Linn), while capsaicin is the principal component of chili pepper (*Capsicum frutescens* Linn) and red peppers (*Capsicum annum* Linn) (*163,165,166*). Piperine and capsaicin are potent inhibitors of the mixed function oxygenase system. Piperine caused a comparable concentration-dependent inhibition of EROD, ethylmorphine *N*-demethylase, AHH and UDP-glucuronyl transferase activities *in vitro* in hepatic postmitochondrial supernatants from control, 3-methylcholanthrene- and phenobarbital-induced rats (*166*). Furthermore, piperine inhibited AHH and ethylmorphine in a noncompetitive manner with similar K_i values from control and 3-methylcholanthrene-induced hepatic microsomes. Piperine appears to be a nonspecific inhibitor of P450 isozymes (*166*). *In vivo*, piperine decreased the level of hepatic cytochrome b_5, and the activities of N,N-dimethylaniline *N*-demethylase, AHH and UDP-glucuronyl transferase in rats (*163,166*). Maximal inhibition of AHH activity was observed 1 h after piperine administration and the activity was completely recovered at 6 h (*166*). Administration of piperine to mice also prolonged hexobarbital-induced sleeping time and zoxazolamine-induced paralysis time (*166*), indicating the metabolism of these drugs were decreased. In rat liver microsomes, capsaicin produced a type I spectral change in a high affinity concentration-dependent manner and was a potent inhibitor of ethylmorphine *N*-demethylase activity (*167*). Capsaicin treatment of rats increased hepatic cytochrome b_5 and aryl hydroxylase activity, as well as pentobarbital-induced sleeping time (*163,167*). Furthermore, a metabolite of capsaicin, dihydrocapsaicin, bound irreversibly to microsomal protein *in vitro* and *in vivo* suggesting that capsaicin is activated to a reactive species which can covalently bind to cellular macromolecules (*167*). Dihydrocapsaicin has been shown to covalently bind to hepatic P450 2E1 (*168*). It has been suggested that P450 2E1 catalyzes the oxidation of dihydrocapsaicin to the phenoxy radical which inactivates P450 2E1 in rats (*168*).

Terpenoids

The monoterpene *d*-limonene is the first cyclic product in the synthesis of terpenes in plants and is found in high concentration in citrus fruit oils (*169*). Limonene is rapidly and extensively metabolized *in vivo*, therefore, one or more of its metabolites may actually be responsible for its anti-cancer activity (*170*). Administration of *d*-limonene one hour prior to *N*-nitrosodiethylamine or NNK exposure inhibited the formation of forestomach and lung tumors in A/J mice (*79,171*). Orange oil and lemon oil, which contain more than 90% *d*-limonene, were also found to be effective in inhibiting NNK-induced lung and forestomach tumors (*171*). Dietary *d*-limonene (5% of the diet) was most effective in reducing the number of DMBA-induced mammary tumors in rats when it was fed during the initiation stage of carcinogenesis (*172*). Furthermore, dietary *d*-limonene inhibited the total DMBA-DNA adducts in the liver, lung, spleen and kidney of rats by 50% (*173*).

    The inhibitory activity of dietary limonene during the initiation phase does not appear to be due to the inhibition of the enzymes important in DMBA metabolism. It has been demonstrated that dietary limonene increased the levels of total hepatic cytochrome P450, P450s 2B and 2C, and epoxide hydratase (*173,174*). These changes in the phase I metabolizing enzymes resulted in an increase (2.6-fold) of *in vitro* microsomal DMBA metabolism (*173*). The lack of inhibition of DMBA metabolism coupled with an increased excretion of DMBA metabolites in *in vivo*-treated rats may indicate that the phase II hepatic conjugating enzymes, such as glutathione *S*-transferase and UDP-glucuronyl transferase, may be important in the anticarcinogenic property of *d*-limonene (*173*).

In addition to limonene, the naturally occurring terpenoids camphor, menthol and pinene also induced hepatic P450 2B (*174*). The induction was mediated by an increase in the amount of mRNA coding for P450 2B. Furthermore, the mRNA species that is induced is highly homologous to the mRNA that codes for P450 2B2. It has been suggested that P450s induced by phenobarbital, a synthetic compound, may have originally evolved in response to terpenoid compounds present in the environment (*174*). It is presently not clear, however, whether terpenoids and phenobarbital act by way of the same mechanism.

Limonene has been found to be an effective inhibitor during the promotion/ progression stage of carcinogenesis. *d*-Limonene and orange oil, when fed as 5% of the diet one week after the direct-acting carcinogen nitrosomethylurea was administered, were found to significantly inhibit rat mammary tumors (*175*). Dietary limonene has also been shown to cause regression of primary, differentiated mammary tumors by DMBA (*176*). Although limonene is capable of increasing P450 enzymes and thereby modifying the activation of a carcinogen (*173,174*), it does not explain limonene's effectiveness in inhibiting nitrosomethylurea-induced mammary tumors or causing tumor regression. It has been suggested that limonene-treated cells may have a decreased capacity to isoprenylate proteins since limonene and menthol have been shown to decrease 3-hydroxy-3-methylglutaryl coenzyme A reductase activity (*177,178*). Terpenoids may act by affecting the farnesylation of *ras* gene product in pre-malignant and malignant cells (*177*).

The citrus limonoids limonin and nomilin are oxidized triterpenes which are responsible for the bitter taste in citrus products. They have been demonstrated to be effective inhibitors of B[*a*]P-induced forestomach tumors in mice and are inducers of glutathione *S*-transferase activity (*179*). Other potent terpenoid inducers of glutathione *S*-transferase activity are kahweol palmitate and cafestol palmitate. Kahweol palmitate and cafestol palmitate are diterpene esters found in green coffee beans (*180*). Kahweol palmitate and cafestol palmitate, as well as their free diterpenes, induced the activity of glutathione *S*-transferase in the small intestine mucosa and liver of mice (*180,181*). Furthermore, diets containing green coffee beans induced glutathione *S*-transferase activity in the liver and small intestine mucosa of mice and rats (*181*). Roasted coffee, instant coffee, and decaffeinated instant coffee are also inducers of glutathione *S*-transferase activity, but their induction activity is less than with green coffee beans (*180*). The furan ring is the structural feature which has been shown to be important in the induction of glutathione *S*-transferase activity by kahweol and cafestol (*181*). The effect of these terpenes on the phase I enzyme system is not known.

Psoralens

Psoralens are naturally occurring tricyclic furocoumarins found in plants such as celery, parsely, parsnips and figs (*182,183*). Psoralen derivatives include methoxsalen (8-methoxypsoralen), bergapten (5-methoxypsoralen) and trioxsalen (trimethylpsoralen) (Figure 5). Methoxsalen is extensively oxidized on its furan ring, possibly into a reactive epoxide. It has been suggested that methoxsalen may be activated to form an unstable species which can covalently bind to the active site of P450 or degrade the epoxide and other reactive metabolites (*184*). The inactivation of P450 by psoralens requires NADPH and oxygen, suggesting that it is mediated by reactive metabolites (*182,184,185*).

Psoralen, methoxsalen and bergapten have been shown to markedly decrease the level of hepatic P450, the activities of aminopyrine *N*-demethylase, benzo[*a*]pyrene hydroxylase, hexobarbital hydroxylase and 7-ethoxycoumarin *O*-deethylase and the metabolic activation of acetominophen in humans (*182,186*) and animals (*184–186*). In addition, psoralen and bergapten inhibited mutagenicity induced by dictamnine and rutacridone in *Salmonella typhimurium* TA98 by inactivation of P450 (*187*). The decrease in the monooxygenase activities by psoralen derivatives and metabolite(s) is due to competitive inhibition and suicide inhibition of the P450 species involved (*184,185*).

The inactivation of P450 by psoralen derivatives may also be accompanied by the induction of microsomal enzymes. It has been demonstrated that methoxsalen markedly increased hepatic AHH and ethylmorphine *N*-demethylase activities in mice. This induction appears to be tissue specific in that these enzyme activities were not affected by methoxsalen in the skin (*188*). Furthermore, repeated administration of methoxsalen or bergapten resulted in both induction and inactivation of hepatic drug-metabolizing enzymes, whereas, psoralen only inactivated the monooxygenase activities (*185*).

The inhibitory action of psoralens appears to be dependent on an unsubstituted furan ring such as in psoralen, methoxsalen and bergapten (Figure 5) (*182,185*). On the other hand, trioxsalen, with a methyl furan structure (Figure 5), had no effect on P450 and monooxygenase activities (*182,185,188*). The presence of this methyl group may sterically hinder its metabolic activation and the inactivation of P450 (*182,185*).

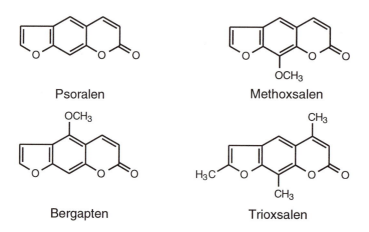

Psoralen   Methoxsalen

Bergapten   Trioxsalen

Figure 5. Chemical structures of some naturally occurring psoralens.

Conclusions

In this review, a wide array of food phytochemicals have been shown to affect the metabolism of various xenobiotics. The mechanism appears to be partly due to altering the levels of the phase I and phase II enzymes. Many dietary compounds are selective in that they can induce certain xenobiotic-metabolizing enzymes while having no effect on or decreasing others. Furthermore, some dietary compounds are tissue-selective in their effect. The enzyme composition and its alteration by the administration of dietary compounds will determine the rate of metabolism of xenobiotics in the tissues of animals and humans. Some of the food phytochemicals cause their effects indirectly, i.e. a metabolite of the compound is actually influencing the enzyme. In addition, the amount of the compound or its metabolite which reaches a particular tissue is not known in many of the cases. Although studies with animals greatly enhanced our understanding of the possible actions of food phytochemicals in affecting xenobiotic metabolism and carcinogenesis, one must apply caution in extrapolating the information obtained in the animal studies to humans because of possible species differences. More detailed studies are needed to clearly define the mechanism(s) of action of food phytochemicals on xenobiotic metabolism.

Acknowledgements

The authors would like to thank Professor Harold Newmark (Rutgers University and Memorial Sloan Kettering Cancer Center) for his helpful suggestions. This work was supported by NIH grants CA46535, CA56673 and ES05693 and NIEHS Center grant ES05022.

Literature Cited

1. Nelson, D. R.; Strobel, H. W. *Mol. Biol. Evol.* **1987**, *4*, 572–593.
2. Nelson, D. R.; Kamataki, T.; Waxman, D. J.; Guengerich, F. P.; Estabrook, R. W.; Feyereisen, R.; Gonzalez, F. J.; Coon, M. J.; Gunsalus, I. C.; Gotoh, O., Okuda, K.; Nebert, D. W. *DNA Cell Biol.* **1993**, *12*, 1–51.
3. Yang, C. S.; Yoo, J.-S. H. In *Nutrition, Toxicology and Cancer*; Rowland, I., Ed.; CRC Press, Inc.: Boca Raton, 1991; pp 53–91.
4. Jakoby, W. B.; Ziegler, D. M. *J. Biol. Chem.* **1990**, *265*, 20715–20718.
5. Mannervik, B.; Alin, P.; Guthenberg, C.; Jensson, H.; Tahir, M. K.; Warholrn, M.; Jornvall, H. *Proc. Natl. Acad. Sci. USA* **1985**, *82*, 7202–7206.
6. Pickett, C. B.; Lu, A. Y. H. *Ann. Rev. Biochem.* **1989**, *58*, 743-764.
7. Prochaska, H. J.; De Long, M. J.; Talalay, P. *Proc. Natl. Acad. Sci. USA* **1985**, *82*, 8232–8236.
8. Zhang, Y.; Talalay, P.; Cho, C.-G.; Posner, G. H. *Proc. Natl. Acad. Sci. USA* **1992**, *89*, 2399–2403.
9. Guengerich, F. P. *Chem. Res. Toxicol.* **1991**, *4*, 391–407.
10. Guengerich, F. P. In *Progress in Drug Metabolism*; Bridges, J. W., Chasseaud, L. F.; Gibson, G. G., Ed.; Taylor & Francis Ltd.: 1987, Vol. 10; Chapter 1, pp. 1–54.
11. Dong, Z.; Jeffery, A. M. *Cancer Invest.* **1990**, *8*, 523–533.
12. Autrup, H. *Drug Metab. Rev.* **1982**, *13*, 603–646.
13. Koga, N.; Inskeep, P. B.; Harris, T. M.; Guengerich, F. P. *Biochem.* **1986**, *25*, 2192–2198.
14. MacFarland, R. T.; Gandolfi, A. J.; Sipes, I. G. *Drug Chem. Toxicol.* **1984**, *7*, 213–227.
15. Guengerich, F. P.; Crawford, W. M., Jr.; Domoradzki, J. Y.; MacDonald, T. L.; Watanabe, P. G. *Toxicol. Appl. Pharmacol.* **1980**, *55*, 303–317.
16. Yang, C. S.; Brady, J. F.; Hong, J.-Y. *FASEB J.* **1992**, *6*, 737-744.

17. Talalay, P.; De Long, M. J.; Prochaska, H. J. *Proc. Natl. Acad. Sci. USA* **1988**, *85*, 8261–8265.
18. Prochaska, H. J.; Talalay, P. *Cancer Res.* **1988**, *48*, 4776–4782.
19. Sones, K.; Heaney, R. K.; Fenwick, G. R. *J. Sci. Food Agric* **1984**, *35*, 712–720.
20. McDanell, R.; McLean, A. E. M. *Food Chem. Toxicol.* **1988**, *26*, 59–70.
21. Bailey, G. S.; Hendricks, J. D.; Shelton, D. W.; Nixon, J. E.; Pawlowski, N. E. *J. Natl. Cancer Inst.* **1987**, *78*, 931–934.
22. Pence, B. C.; Buddingh, F.; Yang, S. P. *J. Natl. Cancer Inst.* **1986**, *77*, 269–276.
23. Morse, M. A.; LaGreca, S. D.; Amin, S. G.; Chung, F.-L. *Cancer Res.* **1990**, *50*, 2613–2617.
24. Bradfield, C. A.; Bjeldanes, L. F. *Food Chem. Toxicol.* **1984**, *22*, 977–982.
25. Baldwin, W. S.; Leblanc, G. A. *Chem.-Biol. Interact.* **1992**, *83*, 155–169.
26. Wortelboer, H. M.; Van Der Linden, E. C. M.; de Kruif, C. A.; Noordhoek, J.; Blaauboer, B. J.; van Bladeren, P. J.; Falke, H. E. *Food Chem. Toxicol.* **1992**, *30*, 589–599.
27. Michnovicz, J. J.; Bradlow, H. L. *J. Natl. Cancer Inst.* **1990**, *82*, 947–949.
28. Bradlow, H. L.; Michnovicz, J. J.; Telang, N. T.; Osborne, M. P. *Carcinogenesis* **1991**, *12*, 1571–1574.
29. Whitty, J. P.; Bjeldanes, L. F. *Food Chem. Toxicol.* **1987**, *25*, 581–587.
30. Salbe, A. D.; Bjeldanes, L. F. *Food Chem. Toxicol.* **1985**, *23*, 57–65.
31. Salbe, A. D.; Bjeldanes, L. F. *Carcinogenesis* **1989**, *10*, 629–634.
32. Vang, O.; Jensen, H.; Autrup, H. *Chem.-Biol. Interact.* **1991**, *78*, 85–96.
33. Chung, F.-L.; Wang, M.; Hecht, S. S. *Carcinogenesis* **1985**, *6*, 539–543.
34. Wortelboer, H. M.; de Kruif, C. A.; van Iersel, A. A. J.; Falke, H. E.; Noordhoek, J.; Blaauboer, B. J. *Biochem. Pharmacol.* **1992**, *43*, 1439–1447.
35. de Kruif, C. A.; Marsman, J. W.; Venekamp, J. C.; Falke, H. E.; Noordhoek, J.; Blaauboer, B. J.; Wortelboer, H. M. *Chem.-Biol. Interact.* **1991**, *80*, 303–315.
36. Bradfield, C. A.; Bjeldanes, L. F. *J. Toxicol. Environ. Health* **1987**, *21*, 311–323.
37. Vang, O.; Jensen, M. B.; Autrup, H. *Carcinogenesis* **1990**, *11*, 1259–1263.
38. Bjeldanes, L. F.; Kim, J.-Y.; Grose, K. R.; Bartholomew, J. C.; Bradfield, C. A. *Proc. Natl. Acad. Sci. USA* **1991**, *88*, 9543–9547.
39. Shertzer, H. G. *Toxicol. Appl. Pharmacol.* **1982**, *64*, 353–361.
40. Bradfield, C. A.; Bjeldanes, L. F. *J. Agric. Food Chem.* **1987**, *35*, 896–900.
41. McDanell, R.; McLean, A. E. M.; Hanley, A. B.; Heany, R. K.; Fenwick, G. R. *Food Chem. Toxicol.* **1987**, *25*, 363–368.
42. Wortelboer, H. M.; de Kruif, C. A.; van Iersel, A. A. J.; Noordhoek, J.; Blaauboer, B. J.; van Bladeren, P. J.; Falke, H. E. *Food Chem. Toxicol.* **1992**, *30*, 17–27.
43. Bradfield, C. A.; Chang, Y.; Bjeldanes, L. F. *Food Chem. Toxicol.* **1985**, *23*, 899–904.
44. Chang, Y.; Bjeldanes, L. F. *Food Chem. Toxicol.* **1985**, *23*, 905–909.
45. Wattenberg, L. W. *Cancer Res.* **1992**, *52(suppl.)*, 2085s–2091s.
46. Salbe, A. D.; Bjeldanes, L. F. *Food Chem. Toxicol.* **1986**, *24*, 851–856.
47. Chung, F.-L.; Morse, M. A.; Eklind, K. I. *Cancer Res.* **1992**, *52(suppl.)*, 2719s–2722s.

48. Eklind, K. I.; Morse, M. A.; Chung, F.-L. *Carcinogenesis* **1990**, *11*, 2033–2036.
49. Wattenberg, L. W. *Carcinogenesis* **1987**, *12*, 1971–1973.
50. Morse, M. A.; Wang, C.-X.; Stoner, G. D.; Mandal, S.; Conran, P. B.; Amin, S. G.; Hecht, S. S.; Chung, F.-L. *Cancer Res.* **1989**, *49*, 549–553.
51. Stoner, G. D.; Morrissey, D. T.; Heur, Y.-H.; Daniel, E. M.; Galati, A. J.; Wagner, S. A. *Cancer Res.* **1991**, *51*, 2063–2068.
52. Morse, M. A.; Amin, S. G.; Hecht, S. S.; Chung, F.-L. *Cancer Res.* **1989**, *49*, 2894–2897.
53. Morse, M. A.; Eklind, K. I.; Amin, S. G.; Hecht, S. S.; Chung, F.-L. *Carcinogenesis* **1989**, *10*, 1757–1759.
54. Morse, M. A.; Eklind, K. I.; Hecht, S. S.; Jordan, K. G.; Choi, C.-I.; Desai, D. H.; Amin, S. G.; Chung, F.-L. *Cancer Res.* **1991**, *51*, 1846–1850.
55. Morse, M. A.; Reinhardt, J. C.; Amin, S. G.; Hecht, S. S.; Stoner, G. D.; Chung, F.-L. *Cancer Lett.* **1990**, *49*, 225–230.
56. Guo, Z.; Smith, T. J.; Wang, E.; Sadrieh, N.; Ma, Q.; Thomas, P. E.; Yang, C. S. *Carcinogenesis* **1992**, *13*, 2205–2210.
57. Guo, Z.; Smith, T. J.; Wang, E.; Eklind, K. I.; Chung, F.-L.; Yang, C. S. *Carcinogenesis* **1993**, (in press).
58. Smith, T. J.; Guo, Z.-Y.; Thomas, P. E.; Chung, F.-L.; Morse, M. A.; Eklind, K.; Yang, C. S. *Cancer Res.* **1990**, *50*, 6817–6822.
59. Doerr-O'Rourke, K.; Trushin, N.; Hecht, S. S.; Stoner, G. D. *Carcinogenesis* **1991**, *12*, 1029–1034.
60. Murphy, S. E.; Heiblum, R.; King, P. G.; Bowman, D.; Davis, W. J.; Stoner, G. D. *Carcinogenesis* **1991**, *12*, 957–961.
61. Smith, T. J.; Guo, Z.; Li, C.; Ning, S. M.; Thomas, P. E.; Yang, C. S. *Cancer Res.* **1993**, (in press).
62. Ishizaki, H.; Brady, J. F.; Ning, S. M.; Yang, C. S. *Xenobiotica* **1990**, *20*, 255–264.
63. Chung, F.-L.; Juchatz, A.; Vitarius, J.; Hecht, S. S. *Cancer Res.* **1984**, *44*, 2924–2928.
64. Huang, Q.; Lawson, T. A.; Chen, S. C.; Mirvish, S. S. *Proc. Amer. Assoc. Cancer Res.* **1992**, *33*, 166.
65. Guo, Z.; Smith, T. J.; Ishizaki, H.; Yang, C. S. *Carcinogenesis* **1991**, *12*, 2277–2282.
66. Guo, Z.; Smith, T. J.; Thomas, P. E.; Yang, C. S. *Arch. Biochem. Biophys.* **1992**, *298*, 279–286.
67. Wattenberg, L. W. *Proc. Nutr. Soc.* **1990**, *49*, 173–183.
68. Vos, R. M. E.; Snoek, M. C.; van Berkel, W. J. H.; Muller, F.; van Bladeren, P. J. *Biochem. Pharm.* **1988**, *37*, 1077–1082.
69. Benson, A. M.; Barretto, P. B.; Stanley, J. S. *J. Natl. Cancer Inst.* **1986**, *76*, 467–473.
70. Benson, A. M.; Barretto, P. B. *Cancer Res.* **1985**, *45*, 4219–4223.
71. Sparnins, V. L.; Chuan, J.; Wattenberg, L. W. *Cancer Res.* **1982**, *42*, 1205–1207.
72. Bogaards, J. J. P.; van Ommen, B.; Falke, H. E.; Willems, M. I.; van Bladeren, P. J. *Food Chem. Toxicol.* **1990**, *28*, 81–88.
73. Talalay, P. In *Cancer Chemoprevention*; Wattenberg, L., Lipkin, M., Boone, C. W.; Kelloff, G. J., Ed.; CRC Press, Inc: Boca Raton, 1992, pp 469–478.

74. Fenwick, G. R.; Hanley, A. B. *CRC Crit. Rev. Tood Sci. Nutr.* **1985**, *22*, 273–377.
75. You, W.-C.; Blot, W. J.; Chang, Y.-S.; Ershow, A. G.; Yang, Z.-Y.; An, Q.; Henderson, B.; Xu, G.-W.; Fraumeni, J. F. J.; Wang, T.-G. *Cancer Res.* **1988**, *48*, 3518–3523.
76. You, W.-C.; Blot, W. J.; Chang, Y.-S.; Ershow, A.; Yang, Z. T.; An, Q.; Henderson, B. E.; Fraumeni, J. F., Jr.; Wang, T.-G. *J. Natl. Cancer Inst.* **1989**, *81*, 162–164.
77. Block, E. *Angew. Chem. Int. Ed. Engl.* **1992**, *31*, 1135–1178.
78. Sparnins, V. L.; Barany, G.; Wattenberg, L. W. *Carcinogenesis* **1988**, *9*, 131–134.
79. Wattenberg, L. W.; Sparnins, V. L.; Barany, G. *Cancer Res.* **1989**, *49*, 2689–2692.
80. Wargovich, M. J.; Imada, O.; Stephens, L. C. *Cancer Lett.* **1992**, *64*, 39–42.
81. Wargovich, M. J.; Woods, C.; Eng, V. W. S.; Stephens, L. C.; Gray, K. *Cancer Res.* **1988**, *48*, 6872–6875.
82. Sumiyoshi, H.; Wargovich, M. *Cancer Res.* **1990**, *50*, 5084–5087.
83. Ip, C.; Ganther, H. E. *Carcinogenesis* **1991**, *12*, 365–367.
84. Ludeke, B. I.; Domine, F.; Ohgaki, H.; Kleihues, P. *Carcinogenesis* **1992**, *13*, 2467–2470.
85. Hong, J.-Y.; Smith, T.; Lee, M.-J.; Li, W.; Ma, B.-L.; Ning, S.-M.; Brady, J. F.; Thomas, P. E.; Yang, C. S. *Cancer Res.* **1991**, *51*, 1509–1514.
86. Brady, J. F.; Li, D.; Ishizaki, H.; Yang, C. S. *Cancer Res.* **1988**, *48*, 5937–5940.
87. Brady, J. F.; Wang, M.-H.; Hong, J.-Y.; Xiao, F.; Li, Y.; Yoo, J.-S. H.; Ning, S. M.; Fukuto, J. M.; Gapac, J. M.; Yang, C. S. *Toxicol. Appl. Pharmacol.* **1991**, *108*, 342–354.
88. Tadi, P. P.; Teel, R. W.; Lau, B. H. S. *Nutr. Cancer* **1991**, *15*, 87–95.
89. Brady, J. F.; Ishizaki, H.; Fukuto, J. M.; Lin, M. C.; Fadel, A.; Gapac, J. M.; Yang, C. S. *Chem. Res. Toxicol.* **1991**, *4*, 642–647.
90. Pan, J.; Hong, J.-Y.; Ma, B.-L.; Xiao, F.; Paranawithana, S. R.; Yang, C. S. **1993**, (in press).
91. Gudi, V. A.; Singh, S. V. *Biochem. Pharm.* **1991**, *42*, 1261–1265.
92. Maurya, A. K.; Singh, S. V. *Cancer Lett.* **1992**, *57*, 121–129.
93. Hayes, M. A.; Rushmore, T. H.; Goldberg, M. T. *Carcinogenesis* **1987**, *8*, 1155–1157.
94. Sparnins, V. L.; Mott, A. W.; Barany, G.; Wattenberg, L. W. *Nutr. Cancer* **1986**, *8*, 211–215.
95. Middleton, E. J. *Trends Pharmac. Sci.* **1984**, *5*, 335–338.
96. Huang, M.-T.; Johnson, E. F.; Mueller-Eberhardt, U.; Koop, D. R.; Coon, M. J.; Conney, A. H. *J. Biol. Chem.* **1981**, *256* , 10897–10901.
97. Siess, M.-H.; Le Bon, A.-M.; Suschetet, M.; Rat, P. *Food Add. Contaminants* **1990**, *7*, S178–S181.
98. Vernet, A.; Siess, M. H. *Food Chem. Toxicol.* **1986**, *24*, 857–861.
99. Siess, M.-H.; Pennec, A.; Gaydou, E. *Eur. J. Drug Metab. Pharmacokinetics* **1989**, *14*, 235–239.
100. Beyeler, S.; Bernard, T.; Perrissoud, D. *Biochem. Pharm.* **1988**, *37*, 1971–1979.
101. Buening, M. K.; Chang, R. L.; Huang, M.-T.; Fortner, J. F.; Wood, A. W.; Conney, A. H. *Cancer Res.* **1981**, *41*, 67–72.

102. Guengerich, F. P.; Kim, D.-H. *Carcinogenesis* **1990**, *11*, 2275–2279.
103. Trela, B. A.; Carlson, G. P. *Xenobiotica* **1987**, *17*, 11–16.
104. Verma, A. K. In *Phenolic Compounds in Food and their Effects on Health II. Antioxidants & Cancer Prevention*; Huang, M.-T.; Ho, C.-T.; Lee, C. Y., Eds.; ACS Symposium Series 507; American Chemical Society: Washington, DC, 1992; Chapter 18, pp 250–264.
105. Brouard, C.; Siess, M. H.; Vernevaut, M. F.; Suschetet, M. *Food Chem. Toxicol.* **1988**, *26*, 99–103.
106. Bokkenheuser, V. D.; Shackleton, C. H. L.; Winter, J. *Biochem. J.* **1987**, *248*, 953–956.
107. Das, M.; Mukhtar, H.; Bik, D. P.; Bickers, D. R. *Cancer Res.* **1987**, *47*, 760–766.
108. Sousa, R. L.; Marletta, M. A. *Arch. Biochem. Biophys.* **1985**, *240*, 345–357.
109. Gryglewski, R. J.; Korbut, R.; Robak, J.; Swies, J. *Biochem. Pharm.* **1987**, *36*, 317–322.
110. Kato, R.; Nakadate, T.; Yamamoto, S.; Suimura, T. *Carcinogenesis* **1983**, *4*, 1301–1305.
111. Deschner, E. E. In *Phenolic Compounds in Food and their Effects on Health II. Antioxidants & Cancer Prevention*; Huang, M.-T.; Ho, C.-T.; Lee, C. Y., Eds.; ACS Symposium Series 507; American Chemical Society: Washington, DC, 1992; Chapter 19, pp 265-268.
112. Lasker, J. M.; Huang, M. T.; Conney, A. H. *J. Pharmacol. Exp. Ther.* **1984**, *229*, 162–170.
113. Huang, M.-T.; Chang, R.; Fortner, J. G.; Conney, A. H. *J. Biol. Chem.* **1981**, *256*, 6829–6836.
114. Kellis, J. T. J.; Vickery, L. E. *Science* **1984**, *225*, 1032–1034.
115. Ibrahim, A.-R.; Abdul-Hajj, Y. J. *J. Steroid Biochem. Molec. Biol.* **1990**, *37*, 257–260.
116. Moochhala, S. M.; Loke, K. H.; Das, N. P. *Biochem. Internat.* **1988**, *17*, 755–762.
117. Chae, Y.-H.; Ho, D. K.; Cassady, J. M.; Cook, V. M.; Marcus, C. B.; Baird, W.; M. *Chem-Biol. Interact.* **1992**, *82*, 181–193.
118. Nagabhushan, M.; Amonkar, A. J.; Nair, U. J.; Santhanam, U.; Ammigan, N.; D'Souza, A. V.; Bhide, S. V. *J. Cancer Res. Clin. Oncol.* **1988**, *114*, 177–182.
119. Steele, C. M.; Lalies, M.; Ioannides, C. *Cancer Res.* **1985**, *45*, 3573–3577.
120. Monteith, D. K. *Mutat. Res.* **1990**, *240*, 151–158.
121. Liu, L.; Castonguay, A. *Carcinogenesis* **1991**, *12*, 1203-1208.
122. Siegers, C.-P.; Larseille, J.; Younes, M. *Res. Commun. Chem. Pathol. Pharmacol.* **1982**, *36*, 61–73.
123. Shi, S. T.; Wang, Z. Y.; Smith, T. J.; Cheng, W. F.; Ho, C. T.; Yang, C.; *Proc. Amer. Assoc. Cancer Res.* **1993**, *34*, 157.
124. Wang, Z. Y.; Bickers, D. R.; Mukhtar, H. *Drug Metab. Dispos.* **1988**, *16*, 93–103.
125. Mukhtar, H.; Wang, Z. Y.; Katiyar, S. K.; Agarwal, R. *Prev. Med.* **1992**, *21*, 351–360.
126. Aldercreutz, H.; Mousavi, Y.; Clark, J.; Hockerstedt, K.; Hamalainen, E.; Wahala, K.; Makela, T.; Hase, T. *J. Steroid Biochem. Molec. Biol.* **1992**, *41*, 331–337.
127. Cassady, J. M.; Zennie, T. M.; Chae, Y.-H.; Ferin, M. A.; Portnondo, N. E.; Baird, W. M. *Cancer Res.* **1988**, *48*, 6257–6261.

128. Chae, Y.-H.; Coffing, S. L.; Cook, V. M.; Ho, D. K.; Cassady, J. M.; Baird, W. M. *Carcinogenesis* **1991**, *12*, 2001–2006.
129. Sariaslani, F. S.; Kunz, D. A. *Biochem. Biophys. Res. Commun.* **1986**, *141*, 405–410.
130. Degan, G. H. *J. Steroid Biochem.* **1990**, *35*, 473–479.
131. Pratt, D. E. In *Phenolic Compounds in Food and their Effects on Health II. Antioxidants & Cancer Prevention*; Huang, M.-T.; Ho, C.-T.; Lee, C. Y., Ed.; ACS Symposium Series 507: Washington, DC, 1992, Chapt. 5; pp 54–71.
132. Ho, C.-T. In *Phenolic Compounds in Food and their Effects on Health II. Antioxidants & Cancer Prevention*; Huang, M.-T.; Ho, C.-T.; Lee, C. Y., Eds.; ACS Symposium Series 507; American Chemical Society: Washington, DC, 1992; Chapter 1, pp 2–7.
133. Das, M.; Bickers, D. R.; Mukhtar, H. *Carcinogenesis* **1985**, *6*, 1409–1413.
134. Mandal, S.; Ahuja, A.; Shivapurkar, N. M.; Cheng, S.-J.; Groopman, J. D.; Stoner, G. D. *Carcinogenesis* **1987**, *8*, 1651–1656.
135. Gali, H. U.; Perchellet, E. M.; Klish, D. S.; Johnson, J. M.; Perchellet, J.-P. *Carcinogenesis* **1982**, *13*, 715–718.
136. Gali, H. U.; Perchellet, E. M.; Perchellet, J.-P. *Cancer Res.* **1991**, *51*, 2820–2825.
137. Lesca, P. *Carcinogenesis* **1983**, *4*, 1651–1653.
138. Athar, M.; Khan, W. A.; Mukhtar, H. *Cancer Res.* **1989**, *49*, 5784–5788.
139. Mandal, S.; Stoner, G. D. *Carcinogenesis* **1990**, *11*, 55–61.
140. Mukhtar, H.; Das, M.; Del Tito, B. J. J.; Bickers, D. R. *Biochem. Biophys. Res. Commun.* **1984**, *119*, 751–757.
141. Del Tito, B. J. J.; Mukhtar, H.; Bickers, D. R. *Biochem. Biophys. Res. Commun.* **1983**, *114*, 388–394.
142. Das, M.; Khan, W. A.; Asokan, P.; Bickers, D. R.; Mukhtar, H. *Cancer Res.* **1987**, *47*, 767–773.
143. Dixit, R.; Teel, R. W.; Daniel, F. B.; Stoner, G. D. *Cancer Res.* **1985**, *45*, 2951-2956.
144. Chang, R. L.; Huang, M.-T.; Wood, A. W.; Wong, C.-Q.; Newmark, H. L.; Yagi, H.; Sayer, J. M.; Jerina, D. M.; Conney, A. H. *Carcinogenesis* **1985**, *6*, 1127–1133.
145. Smart, R. C.; Huang, M.-T.; Chang, R. L.; Sayer, J. M.; Jerina, D. M.; Conney, A. H. *Carcinogenesis* **1986**, *7*, 1663–1667.
146. Castonguay, A.; Allaire, L.; Charet, M.; Rossignol, G.; Boutet, M. *Cancer Lett.* **1989**, *46*, 93–105.
147. Mandal, S.; Shivapurkar, N. M.; Galati, A. J.; Stoner, G. D. *Carcinogenesis* **1988**, *9*, 1313–1316.
148. Barch, D. H.; Fox, C. C. *Cancer Lett.* **1989**, *44*, 39–44.
149. Sayer, J. M.; Yagi, H.; Wood, A. W.; Conney, A. H.; Jerina, D. M. *J. Amer. Chem. Soc.* **1982**, *104*, 5562–5564.
150. Josephy, P. D.; Lord, H. L.; Snieckus, V. A. *Proc. Amer. Assoc. Cancer Res.* **1990**, *31*, 121.
151. Kuenzig, W.; Chau, J.; Norkus, E.; Holowaschenko, H.; Newmark, H.; Mergens, W.; Conney, A. H. *Carcinogenesis* **1984**, *5*, 309–313.
152. Newmark, H. L. *Can. J. Physiol. Pharmacol.* **1987**, *65*, 461–466.
153. San, R. H. C.; Chan, R. I. M. *Mutat. Res.* **1987**, *177*, 229–239.
154. Sivarajah, K.; Lasker, J. M.; Eling, T. E.; Abou-Donia, M. B. *Mol. Pharm.* **1982**, *21*, 133–141.

155. Potter, D. W.; Hinson, J. A. *J. Biol. Chem.* **1987**, *262*, 974–980.
156. Koshihara, Y.; Neichi, T.; Murota, S.-I.; Lao, A.-N.; Fujimoto, Y.; Tatsuno, T. *Biochim. Biophys. Acta* **1984**, *792*, 92–97.
157. Kulkarni, A. P.; Cai, Y.; Richards, I. S. *Int. J. Biochem.* **1992**, *24*, 255–261.
158. Sugiura, M.; Naito, Y.; Yamaura, Y.; Fukaya, C.; Yokoyama, K. *Chem. Pharm. Bull. (Tokyo)* **1989**, *37*, 1039–1043.
159. Huang, M.-T.; Lysz, T.; Ferraro, T.; Abidi, T. F.; Laskin, J. D.; Conney, A. H. *Cancer Res.* **1991**, *51*, 813–819.
160. Newmark, H. L. In *Phenolic Compounds in Food and their Effects on Health II. Antioxidants & Cancer Prevention*; Huang, M.-T.; Ho, C.-T.; Lee, C. Y., Eds.; ACS Symposium Series 507; American Chemical Society: Washington, DC, 1992; Chapter 4, pp 48–53.
161. Azuine, M. A.; Bhide, S. V. *Nutr. Cancer* **1992**, *17*, 77–83.
162. Huang, M.-T.; Ferraro, T. In *Phenolic Compounds in Food and their Effects on Health II. Antioxidants & Cancer Prevention*; Huang, M.-T.; Ho, C.-T.; Lee, C. Y., Eds.; ACS Symposium Series 507; American Chemical Society: Washington, DC, 1992; Chapter 2, pp 8–34.
163. Sambaiah, K.; Srinivasan, K. *Indian J. Biochem. Biophys.* **1989**, *26*, 254–258.
164. Huang, M.-T.; Robertson, F. M.; Lysz, T.; Ferraro, T.; Wang, Z. Y.; Georgiadis, C. A.; Laskin, J. D.; Conney, A. H. In *Phenolic Compounds in Food and their Effects on Health II. Antioxidants & Cancer Prevention*; Huang, M.-T.; Ho, C.-T.; Lee, C. Y., Eds.; ACS Symposium Series 507; American Chemical Society: Washington, DC, 1992; Chapter 27, pp 338–349.
165. Nakatani, N. In *Phenolic Compounds in Food and their Effects on Health II. Antioxidants & Cancer Prevention*; Huang, M.-T.; Ho, C.-T.; Lee, C. Y., Eds.; ACS Symposium Series 507; American Chemical Society: Washington, DC, 1992; Chapter 6, pp 72–86.
166. Atal, C. K.; Dubey, R. K.; Singh, J. *J. Pharmacol. Therap.* **1985**, *232*, 258–262.
167. Miller, M. S.; Brendel, K.; Burks, T. F.; Sipes, I. G. *Biochem. Pharmacol.* **1983**, *32*, 547–551.
168. Gannett, P. M.; Iversen, P.; Lawson, T. *Bioorg. Chem.* **1990**, *18*, 185–198.
169. Shaw, P. E. *Agric. Food Chem.* **1979**, *27*, 246–257.
170. Crowell, P. L.; Kennan, W. S.; Haag, J. D.; Ahmad, S.; Vedejs, E.; Gould, M. N. *Carcinogenesis* **1992**, *13*, 1261–1264.
171. Wattenberg, L. W.; Coccia, J. B. *Carcinogenesis* **1991**, *12*, 115–117.
172. Elson, C. E.; Maltzman, T. H.; Boston, J. L.; Tanner, M. A.; Gould, M. N. *Carcinogenesis* **1988**, *9*, 331–332.
173. Maltzman, T. H.; Christou, M.; Gould, M. N.; Jefcoate, C. R. *Carcinogenesis* **1991**, *12*, 2081–2087.
174. Austin, C. A.; Shephard, E. A.; Pike, S. F.; Rabin, B. R.; Phillips, I. R. *Biochem. Pharmacol.* **1988**, *37*, 2223–2229.
175. Maltzman, T. H.; Hurt, L. M.; Elson, C. E.; Tanner, M. A.; Gould, M. N. *Carcinogenesis* **1989**, *10*, 781–783.
176. Elegbede, J. A.; Elson, C. E.; Tanner, M. A.; Qureshi, A.; Gould, M. N. *J. Natl. Cancer Inst.* **1986**, *76*, 323–325.
177. Russin, W. A.; Hoesly, J. D.; Elson, C. E.; Tanner, M. A.; Gould, M. N. *Carcinogenesis* **1989**, *10*, 2161–2164.
178. Clegg, R. J.; Middleton, B.; Bell, G. D.; White, D. A. *Biochem. Pharm.* **1980**, *29*, 2125–2127.

179. Lam, L. K. T.; Hasegawa, S. *Nutr. Cancer* **1989**, *12*, 43–47.
180. Lam, L. K. T.; Sparnins, V. L.; Wattenberg, L. W. *Cancer Res.* **1982**, *42*, 1193–1198.
181. Lam, L. K. T.; Sparnins, V. L.; Wattenberg, L. W. *J. Med. Chem.* **1987**, *30*, 1399–1403.
182. Tinel, M.; Belghiti, J.; Descatoire, V.; Amouyal, G.; Letteron, P.; Geneve, J.; Larrey, D.; Pessayre, D. *Biochem. Pharmacol.* **1987**, *36*, 951–955.
183. Beir, R. *Rev. Environ. Contam. Toxicol.* **1990**, *113*, 47–137.
184. Fouin-Fortunet, H.; Tinel, M.; Descatoire, V.; Letteron, P.; Larrey, D.; Geneve, J.; Pessayre, D. *J. Pharmac. Exp. Ther.* **1986**, *236*, 237–247.
185. Letteron, P.; Descatoire, V.; Larrey, D.; Tinel, M.; Geneve, J.; Pessayre, D. *J. Pharmac. Exp. Ther.* **1986**, *238*, 685–692.
186. Tinel, M.; Labbe, G.; Pessayre, D. *J. Hepatol.* **1986**, *3*, S42.
187. Schimmer, O.; Kiefer, J.; Paulini, H. *Mutagenesis* **1991**, *6*, 501–506.
188. Bickers, D. R.; Mukhtar, H.; Molica, S. J. J.; Pathak, M. A. *J. Invest. Dermatol.* **1982**, *79*, 201–205.

RECEIVED July 27, 1993

Chapter 3

Micronutrients in Cancer Prevention

Paul A. Lachance

Department of Food Science, Cook College, Rutgers, The State University of New Jersey, New Brunswick, NJ 08903

The human body is composed of sixty three trillion cells plus or minus a few hundred billion. Each day a typical human cell undergoes five thousand mutations. Each mutation of course can lead to the initiation of cancer but it usually does not because a particular mutation does not necessarily occur in both strands of the DNA helix and therefore is repairable. A number of factors contribute to the repair of the mutations, but several of these factors play a significant role in dissipating the initiators which cause the mutation — be they exogenously or endogenously generated free radicals or chemical constituents such as peroxide. This paper discusses the protective effects of micronutrients.

Many nutrient factors are protective against initiators, which cause mutations (*1*). Macronutrients such as energy, proteins, lipids and carbohydrates, as well as fiber have been reported to have effects (*2,3*). The focus of this paper is micronutrients and therefore the macronutrients will not be discussed.

The vitamins listed by Milner (*1*) are riboflavin, folic acid, vitamin B_{12}, vitamins A,C,D,E and the non-vitamin, choline (Table I). Interestingly he did not list β-carotene or other carotenoids. Some carotenoids (50 of nearly 600) are precursors of vitamin A but many carotenoids, even those that are not vitamin A precursors (e.g. lycopene) are absorbed and have very significant antioxidant and free radical scavaging properties which vitamin A does not possess (*4*).

This paper will discuss β-carotene as an example of the "vitamin" category as well as discuss the levels of other vitamin nutrients that may contribute to the prevention of the initiation and possibly the prevention of the promotion and progression of certain cancers. Milner (*1*) also lists calcium, iron, zinc, copper and selenium as minerals known to alter experimental carcinogenesis. These are not specifically discussed in this paper, but an important consideration that needs to be recognized is the interactions between a number of micronutrients. For example, iron nutriture is enhanced by the presence of vitamin C and selenium enhances the antioxidant function of vitamin E.

0097–6156/94/0546–0049$06.00/0

Proper intake — at RDA quantities — of copper, zinc and calcium, along with manganese and vitamin D, are important in the prevention and retardation and possible mitigation of osteoporosis, which, while not related to carcinogenesis, is a very significant disease (5), but which may be applicable in the role of calcium in colon cancer (6). The interactions demonstrate the necessity of identifying and promoting the need for a concert of micronutrients in the prevention of chronic and/or debilitating diseases, including cancer.

Table I. Some Dietary Factors Known to Alter Experimental Carcinogenesis

General	Amino acids	Vitamins	Minerals
Calories	Methionine	Riboflavin	Calcium
Lipids	Cysteine	Folic acid	Zinc
Carbohydrates	Tryptophan	B_{12}	Copper
Proteins	Arginine	A, C, D, E	Selenium
Fiber		Choline	

SOURCE: Milner, 1989 (1).

Antioxidant Nutrients

The relationship between food and oxygen, and the role of food antioxidants in human metabolism is very complex. Namiki (7) has provided an extensive review of antioxidants/antimutagens in food. There are many substances in foods and in human cells that have antioxidant properties; there is substantial evidence of benefits for ascorbic acid (Vitamin C), tocopherol (Vitamin E) and carotenoids (in particular β-carotene).

The data supporting a role for antioxidants in reducing cancer risk has been compiled by Block (8). She reports that of approximately 130 studies that examined vitamin C or β-carotene or vitamin E or food sources thereof in an array of cancers, about 120 have revealed statistically significant reduced risk with increased intakes.

Prasad (10), in reviewing vitamin E in cancer prevention, points out that vitamin E is effective in both the initiation and promotion stages of cancer because of either its antioxidant or immuno-enhancing properties or both. In that review, Prasad concludes that vitamin E appears more effective in conjunction with other nutrients, such as selenium and vitamin C in the prevention of tumor development.

The FDA has cautiously acknowledged that the "publicly available evidence does indicate that diets rich in fruits and vegetables, which are low in fat and high in vitamin A (as β-carotene), vitamin C and dietary fiber are associated with a decreased risk of several types of cancer" (9). The FDA accepts the hypothesis of a protective effect of vitamin E in humans. Block (8) challenges the FDA's reluctance to recognize nutrients *per se* in cancer prevention. The FDA seeks to prove that selected nutrients are effective agents in a medical treatment model or in a pharmacological search for the best therapeutic agent. In contrast, a prevention model acknowledges that all antioxidant nutrients play a role in mitigating cancers and the biochemical evidence is "extremely strong" and the epidemiological evidence is accordingly consistent.

Role of β-Carotene in Cancer Prevention

A substantial number of both retrospective and prospective epidemiological studies that demonstrate that β-carotene is an important factor in the diet relative to the incidence of lung cancer have been conducted. More than 30 retrospective, prospective and biochemical assay studies conducted in Europe, Asia and North America, and which involved more than 250,000 cases of lung cancer, have demonstrated an increased risk with low dietary intakes of carotenoid containing foods and/or low blood levels of carotene. A tabulation of retrospective studies by Gaby and Singh (*11*) (Table II) demonstrates that lung cancer risk is diminished by diets rich in β-carotene and not by diets rich in vitamin A.

This phenomena was extremely well demonstrated in 1981 by Shekelle (*12*) in the Western Electric Study. He reported the nineteen year incidence of bronchiogenic carcinoma, versus quartiles of the the dietary β-carotene intake of the participants in the study. Those who consumed diets in the highest quartile ($\geq$ 4 mg β-carotene daily) — even with a history of smoking up to 30 years — were substantially protected from bronchiogenic lung carcinoma. Nonsmokers were completely protected. An important consideration is that these self same dietaries were also concomitant sources of ascorbic acid, vitamin E and several minerals.

The beneficial role of β-carotene in cancers other than lung is also seen in other retrospective and prospective studies with a significant number of the studies demonstrating benefits of decreased risk for various upper GI cancers and gynecological cancers. A tabulation, again made by Gaby and Singh (*11*), of the positive role of β-carotene in comparison to the lack of correlation to vitamin A in the diet of cancers other than lung is seen in Table III. Again one observes the demonstration of protection when the diet was rich in carotenes and essentially no effect when the diets were screened for vitamin A.

Our work (*13*) has demonstrated that the American dietary has a significant gap in the intake of carotenes or β-carotene equivalents. We calculated that the recommended dietaries of the USDA/DHHS for complying with the DHHS/USDA dietary guidelines of the early 1980's revealed that such dietaries would provide 5.2 mg β-carotene equivalent intake daily. Four years later the National Cancer Institute independently published recommended dietaries to retard the onset of cancer. These dietaries deliver approximately 6 mg β-carotene daily. The nutrient composition of these recommended menu cycles, which I am calling dietaries, has been published (*14*). These dietaries promulgated with two different disease objectives (heart disease and cancer) are remarkably similar in nutrient composition.

The 1985 National Food Consumption Continuing Survey of Individual Intakes (*15*) for 18–50 year old men and women revealed that the mean intake of β-carotene is only 1.5 mg per day. It is therefore obvious that a gap (*13*) exists in the carotene intake of the American population. More recently, Block and Subar (*16*) have reported a daily intake of carotenes of 2.4 to 3.0 mg for career-age adults (Figure 1) using a national probability survey of Americans based on a 60 food frequency survey. Food frequency techniques tend to overestimate, but the results confirm the existence of a significant gap in the intake of carotene containing foods.

Other Nutrient Gaps

We can ask whether there might be nutrient intakes above the RDA that could exert beneficial effects in experimental or human carcinogenesis by inhibiting initiation

**Table II. Dietary Carotenes vs. Vitamin A and Lung Cancer Risk:
Retrospective Studies**

Carotene	Vitamin A
+ (carrots)	+ (milk)
+ (carotene)	0 (retinol)
+ (vegetables)	+ (milk)
+ (carotene)	0 (preformed)
+ (fruits & vegetables)	0 (retinol)
+ (carotene)	0 (retinol)
+ (carotene)	0 (preformed)
+ (carrots)	0 (liver/cheese)
+ (plant foods)	+ (total vitamin A[a])
+ (carotene)	0 (retinol)
0 (β-carotene)	0 (total vitamin A[a])
+ (β-carotene)	0 (retinol)
+ (carrots & green leafy vegetables)	+ (retinol)

SOURCE: Adapted from ref. 11, p. 45.
+ = Increased risk of cancer with lower intake.
0 = No association between the risk of cancer and lower intake.
[a] Total dietary vitamin A, including β-carotene.

**Table III. Dietary Carotenes vs. Vitamin A and Risk of Cancers
Other Than in the Lung: Retrospective Studies**

Site	Carotene	Vitamin A
bladder	+ (vegetables)	+ (milk)
bladder	+ (carotenoids)	0 (retinoids)
breast	+ (carotenes)	+ (retinol)
breast	+ (β-carotene)	0 (retinol)
breast	+ (β-carotene)	0 (retinol)
colon	0 (carotene)	+ (retinol)
cervix	+ (β-carotene)	0 (retinol)
cervix	+ (β-carotene)	0 (retinol)
endometrium	+ (β-carotene)	0 (retinol)
esophagus	+ (carotene)	0 (dairy/eggs)
esophagus	+ (β-carotene)	0 (retinol)
esophagus	+ (β-carotene)	0 (retinol[a])
larynx	+ (carotene)	0 (retinol)
oropharynx/head and neck	+ (fruits &vegetables)	0 (dairy/eggs)
stomach	+ (green vegetables)	0 (retinol)
stomach	+ (carotene)	0 (retinol)

SOURCE: Adapted from ref. 11, p. 44.
+ = Increased risk of cancer with lower intake.
0 = No association between the risk of cancer and lower intake.
[a] Retinol from butter and offal was associated with an increased risk with
higher intake; retinol from other foods (e.g., eggs and milk) was not.

of carcinogenesis. I have selected a few of the nutrients identified by Milner (*1*) for purposes of discussing safe ranges of intakes and of more substantial intakes of these micronutrients which could prove to be beneficial.

The concept of determining safe ranges of intake assumes that there is individual variability of both requirement for a nutrient and tolerance for high intake. The safe range of intake is associated with a very low probability of either inadequacy or excess for an individual.

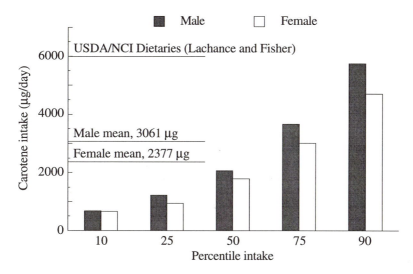

Figure 1. Carotene intake distribution for Americans 18–49 years old. The mean intake of dietary carotene is substantially below the intake that would be provided by ideal diets. Over 50% of career-age adults even fail to obtain the mean intake. (Data are from reference 16.)

Riboflavin

Little information is available as to the effects of riboflavin as an anticarcinogenic compound; the flavin molecule, however, has antioxidant potential (*17,18*). Therefore riboflavin might be expected to reduce free radical damage. Figure 2 illustrates the safe range of riboflavin intakes. The intake value that is considered inadequate below which actual riboflavin deficiency occurs is 0.8 mg per day. The 1989 RDAs for riboflavin are 1.2–1.8 mg per day depending on sex. The USDA/NCI dietaries would deliver 1.8–1.9 mg/day. It is also noteworthy that riboflavin absorption is self limiting at 1,000 mg per day and the compound is non-toxic.

Folate

Figure 3 describes the safe range of folate intakes. Folic acid is important in DNA synthesis. The requirement for folic acid for cellular replication is the rationale for

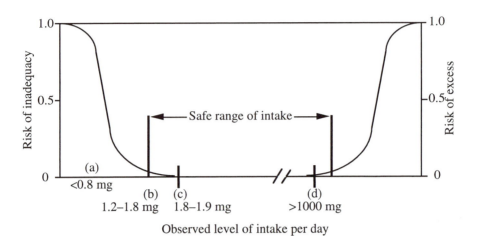

Figure 2. The safe range of riboflavin intake. Riboflavin is non-toxic therefore the safe range of intake is broad. The current intake/day approximates the intake that would be provided by ideal diets. (a) Dietary deficiency; (b) 1989 RDA; (c) USDA/NCI menus (*14*); (d) absorption saturated, nontoxic.

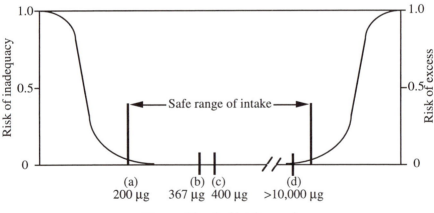

Figure 3. The safe range of folate intake. Folacin is a limiting nutrient in the American dietary. Intakes of less than 250 μg/day by women has been associated with neural tube defects occurring in the fetus in early pregnancy. (a) 1989 RDA; (b) USDA/NCI menus (*14*); (c) 1980 RDA; (d) mask vitamin B_{12} deficiency.

the use of folic acid antagonists in cancer chemotherapy. A classic example of such an antagonist is the drug Methotrexate®. Folic acid also has antioxidant properties.

The current 1989 RDA for folate is 180–200 μg per day, down from the 1980 RDA of 400 μg per day. The USDA/NCI dietaries would provide 367 mg per day. One wonders why the 1989 RDA was decreased by one half that of the 1980 RDA. We do know that the mean intake of folate is in the vicinity of 180 μg per day, and that the RDA of 400 μg was rarely ever consumed by career-age Americans. Attaining this value would require substantial improvements in the daily intake of dark green vegetables. Folate levels in supplements and in food fortification are restricted by law to 400 μg per day. A vitamin product containing more than 400 μg per day requires a prescription. The purpose of the law was to prevent the masking of B_{12} deficiency by intakes of folic acid that exceeded 1,000 μg or so.

Today, we know the critical role that folic acid plays in the prevention of neural tube defects (*19*). More importantly, folate at 10,000 μg (10 mg) per day administered to women at risk for cervical dysplasia has been reported by Butterworth *et al.* (*20*) to not only have no adverse effect but, in the presence of the HPV (human papilloma virus type 16) infection, to decrease the risk of the precancerous lesions by 5 fold. The argument that folic acid ingestion above 400 μg/day may mask B_{12} deficiency is becoming a conundrum in view of today's diagnostic technology. The FDA (*9*) has recently proposed that the RDI for folic acid be 400 μg in order to prevent neural tube defects. This level may also thwart certain cancers.

Safe Intake of Antioxidant Vitamins

It is well documented that the antioxidant vitamins, ascorbic acid, vitamin E and the previously discussed β-carotene have antioxidant capacities at different locations within the cell and exert an important role in preventing oxidation damage, free radical damage and thus carcinogenesis.

Vitamin C. The safe intakes of ascorbic acid spans a considerable range (Figure 4). The current RDA is 60 mg per day which adequately prevents the deficiency disease of scurvy, but scurvy is rarely seen in developed society. Both the Canadian and American dietary allowances recognize that smokers should consume at least 100 mg per day. A number of investigators have long promoted substantially higher intakes of ascorbic acid.

The USDA/NCI recommended dietaries (*14*) deliver at least 220 mg vitamin C per day. A recent study (*21*) indicates that the long term ingestion of 250 mg per day can lead to tissue saturation. One thousand milligrams per day can interfere with clinical tests such as the guiac test for presence of occult blood in the stool. Individuals have consumed ten or more grams per day for years without untoward consequences, and possible, but poorly documented benefit.

Vitamin E. The safe range of vitamin E intake is given in Figure 5. The RDA is 10 mg per day. The USDA/NCI dietaries would provide approximately 25 mg per day. A literature review prepared for the FDA in 1980 states that 1000 mg per day of vitamin E does not lead to toxicity (*22*).

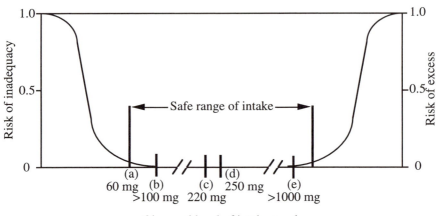

Figure 4. The safe range of ascorbic acid intake. Ascorbic acid intakes approaching 250 mg/day appear necessary to saturate body tissues, and approximates that provided by ideal diets. (a) 1989 RDA; (b) 1989 RDA for smokers; (c) USDA/NCI menus (*14*); (d) for saturation (*21*); (e) interference with clinical test results.

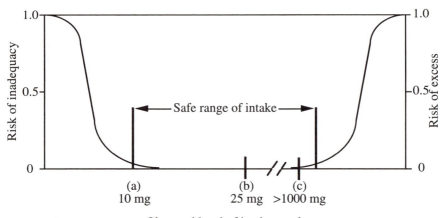

Figure 5. The safe range of vitamin E intake. Vitamin E intakes from ideal diets would provide 2.5 times the RDA for career-age adults. Vitamin E is non-toxic. (a) 1989 RDA; (b) USDA/NCI menus (*14*); (c) Bendich and Machlin, 1992 (*23*).

Vitamin A. The safe range for carotene intake is given in Figure 6. There is no RDA for carotenes but several are precursors for vitamin A. Whereas vitamin A intake should not exceed 25,000 IU per day, β-carotene is considered nontoxic. We have previously discussed β-carotene intakes. Patients with the genetic disease erythropoietic protoporphyria consume 300 mg β-carotene daily for a lifetime to prevent the condition being exacerbated in the presence of sunlight (*24*).

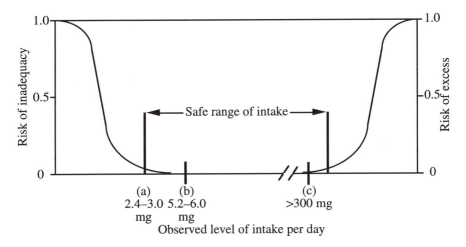

Figure 6. The safe range of carotene intake. Carotene intakes should contribute 90% of the diet but the RDA assumes 50%. Ideal diet intakes are at least double current intakes. B-carotene is non-toxic. (a) 50% of the 1989 RDA for vitamin A; (b) USDA/NCI menus (*14*); (c) daily dose for erythropoietic protoporphyria.

Interaction of Antioxidants

Figure 7 is taken from the work of Estabauer *et al.* (*25*). He and his coworkers have studied the oxidation *in vitro* of the human low density lipoprotein fractions. The work readily demonstrates that in the presence of ascorbic acid, no lipid oxidation occurs when copper chloride is added. Once the ascorbic acid is dissipated, vitamin E, as α- and γ-tocopherols, becomes protective. As the effects of vitamin E are dissipated, carotenoids such as lycopene and a compound (which has not been identified) related to retinyl stearate, play roles in preventing the induced oxidation from accelerating. The last defense appears to be β-carotene. Once these antioxidants have been dissipated, lipid oxidation proceeds very rapidly.

A very critical lesson that this work demonstrates is the importance of the stepwise series of interactions of preventative nutrients in the phenomenon of oxidation, which leads to oxidative damage, which in turn leads to mutations that can subsequently lead to the initiation of carcinogenesis. In this era, we must now question restricting the concept of nutrients as we have previously perceived them, namely as being only relevant to deficiency diseases. In other words, there is no question that ascorbic acid is critical for the prevention of the disease scurvy. This type of one nutrient-one disease relationship is classical and was critical to the

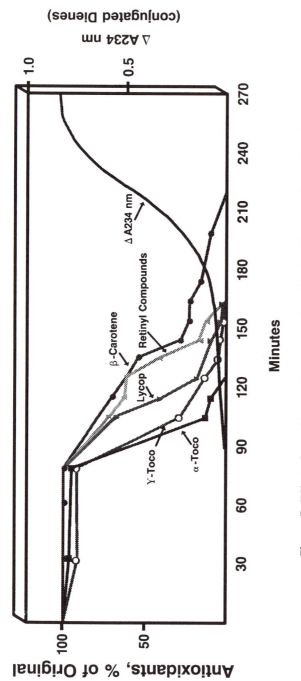

Figure 7. Effect of ascorbate on the consumption of antioxidants and formation of conjugated dienes during oxidation of LDL. The LDL solution was supplemented with 10 μM ascorbate prior to addition of $CuCl_2$. The key antioxidants available from the diet appear to exert their function in series commencing with water soluble ascorbic acid with final protection provided by fat soluble β-carotene (Reproduced with permission from reference 30. Copyright 1989 N.Y. Academy of Science).

discovery of the vitamins and other micronutrients. As a concept for the role of nutrients in the prevention of cancer and other chronic debilitating or killer diseases of the 21st century, however, tunnel vision nutrition does not serve us well.

If one reflects upon the findings of Estabauer and others (*17,25,26*) and realizes that the human body is 70% water, it is understandable that ascorbic acid, which is water soluble, is a critical antioxidant and electron transfer agent in the intra- and extra-cellular fluid spaces of the body. If one recognizes that the cell membrane and the membranes of all the inclusion bodies of each cell (the nuclear membrane, the plasma membrane, the membranes of the endoplasmic reticulum, golgi apparatus, mitochondria etc.) are composed of lipid bilayers, it becomes evident that fat soluble antioxidants and free radical scavengers interspersed in this lipid bilayer become critical in preventing the oxidation of lipids. It is noteworthy that in terms of relative structure and function dimensions, the structure of vitamin E spans approximately the same number of carbons as one half a bilayer (that is one layer of lipid) and that the β-carotene or other carotenoid molecules are capable of spanning the entire bilayer. Folic acid and niacin are critical in the repair of mutations in the nucleus and in the transcription and translation of gene directed instructions.

American Intakes

The figures on micronutrient intake are based upon a recent national probability survey by Block and Subar (*16*) of the nutrient intakes of Americans by age categories based upon a very large sampling and the use of a sixty item food frequency survey. First (Figure 8), we see that folate intakes are not adequate. Second (Figure 9), one can see that ascorbic acid intakes appear to meet the mean for the RDA but the intakes are otherwise below that which would be provided by the ideal dietary promoted to prevent heart disease as was the intention of the DHHS/USDA dietary guidelines, or the prevention of cancer as was the intention of the NCI dietary guidelines. The same can be said of the data on vitamin E intakes (Figure 10) as well as β-carotene intakes (Figure 1).

Given these insights and even with a possible error of food intake assessment by as much as 25 percent due to under-reporting or overestimation, one readily perceives the seriousness of the current situation relevant to the probability of realizing the benefits of micronutrients in cancer prevention.

Significance of Micronutrients in Disease Prevention

A recent report of Enstrom *et al.* (*27*) which is based on a ten year follow-up study of the 1971–1974 NHANES survey demonstrates that the longevity of individuals, both male and female, were substantially enhanced when individuals were users of supplements; particularly if their dietary plus supplemental ascorbic acid intake exceeded 150 mg per day. These data indicate that longevity was enhanced by several years by these self-initiated preventative practices on the part of a segment of the population. The behavior of ingesting the higher intakes of ascorbic acid in the diet plus supplemental sources was most prevalent in individuals who were better educated and of higher income. Considering the various factors we have discussed, the stance of the regulatory agencies and even of selected nutrition professional organizations against the practice of supplementation should be called into question. Table IV lists a number of clinical conditions that plague Americans

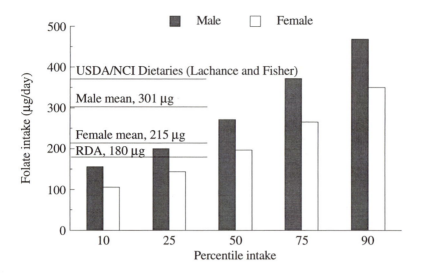

Figure 8. Folate intake distribution for Americans 18–49 years old. Mean folate intake is below the 1989 RDA for nearly 50% of career-age women and rarely meets the intake that would be provided by ideal diets or the Canadian Reference Daily Intake of 400 µg/day. (Data are from reference 16.)

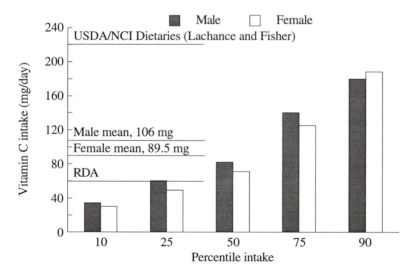

Figure 9. Vitamin C intake distribution for Americans 18–49 years old. The mean vitamin C intake of approximately 25% by career-age women falls below the 1989 RDA, and all women rarely consume the daily intake that would be provided by ideal diets. (Data are from reference 16.)

and the percent of the population that is estimated to be at risk for selected vitamins that are related to these conditions.

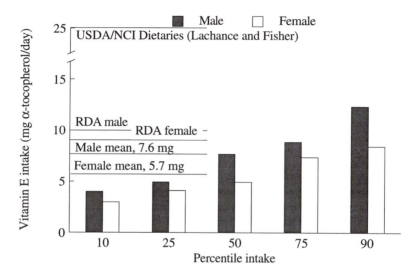

Figure 10. Vitamin E intake distribution for Americans 18–49 years old. The mean vitamin E intake of approximately 75% by career-age women falls below the 1989 RDA, and all women rarely consume the intake that would be provided by ideal diets. (Data are from reference 16.)

If the results of various animal studies, human clinical trials and epidemiological studies are credible, one has difficulty denying the importance of intakes of micronutrients that are substantially greater than the RDAs, which are based on thwarting nutrient deficiency diseases, as currently promulgated. Of greater consequence to all consumers is the use of the RDA by regulatory agencies for purposes of labeling and thus purveying to the public the false sense of the adequacy of such intakes to optimal health. Quality of life and optimal intakes are necessarily related, so the American public deserves to be better informed as to sources of nutrients — be they food, fortified foods or supplements.

The greatest potential risks are with supplements since these can be taken indiscriminantly or by accident and in dosages that, for certain nutrients, can lead to severe toxicity and, in rare occasions, death. There is no evidence, however, that fortified foods can lead to such risks. The reason for the safety of nutrients in food fortification is that the volume of food that one must consume to reach at-risk levels is self-controlling.

It is also evident that adopting dietary practices that follow the recommended dietaries as promulgated by the USDA/DHHS in the case of the dietary guidelines for Americans intended to thwart heart disease, and by the National Cancer Institute to thwart cancer risks are to be highly recommended. A very substantial subsidy to guarantee the lower costs of vegetables, especially dark green and yellow vegetables and fruit could have a very significant impact on the dietary intake of micronutrients and therefore the health of Americans. The transfer of subsidies from tobacco and sugar could readily cover the needed subsidies for dark green and yellow vegetable and fruits.

**Table IV. Selected Groups in the United States Population
at Risk for Poor Vitamin Status**

Group	Estimated number (% of population[a])	Limiting nutrients
Adolescents (10–19 years old)	14.6	Folate, vitamin A
Alcohol users (3 or more drinks/week)	22*	Thiamine, vitamins A, B_6, and D, folate, β-carotene
Cigarette smokers	28.6**	Vitamins B_6, E, and C, β-carotene, folate
Diabetics[b]	4.6	Vitamins B_6, C, and D
Dieters[c]	20	Potentially any or all
Elderly (≥65 years old)[d]	12.1	Vitamins C and D, folate
Oral contraceptive users[e]	10.4***	Vitamins B_6, folate, β-carotene
Pregnant women[f]	3.9***	All, especially folate
Strict vegetarians[g]	>0.9	Vitamins B_6 and B_{12}

SOURCE: Adapted from ref. 11, p. 7.

[a] U.S. Bureau of the Census (1987). Asterisks indicate percentage of the population that was: * ≥ 18 years old; ** ≥ 20 years old; *** women ≥ 16 years old.

[b] The Surgeon General's Report on Nutrition and Health (1988); based on an estimated 11 million diabetic individuals in a total population of 240 million.

[c] Fisher and Lachance, 1985 (28).

[d] Projected for 1990, *Current Population Reports*, Series P-25, No. 704, "Projections of the population of the United States: 1977–2050."

[e] The Surgeon General's Report on Nutrition and Health (1988); based on an estimated 10 million women taking oral contraceptives in a female population (over the age of 15) of 96 million.

[f] Estimate based on the number of live births. The U.S. RDA is the same for pregnant and lactating women; in 1980–81, 32.6% of infants were breast fed 3 or more months (U.S. Bureau of the Census, 1987).

[g] Estimate from Smith, 1988 (29).

Summary

The classic rationale of cancer development, namely, initation, promotion and progression actually limits the realization of the full potential of the role of micronutrients in cancer prevention. Although some micronutrients can thwart initation and promotion, a concert of micronutrients, at intakes usually exceeding the RDA, has the potential to (a) nullify the accidents (e.g., trap oxygen and free radicals required for initiation to occur) and (b) repair mutations and modulate gene expressions that decrease inheritable risks. A new and updated appreciation of micronutrients is in order.

Acknowledgment

This paper is a contribution of the New Jersey Agricultural Experiment Station, Project 10405-93-1.

Literature Cited

1. Milner, J. A. In *Nutrition and Cancer Prevention: Investigating the Role of Micronutrients*; Moon, T.E.; M.S. Micozzi, Eds.; Marcel Dekker, Inc.: New York, 1989; pp 13–32.
2. Shils, M. E. In *Modern Nutrition in Health and Disease*; Shils, M. E.; Young, V. A., Eds.; Lea and Febiger: Philadelphia, 1988; pp 1380–1422.
3. Micozzi, M. S.; Moon, T. E., Eds. *Macronutrients: Investigating Their Role in Cancer*; Marcel Dekker, Inc.: New York, 1992.
4. Krinsky, N. Y. *Free Rad. Biol. Med.* **1989**, *7*, 617–35.
5. Strauss, L.; Saltman, P.; Smith, K.; Andon, M. *J. Am. College Nutri.* **1993**, in press.
6. Sorenson, A. W.; Slattery, M. L.; Ford, M. M. *Nutr. and Cancer* **1988**, *11*, 135–145.
7. Namiki, M. *CRC Critical Rev. Food Sci. Nutrition* **1990**, *29*(4), 273–300.
8. Block, G. *Nutr. Revs.* **1992**, *50*, 207–213.
9. *Fed. Regist.* 58(3), January 6, 1993, p 2622.
10. Prasad, K.; Edwards-Prasad, J. *J. Am. Col. Nutr.* **1992**, *11*, 487–500.
11. Gaby, S. K.; Singh, V. N. In *Vitamin Intake and Health: A Scientific Review*; Gaby S. K.; Bendich, A; Singh, V. N.; Machlin, L. J., Eds.; Marcel Dekker, Inc.: New York, 1991; pp 29–58.
12. Shekelle, R.; Liu, S.; Raynor, W.; Lepper, M.; Maliza, C.; Rossof, A.; Paul, O.; Shryock, A.; Stamler, J. *Lancet* **1981**, *2*, 1185–1190.
13. Lachance, P. A. *Clin. Nutrition* **1988**, *7*, 118–122.
14. Lachance, P. A. In *Food Safety Assessment*; Finley, J. W.; Robinson, S.F.; Armstrong, D.J., Eds.; ACS Symposium Series No. 484; American Chemical Society: Washington, DC, 1992, pp 278–296.
15. *Continuing Survey of Food Intake by Individuals* (reports no 85-3; 86-3), U.S. Department of Agriculture, U.S. Government Printing Office: Washington, DC, 1987/88.
16. Block, G.; Subar, A. *J. Am. Dietetic Assoc.* **1992**, *92*, 969–977.
17. Merrill, A. H., Jr.; Lambeth, J. D.; Edmonson, D. E.; McCormick, D. B. *Ann. Rev. Nutr.* **1981**, *1*, 281–317.
18. Muller, F. *Free Rad. Biol. Med.* **1987**, *3*, 215–30.
19. Schorach, C. J.; Smithells, R. W. In *Micronutrients in Health and in Disease Prevention*; Bendich, A; Butterworth, C.E., Eds.; Marcel Dekker Inc.: New York, 1991; pp 263–285.
20. Butterworth, C. E., Jr.; Hatch, K. D.; Macaluso, M.; Cole, P.; Sauberlich, H. E.; Soong, S.-J.; Borst, M.; Baker, V. V. *J. Am. Med. Assoc.* **1992**, *267*, 528–533.
21. Simon, J. A. *J. Am. College Nutr.* **1992**, *11*, 107–125.
22. Miller, D. R.; Hayes, K. C. In *Nutritional Toxicology*; Hathcock, J.N., Ed. Academic Press: New York, 1982; Vol. 1, pp 81–133.
23. Bendich, A.; Machlin, L. J. In *Vitamin E in Health and Disease* Packer, L.; Fuchs, J., Eds.; Marcel Dekker, Inc.: New York, 1992; pp 411–416.
24. Mathews-Roth, M. M. *Biochimie* **1986**, *68*, 875–84.

25. Esterbauer, H.; Gebicki, J.; Puhl, H.; Jurgens, G. *Free Rad. Biol. Med.* **1992**, *13*, 341–390.
26. Borek, C. In *Vitamins and Minerals in the Prevention and Treatment of Cancer*; Jacobs, M. M., Eds.; CRC Press, Inc.: Boca Raton, FL, 1991.
27. Enstrom, J. E.; Kamin L. E.; Klein, M. A. *Epidemiology* **1992**, *3*, 194–202.
28. Fisher, M. C.; Lachance, P. A. *J. Am. Diet. Assoc.* **1985**, *85*, 450–454.
29. Smith, M. V. *Am. J. Clin. Nutr.* **1988**, *48*, 906–909.
30. Esterbauer, H.; Striegl, G.; Puhl, H.; Oberreither, S.; Rotheneder, M; El-Saadani, M.; and Jurgens, G., *Annals N.Y. Acad. Sci.* **1989**, *570*, 254-267.

RECEIVED October 4, 1993

Chapter 4

Antimutagen and Anticarcinogen Research in Japan

Mitsuo Namiki[1]

Tokyo University of Agriculture, Setagaya-ku, Tokyo 156, Japan, and Nagoya University, Chikusa, Nagoya 464–01, Japan

Marked changes in Japanese food and diet over the past 40 years have resulted in significant elongation of the average life expectancy along with an increase in cancer deaths involving changes in incidence distribution. Following progress in studies on mutagenesis, carcinogenesis and oxygen diseases, studies on antioxidants, antimutagens and anticarcinogens in food are being promoted in Japan as an important part of research in the physiological functionality of food. Various novel physiological activities are being elucidated in Japanese traditional foods, e.g. antimutagenicity of tea catechins, sea weeds and flavoring compounds, antioxidative and antiaging activities of sesame, spices, tea, fermented foods and others. Various oligosaccharides and dietary fibers are being developed to improve colonic microflora correlated to carcinogenesis.

Due to the mild and oceanic climate of the Japanese islands, and perhaps also to Japanese religious and racial features, the Japanese people have developed a unique food culture. Their diet consists of a variety of plants and seafood with rice as the staple food. The use of raw foods (such as raw fish, sushi, seaweeds), many fermented products (including soy sauce, miso, natto, sake, and others), and various salty pickles is particular to the Japanese diet.

The Japanese food culture was originally affected by the Chinese belief that daily diet controls health, disease and aging, and has traditionally used various medicinal foods, for example, green tea, sesame seed and oil, sake and rice vinegar, mushroom (shiitake), various seaweeds and other foods. They are believed to chronically affect health and aging, but until recently little work had been done to elucidate their effects using modern chemical and medical methods.

Food and Health, Cancer and Aging

After the Second World War, the Japanese diet greatly changed in quality and quantity from very poor to saturation. Thus, as shown in Figure 1, the average life

---

[1]Current address: Meito-ku, Yashirodai 2–175, Nagoya 465, Japan

0097–6156/94/0546–0065$06.00/0

expectancy increased rapidly over the past 40 years (1950–1990), from 50 to 78 years of age for males and 54 to 80 years of age for females. It is now the longest in the world (*1*).

Among many factors to be considered as a reason why such rapid elongation of life expectancy was achieved, first of all, improvement of diet in nutrition and sanitation should be pointed out. The Japanese diet changed from the traditional one to become closer to the U.S.-European type, resulting in increased intake of animal protein and fat and decreased intake of carbohydrate and dietary fiber (Figure 2).

At the same time, the pattern of cause of mortality in Japan changed, and since 1981 cancer has become number one (Figure 3). In Japan in 1955, the cancer mortality rate by type was by far the highest in stomach, over 50% for males, then in decreasing order liver, lung, esophagus, colon and other cancers. This is quite different from the U.S.-European countries, where lung, colon, breast and liver cancers are dominant and the stomach cancer rate is low. This apparent difference between Japan and Western countries is thought to be due mainly to differences in diet, i.e. high cereal, fish and salt intake in Japan and high animal fat and meat intake in the U.S.-European countries. In recent years, the mortality ratio in Japan has tended toward the U.S.-European type, that is, a gradual decrease in stomach cancer and increase in lung and colon cancers (Figure 4) (*1*), probably due to changes in the Japanese diet making it more closely resemble the U.S.-European type.

Research Projects on Physiological Functionality of Food

Given these circumstances, starting in the 1970's, a new research field has developed in food science in Japan to elucidate the unknown physiological activities of our daily diets, especially those concerning immune response, circulatory diseases, carcinogenesis and aging. These comprehensive studies are conducted in conjunction with advanced medicinal, biochemical and food-chemical investigations.

One large project, "Systematic Analysis and Development of Food Functionalities," was started in 1984 with the support of a grant-in-aid from the Ministry of Education, Science and Culture of Japan to Professor M. Fujimaki and 81 collaborating professors (1984–1986) (*2*). This was continued in two other projects, "Analyses of Body-modulating Functions of Foods" (Professor H. Chiba, 1988-1990) (*3*) and "Analysis and Molecular Design of Physiologically Functional Foods" (Professor S. Arai, 1992–). Another project, "Research on Cancer Chemoprevention" was started 1983 within the framework of a Japanese cancer research program supported by the Ministry of Health and Welfare of Japan (*4*). In these projects and others, research on antioxidants, antimutagens and anti-carcinogens in food and related materials are being conducted as important fundamental studies on prevention of cancer and aging.

Progress in Environmental Mutagen Research

Studies on the effects of food components on mutagenesis and carcinogenesis greatly advanced in the 1970's with the development of rapid and sensitive methods to assay mutagens using microbial mutants, e.g., the rec assay using *Bacillus subtilis* mutant by Kada (*5*) and the Ames test using *Salmonella typhimurium* mutants by Ames (*6*). Many studies using these assays demonstrated the presence

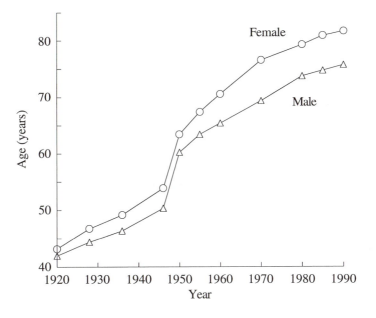

Figure 1. Average life expectancy in Japan (1920–1990).

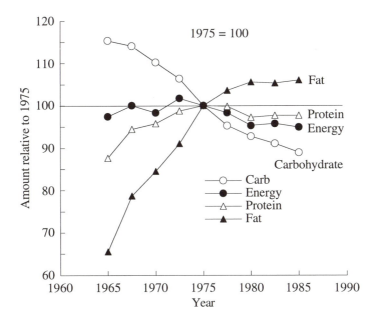

Figure 2. Changes in daily intake of major nutrients in Japan.

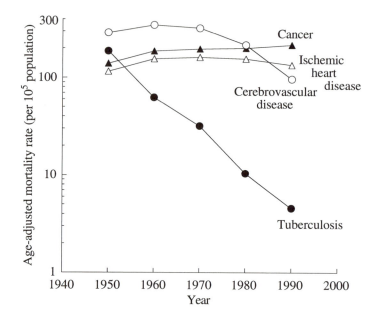

Figure 3. Age-adjusted mortality rates of leading causes of death in Japan.

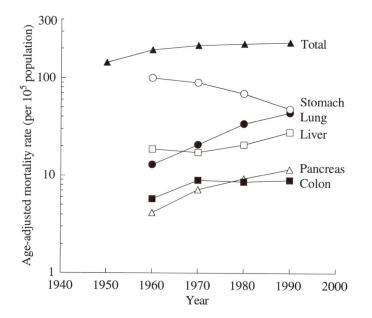

Figure 4. Age-adjusted mortality rates of malignant neoplasma in Japan.

of a number of mutagens in our environment and gave valuable information on their role in carcinogenesis, leading to the conclusion that food and diet may be the most important factors in chemical carcinogenesis (*7*).

One of strongest impacts caused by this mutagen research concerned the mutagenicity of food additives, especially that of nitrofuran derivatives which have very strong antiseptic activity. They have been widely used as preservatives for various foods in Japan, such as fish, meat sausages and ham, soybean tofu, and red bean jam, especially after the Second World War when the cold chain system of food supply was not yet established. In the early 1970's, Kada pointed out the strong mutagenicity of these additives by microbioassay (*8*), and after further research on their carcinogenicity, most came to be excluded from the list of food additives permitted by the government. After this, use of food additives became a strong public concern and the trend now is to use as little additives as practical for food preservation.

Another impact was the discovery of the so called pyrolysate mutagens, or heterocyclic amine mutagens, by Sugimura *et al.* (*9,10*) in pyrolysates of amino acids (Trp-P-1) and in various roasted foods (MeIQ). These mutagens showed extraordinarily strong mutagenicity in the Ames test, and it was demonstrated that they are not contaminants but Maillard reaction products formed during roasting and broiling of various foods. These facts provided not only an important subject of chemical carcinogenesis research but also had a strong social impact in the daily diet once it was suggested that they are an important factor in the high incidence of stomach cancer among the Japanese people.

Antimutagens in Food and Related Materials

At about the same time, Kada discovered suppressive activity against heterocyclic amine mutagens as Trp-P-1 in various vegetable juices, including cabbage, broccoli and burdock (*11*). After investigations on the action mechanisms of various agents suppressing mutagenicity, Kada proposed that antimutagens be divided into two types, "desmutagens" and "bio-antimutagens" (*12*). As shown in Figure 5, the former are concerned in various ways with the prevention of DNA damage in cells, as explained in the following subsection. The latter act in the processes of mutagenesis of the DNA-damaged cells, and the effect mostly involves enhancement of repair and suppression of repair error.

Desmutagenesis Research in Japan.   Recent studies on desmutagens in food system in Japan can be summarized according to their assumed mode of action.

Inhibition of Mutagen Formation.   Inhibition of mutagen formation is the first step of desmutagenesis and the effects of ascorbic acid, α-tocopherol and cysteine against nitrosamine formation are well known (*13–15*). Recently, scorbamic acid, an amino-carbonyl reaction product of dehydroascorbic acid with amino acid, was shown to eliminate nitrite more effectively than ascorbic acid (*16*). Phenolic compounds are also known to inhibit nitrosamine formation, as observed in the cases of sesamol, an antioxidative phenolic compound formed from sesamolin (*17*), and BHA (*18*). When considering these eliminating effects, however, it must be noted that they are markedly influenced by pH and the compounds sometimes act as promoters of the formation and also provide other types of mutagens.

As mentioned above, heterocyclic amine mutagens such as IQ are a kind of Maillard reaction product but its common product melanoidin was shown to inhibit nitrosamine formation, probably due to reduction of nitrite by melanoidin (*19*).

Direct Chemical Inactivation of Mutagens. Chemical inactivation has been observed in the case of the metabolically active products of MNNG and Trp-P-1 by green and black tea extracts as well as tea catechins (*20*). This result is of interest because of an epidemiological report showing that stomach cancer incidence in a tea producing area is lower than in other areas in the Shizuoka prefecture of Japan (*21*). Oolong tea extract has also shown to suppress chromosome aberrations induced by benzo[*a*]pyrene (B[*a*]P) and Trp-P-2 in CHL cells (*22*). Inactivation of a metabolic activation product of Trp-P-2 [Trp-P-2(NHOH)] due to hemin, chlorophyll, and Cu-chlorophyllin is also assumed to be caused by complex product formation (*23*). Medicinal plant tannins, such as geraniin from the widely used drug plant *Geranium thunbergii*, effectively inactivate B[*a*]P epoxide by promotion of hydrolysis to inactive the tetraol compound (*24*). Strong inhibitory activity against Trp-P-2 mutagenicity was also observed in the water extracts of various herbs, especially bay, berfamont, English lavender, Florence dernel, peppermint, thyme (*25*).

Enzymatic Inactivation of Mutagens. As noted above, Kada *et al.* first discovered a suppressive effect of cabbage, broccoli and other vegetable juices against the mutagenicity of Trp-P-1 and others. They pursued the active principle of cabbage juice by fractionation and purification and confirmed that the active component is a heat sensitive heme-protein substance like horseradish peroxidase (*26*). A similar inactivating effect on Trp-P-1 and other mutagens was observed in human saliva (*27*) The activity was shown to be influenced by health condition, age and other factors, and well correlated with peroxidase and catalase activities in saliva.

Physical Inactivation. Fractionation and purification of the component of burdock that is active against Trp-P-1 and others led to a lignin-like polymer compound. The inactivating mechanism is thought to form a complex with mutagens followed by their excretion (*28*). The inactivating effects of vegetable fibers (*29*) and humic acid (*30*) are considered to be caused in a similar manner.

Inhibition of Metabolic Activation of Promutagens. Some heterocyclic amine mutagens require enzymatic activation in the liver for mutagenicity and the inhibition of that process is assumed result in desmutagenesis. Suppression of the mutagenicity of Trp P-1 and nitrosamines by oleic acid (*31*), and by extracts of enokitake (a kind of mushroom) are assumed to be due to the inhibition of the activating enzyme by complex formation with unsaturated fatty acids (*32*). This type of suppression of mutagenicity is also reported for hemin against Trp-P-1 and Trp-P-2 and aflatoxin B_1 (*33*), hemin against B[*a*]P in Chinese hamster V79 cells (*34*), chlorophyll and Cu-chlorophyllin against various mutagens (*35*), some naphthol and quinone derivatives in vegetables against B[*a*]P (*36*), tannins in medicinal plants (*37*) and polyphenols in green and black tea against chromosome aberration induced by B[*a*]P and aflatoxin B_1 in rat bone marrow cells (*38*).

Bio-antimutagens in Food and Related Materials. As mentioned above, bio-antimutagens are agents that reduce the yield of mutations from cells that have already been exposed to mutagens and have some damage to their DNA. In the commonly used screening test for bio-antimutagens, *E. coli* WP2 B/r cells are exposed to X-ray, UV, MNNG, or other mutagens, then incubated with or without the test compound, and the frequencies of induced mutations are compared. In 1978, Kada found that the mutation frequency was markedly reduced by the presence of cobaltous chloride (*39,40*). Following this discovery, a survey of effective bio-antimutagens in food and related materials and molecular genetic studies on the mechanism of the active substances were developed in Japan. Ohta *et al.* found marked bio-antimutagenic activity by cinnamaldehyde (*41*) and vanillin, coumarin, umbelliferone and their related compounds (*42,43*). Interestingly, they are widely used food-flavoring compounds, and commonly possess an α-β-unsaturated aldehyde structure which is assumed to be the active site (*44*).

Shimoi *et al.* also screened various phytochemicals for bio-antimutagenic activity against UV-induced mutagenesis. They showed that components which have a pyrogallol moiety, like gallic acid, (-)-epicatechin gallate (ECg), (-)-epigallocatechin (EGC), and (-)-epigallocatechin gallate (EGCg), reduced mutation, while polyphenol compounds — caffeic acid, chlorogenic acid and quercetin — did not (*45,46*).

Molecular genetic studies on these bio-antimutagens indicate that most of them involve the repair system of damaged DNA cells. As shown in Figure 6, vanillin, coumarin and cinnamaldehyde are assumed to accelerate recombinational repair of damaged DNA induced by 4-NQO and UV in *E. coli* as well as in mammalian cells (*47*). Tannic acid is assumed to contribute to the promotion of excision repair and tea catechins such as EGC may act in some way to improve the fidelity of DNA replication attributable to altered DNA polymerase III (*46*).

Antipromoters in Food and Related Materials

Recently, chemical carcinogenesis has been interpreted to be a two-stage process of initiation and promotion. Inhibition of promotion is considered to be very important in cancer prevention. Wide screening of various food plants, seaweeds and related phytochemicals has been done by Koshimizu *et al.* employing an *in vitro* method using inhibitory activity against tumor promoter-induced activation of Epstein-Barr virus (*48*). Among the tested materials, a considerable part — over 120 — showed positive activity, but the active principle was often difficult to isolate, suggesting that the effect is caused by the coordinated action of several components. Several active factors, however, have been isolated and identified including oleanolic acid from green perilla, mokkolactone from burdock, gingerol from ginger, among others (*49*). It is interesting that many seaweeds, especially the edible brown algae such as wakame (*Undaria pinnatifida* S.), which is common in the Japanese daily diet showed significant antipromoter activity (*50,51*).

Antioxidants in Food and Related Materials

Important research in antimutagenesis is in studies on active oxygen damage *in vivo*; it is assumed that active oxygen species are important as direct and indirect initiators as well as promoters of mutagenesis and carcinogenesis (*52*). Living systems have effective endogenous defense systems involving enzymatic and

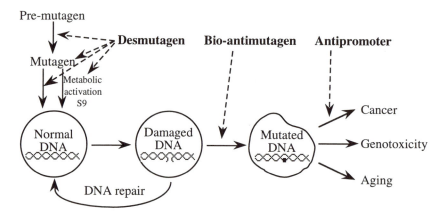

Figure 5. Schematic representation of antimutagenic and anticarcinogenic processes. (Modified after Figure 5.1 in Reference 12.)

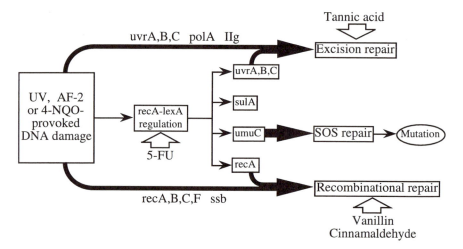

Figure 6. Repair processes of DNA damage and bio-antimutagenesis mechanisms (*43–46*).

nonenzymatic processes, but in addition, it has been shown that dietary antioxidants can act to reduce tumor incidence in animals. Therefore, antioxidants in food have drawn much attention as agents to scavenge and eliminate active oxygen species and act as potential agents to prevent mutagenesis, carcinogenesis and aging (*53–55*).

In this respect, many studies on the effects of vitamins A, C and E as well as on the isolation and identification of natural antioxidants from various foods and related materials are being conducted in Japan (*56*). The main antioxidants are listed in Table I according to the food and materials they are found in.

Table I. Materials of Natural Antioxidants in Recent Studies in Japan

Group	Example	Reference
I. Plants		
1) Oil seeds	sesame, sunflower	(*64, 86–91*)
2) Grains	rice hull, rice bran	(*57,58*)
3) Beans and Nuts	soybean, red bean	(*61*)
	peanut shell	(*59*)
	lotus seed	(*60*)
4) Germ	sesame	(*64*)
5) Sprout and leaves	barley	(*65*)
	tea	(*93*)
6) Leaf waxes	*Eucalyptus*	(*66*)
	Prunus	(*67*)
7) Barks and roots	*Eucalyptus etc.*	(*66,68*)
8) Vegetables and fruits	carrot, parsley, wild grapes	(*70*)
9) Spices and herbs	rosemary, sage, thyme	(*71-75*)
10) Medicinal plant	Osbeckia chinesis L	(*68,69*)
11) Algae	ananori	(*76*)
II. Marine products	short-necked clam	(*77*)
	oyster, scallop	(*78*)
III. Microbial products	*Streptomyces* metabolites	
IV. Fermentation products	*Tempeh*	(*62*)
	Natto, Miso	(*63*)
V. Protein hydrolysates		
VI. Maillard reaction	Melanoidin	(*82,83*)
VII. Others		

Seeds, Beans and Nuts. Oil and crop seeds, beans and nuts are seeds of plants and they must retain their germinating ability for long term preservation. Therefore, they usually contain effective antioxidants such as tocopherols in or around their germs, which are usually rich in lipids, for protection against oxidative damage. Sesame is one representative antioxidative oil seed, and recent studies on the antiaging effect of sesame and its antioxidative components are introduced in the next section.

The fact that effective antioxidants were isolated from rice hull (57,58) and peanut shell (59) presents a new perspective on the role of seed shells as an antioxidative defense system.

Lotus seed is famous in Japan and elsewhere as a seed that has been demonstrated to retain its germination potential for over 2000 years. To elucidate this extraordinarily long germination potential, the antioxidative components of lotus seed were investigated, and it was demonstrated that the seed germ inside its very hard shell is rich in tocopherols, tocotrienols and chlorophylls as strong antioxidants (60).

Among legumes, antioxidants are found in the beans and, interestingly, in the fermentation products of soybean. Antioxidants more effective than α-tocopherol have been isolated from azuki (red bean, important ingredient used in Japanese sweet) and identified as pyrocianidine dimers (61). Among the various Japanese fermentation products, those produced from soybean, such as miso, tempeh and natto have been noted as foods having antioxidative potential, because lipid components, e.g. vitamin A, retain their activity over a long period of time. Chemical investigation of their antioxidative components identified isoflavones such as daid zein and genistein from tempeh (62) and natto (63), which are formed from their glucosides in soybean by fermentation. The studies also suggested the presence of other water-soluble antioxidative substances (63).

Sprouts, Leaves, Bark and Roots. Sprout and young leaves are usually rich in polyphenol and tocopherol antioxidants. Tea is representative and a number of studies on its physiological activities are explained in the section on tea. The presence of strong antioxidant potential in sprouts of sesame seed (64), and in young barley leaves (65) has been reported as increase in phenolic components such as flavones.

New β-diketone type antioxidants have been isolated from leaf waxes of Eucalyptus (66). Studies of Prunus identified new tocopherol derivatives as the main antioxidants (67).

Bark and root are usually abundant in polyphenolic antioxidants, among them, ellagic acid is widely distributed in bark and medicinal plants (66,68), and shown to act as a very strong antioxidant as well as effective antimutagen (69,54).

Juices. Much attention is being focused on vegetable juices, e.g., carrot, parsley, celery, and others, especially those rich in β-carotene, ascorbic acid, and tocopherols for prevention of oxidative diseases related to cancer and aging. A kind of anthocyanin antioxidant has been isolated and identified from a wild grape in Japan (70).

Spices and Herbs. Spices and herbs are also abundant sources of effective antioxidants. Plants of the Perilla family (rosemary, sage, thyme, etc.) are especially abundant in various antioxidants from petroleum ether soluble and insoluble extracts (56). Recently, many new antioxidants have been isolated from various kinds of spices by Nakatani et al. (71–73). Components which are structurally identified are mostly a kind of terpenephenol or lactone compound and compounds such as rosemanol showed antioxidative activity stronger than BHA. Characteristic taste or flavor components of spices are sometimes antioxidative, though not suitable for general food use. Interestingly, the presence of effective and tasteless antioxidants was shown in capsaicin (*74*), pepper (*73*), and turmeric (*75*).

Marine Products. Antioxidants isolated from some algae (*76*) and, interestingly, those isolated from shellfish such as the short-necked clam and oysters (*77,78*) are chlorophyll derivatives. Chlorophyll and its derivatives were once said to be strong oxidation promoters (*79*), but has been shown that they act as oxidation promoters in light conditions but as strong antioxidants under dark conditions (*76,80*).

Related to the antioxidative activity of chlorophyll derivatives, an interesting SOD-like activity of metal-chlorophyllin derivatives was observed, i.e., in biochemical and ESR determination, Fe-chlorophyllin (especially chlorin e6-Fe) showed strong SOD-like antioxidative activity in comparison with authentic bovine RBC-SOD. Co-chlorophyllin was also effective, while Cu-chlorophyllin was very weak. The order of derivatives was Fe>Co>Ni>Cu>Zn>Mg. Fe-chlorophyllin is permitted for use as food colorant, is a green pigment like Cu-chlorophyllin, and is stable, water soluble, and more nonspecific than proteins like SOD enzymes. At present, *in vivo* SOD-like activities involving anticancer activity are being examined by our research group (*81*).

Malliard Products. Strong antioxidative activities of prepared melanoidin in a model system (*82*) as well as in practical foods such as roasted sesame oil (*83*) are known but the active principles have not been isolated and identified.

Antioxidants and Antiaging Activity of Sesame

Sesame seed and its oil have long been evaluated as representative health foods that increase energy and prevent aging (*84*). Sesame oil has been traditionally known as being extraordinarily stable against oxidative deterioration. In serial studies on antioxidative and antiaging effects of sesame seed and oil (*85–88*), we have identified several antioxidants — like new lignan phenol compounds such as sesaminol and sesamolinol.

Moreover, we examined the effects of sesame and sesaminol on senescence in mice using the Senescence Accelerated Mouse (*89*), and showed that long term feeding of sesame seed suppressed the advance to various senescence grades (*90*). In relation to this antiaging effect, an interesting synergistic effect of sesame and its lignans with γ-tocopherol was elucidated. The experiment was conducted by feeding rats the following four kinds of diet for 8 weeks; vitamin E-free control, α-tocopherol-containing diet, γ-tocopherol-containing diet, and sesame seed-containing diet. Using changes in red blood cell hemolysis, plasma pyruvate kinase activity, and peroxides in plasma and liver as indices of vitamin E activity, the sesame seed diet showed high vitamin E activity comparable to the α-tocopherol diet, while this activity was very low in the γ-tocopherol diet. Here it should be noted that sesame seed contains only γ-tocopherol and is negligible in α-tocopherol content. Thus this result indicates that there exist some component(s) in sesame seed that enhances the vitamin E activity of γ-tocopherol to a level equivalent to that of α-tocopherol. It was further demonstrated experimentally that the sesame lignans sesaminol and sesamin also exert this novel synergistic effect (*91*).

It should be noted that food and daily diets comprise very complicated range of many components and their coordinated entities, so physiological activities of food and related materials should be investigated taking this into account.

Physiological Functionalities of Tea

Since ancient times tea has been considered to be effective in keeping body and soul in good condition. Recently, due to the progress in large-scale isolation and purification techniques of tea catechins, the promotion of studies on new physiological functionalities such as the antimutagenicity of tea by Kada *et al.*, and the inauguration of the Japan Tea Science Research Association in 1985, many studies on various biological activities of tea polyphenols have come to be conducted in Japan. The main physiological studies on tea [listed primarily from the Proceedings of the International Symposium on Tea Science at Shizuoka, Japan in August 1991 (*92*)] are as follows:

1. Antioxidative activities of purified tea catechins were shown to be stronger than BHA or tocopherol in a lard autoxidation system in the order (-)-epicatechin < ECg < EGC < EGCg, and have synergistic effects with tocopherol and ascorbic acid (*93*). Antioxidative activities of green and black tea extracts were also observed in lipid peroxidation induced in rat organ and blood plasma (*92*).
2. Antibacterial activities on food-borne, cariogenic and phytopathologic bacteria (*94,95*), and on *Vibrio cholerae* 01 (*92*) have been reported.
3. Inhibitory effect on dental caries is expected with tea catechins based on antibacterial activity against cariogenic bacteria (*96*), and inhibition of glucosyltransferase induced dental plaque formation (*92*).
4. Effects of tea polyphenol on intestinal microflora and metabolites (*97*), and enhancement of proliferation of *Bifidobacteria* by tea extracts (*98*) are expected to be favorable for colon cancer prevention.
5. Tea extract inhibits influenza virus infection (*92*).
6. Antimutagenic and anticarcinogenic activities of tea extracts are raising much interest not only in Japan but also in the USA and China (*92,99,100*). Green tea extracts were shown to exhibit antimutagenic activities against direct-acting mutagens (*101*) and MNNG-induced mutagenesis (*20*), antitumor activity against the growth of Sarcoma 180 and 20-methylcholanthrene-induced carcinogenesis (*102*). Studies on inhibition of azoxymethane-induced colon carcinogenesis in rat by tea polyphenols, inhibition of DMBA-induced carcinogenesis by green tea extracts, and inhibition of HIV-reverse transcriptase and DNA polymerase activity were also reported at the Symposium (*92*).
7. Antipromoter effects by EGCg against DMBA plus teleocidin in mice (*92,102*) and by tea extracts against pancreatic cancer induced by nitrosamines and mouse epidermal JB 6 cells (*92*) have been reported.
8. Antilipidemic effects by tea are expected because of it's reduction of cholesterol absorption and suppression of accumulation of body and liver fat (*92*).
9. Hypotensive activities of tea are expected because of tea's inhibition of α-amylase, reduction of hypertensive enzyme, and suppression of NaCl-induced blood pressure increase (*92*).
10. The inhibitory effects of tea extracts and catechins on platelet aggregation are interesting from the viewpoint of inhibition of thrombosis (*103*).
11. Tea extracts had antiaging effects in Senescence Accelerated Mouse (*92*).

12. Antiallergic effects of tea, an inhibitory effect on halitosis by tea catechins, and other physiological activities of tea have also been reported (*92*).

Dietary Fibers and Oligosaccharides in Japan

Another interesting advance in food science in Japan related to the prevention of cancer by the diet concerns the development of effective dietary fibers and oligo-saccharides to prevent colon cancer through favorable control of fecal flora (intestinal bacteria). As recently recognized, the large intestine and colon have a vast number of intestinal bacteria, and among them, clostridia, represented by *Clostridium perfringens*, is considered to participate in the production of carcino-gens and other toxins, while *Bifidobacteria* suppress growth of *Clostridium* and production of toxic substances (*104*).

Recently, industrial production of various kinds of oligosaccharides, such as fructo-oligosaccharides, isomalto-oligosaccharides, galacto-oligosaccharides, xylo-oligosaccharides and others, has progressed in Japan using microbial enzymes such as transglucosidase. They were first developed as less sweet sugars with low caloric content and/or low dental plaque formation potential. In addition to these intended effects, however, they were shown to act as growth enhancers of *Bifidobacteria* while suppressing growth of *Clostridium* in fecal microflora (*105*).

As a result of the marked progress in research on the functionalities of food in Japan and the discovery of various new physiological activities of food components involving those mentioned here, there has been social activity to promote the designation of certain foods as a new category of food. After careful deliberations at the Government offices concerning food administration, such foods are now being designated as "food for specified health use." Those foods containing dietary fibers and oligosaccharides as well as tea catechins are considered representative and expected to be the first designated as foods of this new classification carrying the approval of the Ministry of Health and Welfare of Japan in the near future.

Literature Cited

1. *J. Health and Welfare Statistics* Ministry of Health and Welfare Statistics Association; Tokyo, 1990.
2. Fujimaki, M. *Reports of Systematic Analysis and Development of Food Functionalities* 1988; pp 1–560.
3. Chiba, H. *Analyses of Body-modulating Functions of Foods* Gakkai Shuppan Center: Tokyo, 1992; pp 1–562.
4. Muto, Y.; Ninomiya, M.; Fujiki, H. *Jap. J. Clin. Oncol.* **1990**, *20*, 219.
5. Kada, T.; Tsuchikawa, K.; Sadaie, Y. *Mutation Res.* **1972**, *16*, 165.
6. Ames, B. N.; McCann, J.; Yamazaki, E. *Mutation Res.* **1975**, *31*, 347.
7. Kee, M. *Nature* **1983**, *303*, 648.
8. Kada, T. *Jap. J. Genet.* **1973**, *48*, 301.
9. Sugimura, T.; Kawachi, K.; Nagao, M.; Yahagi, T.; Seino, Y.; Okamoto, T.; Shoda, K.; Kosuge, T.; Tsuji, K.; Watanabe, K.; Iitaka, Y.; Itai, A. *Proc. Jpn. Acad.* **1977**, *53*, 58.
10. Nagao, M.; Sato, S.;Sugimura, T. In *The Maillard Reaction in Foods and Nutrition*; Waller, G.R.; Feather, M.S., Eds.; ACS Symposium Series No. 215; American Chemical Society: Washington, DC, 1983; p 521.

11. Kada, T.; Morita, K.; Inoue, T. *Mutation Res.* **1978**, *53*, 351.
12. Kada, T.; Inoue, T.; Namiki, M. In *Environmental Mutagenesis, Carcino-genesis, and Plant Biology* Klelowski, E.J. Ed.; Praeger Publishers: N.Y. 1982; Vol. I, pp 133–151.
13. Mirvish, S. S.; Wallcave, L.; Eagen, M.; Shubik, P. *Science* **1972**, *177*, 65.
14. Mergens, W. J.; Chau, J.; Newmark, H. L. In *Nitrosocompounds: Analysis, Formation, and Occurrence* IARC Scientific Publications; Lyon, 1980; No. 30, 259–267.
15. Gray, J. I.; Dugan, L. R. J., Jr. *Food Sci.* **1975**, *40*, 981.
16. Sakashita, Y.; Otsuka, M., Ochanomizu University, personal communication, 1991.
17. Kurechi, H.; Kikukawa, K.; Kato,T. *Chem. Pharm. Bull.* **1979**, *27*, 2442.
18. Kurechi, H.; Kikukawa, K.; Kato, T. *Chem. Pharm. Bull.* **1980**, *28*, 1314.
19. Kato, H.; Lee, I. E.; Chuyen, N. V.; Kim, S. B.; Hayase, F. *Agric. Biol. Chem.* **1987**, *51*, 1333.
20. Jain, A. K.; Shimoi, K.; Nakamura, Y.; Kada, T.; Hara, Y.; Tomita, I. *Mutat. Res.* **1989**, *210*, 1.
21. Oguni, I.; Nasu, K.; Kanaya, S.; Ota, Y.; Yamamoto, S.; Nomura, T. *Jpn J. Nutr.* **1989**, *47*, 93.
22. Kojima, H.; Miwa, N.; Mori, M.; Ohsaki, M.; Konishi, H. *Jpn. J. Food Hygiene* **1989**, *30*, 233.
23. Arimoto, S.; Hayatsu, H. *Mutat. Res.* **1987**, *192*, 253, and **1989**, *213*, 217.
24. Okuda, T.; Mori, K.; Hayatsu, H. *Chem. Pharm. Bull.* **1984**, *32*, 716.
25. Natake, M.; Kanzawa, K.; Mizuno, M.; Ueno, N.; Kobayashi, T.; Danno, G.; Minomoto, S. *Agric. Biol. Chem.* **1989**, *53*, 1423.
26. Inoue, T.; Morita, K.; Kada, T. *Agric. Biol. Chem.* **1981**, *45*, 345.
27. Nishioka, H. In *Antimutagenesis and Anticarcinogenesis Mechanisms*; Shankel, D. M.; Hartman, P. E.; Kada, T.; Hollaender, A., Eds; Basic Life Sciences No. 39; Plenum Press: New York, 1986; pp 143-151.
28. Morita, K.; Nishijima, Y.; Kada, T. *Agric. Biol. Chem.* **1985**, *49*, 925.
29. Kada, T.; Kato, M.; Aikawa, K.; Kiriyama, S. *Mutat. Res.* **1984**, *141*, 149.
30. Sato, T.; Ose, Y.; Nagase, H.; Hayase, K. *Mutat. Res.* **1989**, *176*, 19931.
31. Hayatsu, H.; Inoue, K.; Ohta, H.; Namba, T.; Togawa, K.; Hatatsu, T.; Makita, T.; Wataya, Y. *Mutat. Res.* **1981**, *91*, 437.
32. Ohnishi, Y.; Kinouchi, T.; Tsutsui, H.; Uejima, M.; Nishifuji, T. Proc. 16th Int. Symp. Princess Takamatsu Cancer Res. Fund, 1985 *Diet, Nutrition and Cancer* Hayatsu Ed. Japan Sci. Soc. Press: Tokyo, 1986; pp 107–118.
33. Hayatsu, H.; Arimoto, S.; Negishi, T. *Cancer Lett.* **1980**, *11*, 29.
34. Katoh, Y.; Nemoto, N.; Tanaka, M.; Takayama, S. *Mutat. Res.* **1983**, *121*, 153.
35. Arimito, S.; Ohta, Y.; Namba, T.; Negishi, T.; Hayatsu, H. *Biochem. Biophys. Res. Commun.* **1980**, *92*, 662.
36. Michioka, M.; Kamiyama, S. *J. Cancer Res. Clin. Oncol. (Suppl.)* 1990, *116S*, 76.
37. Sakai, K.; Kose, H.; Nagase, H.; Kito, H.; Sato, T.; Kawai, S.; Mizuno, T. *Jpn J. Sanitory Chem.* **1989**, *35*, 433.
38. Ito, Y.; Ohnishi, S.; Fujie, K. *Mutat. Res.* **1989**, *222*, 253.
39. Inoue, T.; Ohta, Y.; Sadaie, Y.; Kada, T. *Mutat. Res.* **1981**, *91*, 41.

40. Kada, T.; Inoue, T.; Ohta, T.; Shirasu, Y. In *Antimutagenesis and Anticarcinogenesis Mechanisms*; Shankel, D. M.; Hartman, P. E.; Kada, T.; Hollaender, A., Eds; Basic Life Sciences No. 39; Plenum Press: New York, 1986; p 181.
41. Ohta, T.; Watanabe, K.; Moriya, M.; Shirasu, Y.; Kada, T. *Mutat. Res.* **1983**, *107*, 219.
42. Ohta, T.; Watanabe, K.; Moriya, M.; Shirasu, Y.; Kada, T. *Mol. Gen. Genet.* **1983**, *192*, 309.
43. Ohta, T.; Watanabe,M.; Watanabe, K.; Shirasu, Y.; Kada, T. *Fd. Chem. Toxic.* **1986**, *24*, 51.
44. Watanabe, K.; Ohta, T.; Shirasu, Y. *Agric. Biol. Chem.* **1988**, *52*, 1041.
45. Shimoi, K.; Nakamura, Y.; Tomita, I.; Kada, T. *Mutat. Res.* **1985**, *149*, 17.
46. Shimoi, K.; Nakamura, Y.; Tomita, I.; Hara, Y.; Kada, T. *Mutat. Res.* **1986**, *173*, 239.
47. Ohta, T.; Watanabe, M.; Shirasu, Y.; Inoue, T. *Mutat. Res.* **1988**, *201*, 107.
48. Tokuda, H.; Ohigashi, H.; Koshimizu, K.; Ito, Y. *Cancer Lett.* **1986**, *33*, 279.
49. Ohigasji, H.; Tokuda, H.; Koshimizu, K.; Ito, Y. *Cancer Lett.* **1986**, *30*, 143.
50. Koshimizu, K.; Ohigashi, H.; Tokuda, H.; Kondo, A.; Yamaguchi, K. *Cancer Lett.* **1988**, *39*, 247.
51. Ohigashio, H.; Sakai, Y.; Yamaguchi, K.; Umegaki, I.; Koshimizu, K. *Biosci. Biotech. Biochem.* **1992**, *56*, 994.
52. Ames, B. N. In *Antimutagenesis and Anticarcinogenesis Mechanisms*; Shankel, D. M.; Hartman, P. E.; Kada, T.; Hollaender, A., Eds; Basic Life Sciences No. 39; Plenum Press: New York, 1986; p 7–36.
53. Namiki, M.; Osawa, T. In *Antimutagenesis and Anticarcinogenesis Mechanisms*; Shankel, D. M.; Hartman, P. E.; Kada, T.; Hollaender, A., Eds; Basic Life Sciences No. 39; Plenum Press: New York, 1986; pp 131-142.
54. Osawa,T.; Namiki, M.; Kawakishi, S. In *Antimutagenesis and Anticarcinogenesis Mechanisms II*; Kuroda, Y.; Shankel, D. M.; Waters, M. D., Eds; Basic Life Sciences No. 52; Plenum Press: New York, 1990; pp 139–153.
55. Cutler, R. G. In *Free Radicals in Biology*; Proyor, W. A., Ed; Academic Press: New York, 1984; Vol. IV, pp 371–428.
56. Namiki, M. *Critical Reviews in Food Science and Nutrition* **1990**, *29*, 273–300.
57. Ramarathnam, N.; Osawa, T.; Namiki, M.; Tashiro, T. *J. Sci. Food Agric.* **1986**, *37*, 719.
58. Ramarathnam, N.; Osawa, T.; Namiki, M.; Kawakishi, S. *J. Agric. Food Chem.* **1989**, *37*, 316.
59. Namiki, K.; Yamanaka, M.; Namiki, M. *Nippon Shokuhinn Kogyo Gakkai Koenyokoushi* **1990**, 68.
60. Achiwa, Y.; Namiki, K.; Namiki, M. *Nippon Nogeikagaku Kaishi* **1991**, *66(3)*, 181.
61. Ariga, T.; Koshiyama, I.; Fukushima, D. *Agric. Biol. Chem.* **1988**, *52*, 2717.
62. Ikehata, H.; Wakaizumi, H.; Murata, S. *Agric. Biol. Chem.* **1968**, *32*, 740.
63. Ezaki, H.; Nohara, Y.; Onozaki, H.; Osawa, T. *Nippon Shokuhin Kogyo Gakkaishi* **1990**, *37*, 474.
64. Fukuda, Y.; Osawa, T.; Namiki, M. *Nippon Shokuhin Kogyo Gakkaishi* **1985**, *32*, 407.
65. Osawa, T.; Katsuzaki, H.; Hagiwara, Y.; Hagiwara, H.; Shibamoto, T. *J. Agric. Food Chem.* **1992**, *40*, 1135.
66. Osawa, T.; Namiki, M. *J. Agric. Food Chem.* **1985**, *33*, 777.

67. Osawa, T.; Kumazawa, S.; Kawakishi, S. *Agric. Biol. Chem.* **1991**, *55*, 1727.
68. Okuda, T.; Kimura, Y.; Yoshida, T.; Hatano, T.; Okuda, H.; Arichi, S. *Chem.Pharm. Bull.* **1984**, *119*, 751.
69. Osawa, T.; Ide, A.; Su J-D.; Namiki, M. *J. Agric. Food Chem.* **1987**, *35*, 808.
70. Igarashi, K. Takanashi, K.; Makino, M.; Yasui, T. *J. Jap. Soc. Food Sci. Tech.* **1989**, *36*, 852.
71. Inatani, R.; Nakatani, N.; Fuwa, H.; Seto, H. *Agric. Biol. Chem.* **1982**, *46*, 1661.
72. Nakatani, N.; Inatani, R. *Agric. Biol. Chem.* **1984**, *48*, 2081.
73. Nakatani, N.; Inatani, R.; Ohta, H.; Nishioka, A. *Environ. Health Perspect.* **1986**, *67*, 135.
74. Nakatani, N.; Tachibana, Y.; Kikuzaki, H. *Proc. 4th Biennial General Meeting of the Soc. Free Radical Res., Kyoto* **1988**, 453.
75. Toda, S.; Miyase, T.; Arichi, H.; Tanizawa, H. Takino, Y. *Chem. Pharm. Bull.* **1985**, *33*, 1725.
76. Nishibori, S.; Namiki, K. *Nippon Kaseigaku kaishi* **1988**, *39*, 1173.
77. Saklata, K.; Yamamoto, H.; Ishikawa, A.; Yai, A.; Etoh, H.; Ina, K. *Tetrahedron Lett.* **1990**, *31*, 1165.
78. Yamamoto, K.; Sakata, K.; Watanabe, N.; Yagi, A.; Brinen, L.S.; Clardy, J. *Tetrahedron Lett.* **1992**, submitted.
79. Endo, Y.; Usuki, R.; Kaneda, T. *J. Am. Oil Chem. Soc.* **1984**, *61*, 781.
80. Endo, Y.; Usuki, R.; Kaneda, T. *J. Am. Oil Chem. Soc.* **1985**, *62*, 1387.
81. Kariya, K.; Nomoto, K.; Nakamura, T.; Dong, G-S.; Kobayashi, Y.; Namiki, M. In *Active Oxygen, Lipid Peroxides and Antioxidants, Proc. 5th Intern. Cong. on Oxygen Radicals*; Yagi, K., Ed.; Japan Sci. Societies Press: Tokyo, 1992.
82. Namiki, M. *Advances in Food Research* **1988**, *32*, 115–184.
83. Fukuda, Y.; Matsumoto, K.; Yanagida, T.; Koizumi, Y.; Namiki, M. *Nippon Shokuhin Kogyo Gakkai Koenyokoshu* **1992**, 91.
84. Namiki, M.; Kobayashi, T., Eds. *Goma no Kagaku (Science of Sesame)* Asakura Shoten: Tokyo, 1989; 1–41.
85. Fukuda, Y.; Namiki, M. *Nippon Shokuhin Kogyo Gakkaishi* **1988**, *35*, 552.
86. Fukuda, Y.; Osawa, T.; Namiki, M.; Ozaki, T. *Agric. Biol. Chem.* **1985**, *49*, 301.
87. Osawa, T.; Nagata, M.; Namiki, M.; Fukuda, Y. *Agric. Biol. Chem.* **1985**, *49*, 3351.
88. Nagata, M.; Osawa, T.; Namiki, M.; Fukuda, Y.; Ozaki, T. *Agric. Biol. Chem.* **1987**, *51*, 1285.
89. Takeda, H.; Hoshikawa, M.; Takeshita, S.; Irino, M.; Higuchi, K.; Matsushita, T.; Tomoita, Y.; Yasuhira, K.; Hamamoto, H.; Shimizu, K.; Ishii, M.; Yamamuro, T. *Mech. Ageing Dev.* **1981**, *17*, 183.
90. Yamashita, K.; Kawagoe, Y.; Nohara, Y.; Namiki, M.; Osawa, T.; Kawakishi, S. *J. Jpn. Soc. Nutr. Food Sci.* **1990**, *43*, 445.
91. Yamashita, K.; Nohara, Y.; Katayama, K.; Namiki, M. *J. Nutrition* **1992**, *122*, 2440.
92. *Proc. Internat. Symp. on Tea Science (ISTS) Shizuoka, Japan* 1991, 1–785.
93. Matsuzaki, T.; Hara, M. *Nippon Nougeikagaku Kaishi* **1985**, *59*, 129.
94. Hara, Y.; Ishigami, T. *Nippon Shokuhin Kogyo Gakkaishi* **1989**, *36*, 996.
95. Hara, Y.; Watanabe, M.; Sakaguchi, G. *Nippon Shokuhin Kogyo Gakkaishi* **1989**, *36*, 375.

96. Sakanaka, S.; Kim, M.; Taniguchi, G.; Yamamoto, T. *Agric. Biol. Chem.* **1989**, *53*, 2307.

97. Ahn, Y. J.; Kawamura, T.; Kim, M.; Yamamoto, T. Mitsuoka, T. *Agric. Biol. Chem.* **1991**, *55*, 1425.

98. Okubo, T.; Ishihara, N.; Oura, A.; Serit, M.; Kim, M.; Yamamoto, T.; Mitsuoka, T. *Biosci. Biotech. Biochem.* **1992**, *56*, 588.

99. Chen, Z. *Proc. Internat. Seminar on Green Tea* Korean Soc. Fd. Sci. Tech. **1989**, 11–27.

100. Cheng, S-J.; Gao Y-N.; Ho, C-T., Wang, Z-Y. *Proc. Internat. Seminar on Green Tea* Korean Soc. Fd. Sci. Tech. **1989**, 45–50.

101. Okuda, T.; Mori, K.; Hayatsu, H. *Chem. Pharm. Bull.* **1984**, *32*, 3755.

102. Yoshizawa, S.; Horiuchi, T.; Fujiki, H.; Yoshida, T.; Okuda, T.; Sugumura, T. *Phytotherapy Res.* **1987**, *1*, 44.

103. Namiki, K.; Yamanaka, M.; Tateyama, C.; Igarashi, M; Namiki, M. *Nippon Shokuhin Kogyo Gakkaishi* **1991**, *38*, 189.

104. Mitsuoka, T. J. *Industrial Microbiology* **1990**, *6*, 263.

105. Terada, A.; Hara, H.; Kataoka, M.; Mitsuoka, T. *Microbial Ecology in Health and Disease* **1992**, *5*, 43.

RECEIVED May 17, 1993

SULFUR-CONTAINING PHYTOCHEMICALS IN GARLIC AND ONIONS

Chapter 5

Flavorants from Garlic, Onion, and Other Alliums and Their Cancer-Preventive Properties

Eric Block

Department of Chemistry, State University of New York at Albany, Albany, NY 12222

HPLC, cryogenic GC-MS and proton NMR spectroscopy are employed in the analysis of room temperature vacuum distillates and extracts of onion, garlic, and related members of the genus *Allium* [wild garlic, leek, scallion, shallot, elephant (or great-headed) garlic, chive, Chinese chive] using authentic samples of suspected thiosulfinate components to evaluate the methods. Eight or more thiosulfinates (RS(O)SR′) and related organosulfur compounds ("zwiebelanes," *cis*- and *trans*-2,3-dimethyl-5,6-dithiabicyclo[2.1.1]-hexane 5-oxides; (Z,Z)-*d,l*-2,3-dimethyl-1,4-butanedithial S,S'-dioxide) can be separated and identified in each plant extract, several for the first time. Research on the cancer preventative properties of the *Allium* flavorants is also summarized.

Among the earliest of cultivated spices and foods, garlic (*Allium sativum*) and onion (*Allium cepa*) were easily identified by primitive food-seekers by their distinctive smell, now known to be associated with organosulfur compounds. Through the centuries garlic and onion were used in folk medicine for treatment of such varied disorders as dog bites, insect stings, earaches, burns and wounds, baldness, headaches, chest colds, respiratory ailments, asthma, pneumonia, diabetes, cardiovascular disorders, and rheumatism, among others. The popularity of these plants can be attributed to their pungent aroma, strong taste and, in the case of onion, its potent lachrymatory effect. These plants are members of the well known and widely appreciated genus *Allium*, consisting of more than 600 different species, other common members being leek, chive, shallot, and scallion. *Allium* species were valued by early civilizations both as important dietary constituents and as medicinals for the treatment of many disorders, for example as recorded in the Bible, "…[the Jews who fled Egypt to wander the Sinai wilderness for forty years fondly remembered] the fish which we did eat in Egypt so freely, and the pumpkins and melons, and the leeks, onions and garlic," and as documented by discovery of dried garlic cloves and wooden models of onions among the relics found in the burial chambers of the pharaohs, and by references to medicinal uses

0097–6156/94/0546–0084$06.00/0

of these plants in the writings of Aristotle, Hippocrates, Aristophanes, and Pliny the Elder (*1,2*).

The popularity of *Allium* plants as foodstuff and the reputation of garlic and onion as "cure-alls" stimulated scientific investigations, such as the early work by Pasteur in France (ca. 1858) into garlic's antibacterial activity and by Wertheim (1844) and Semmler (1892) in Germany into the composition of distilled garlic and onion oils (mainly diallyl disulfide, $CH_2=CHCH_2SSCH_2CH=CH_2$, **1**, in the former case, and propenyl propyl disulfide in the latter case). With the advent of modern spectroscopic and chromatographic techniques it became possible to determine the molecular basis for the odor, taste and biological activity of the fresh or processed plants. Key discoveries on this subject were made by Cavallito (1944) and Wilkens (1962) in the United States, by Stoll and Seebeck in Switzerland (1948) and by Virtanen in Finland (1960) (*3,4*). Cavallito discovered an unstable, odoriferous liquid substance in extracts of fresh garlic he termed allicin [2-propene-1-sulfino-thioic acid S-2-propenyl ester, **2**, $CH_2=CHCH_2S(O)SCH_2CH=CH_2\equiv AllS(O)SAll$] which possessed antibacterial properties. Stoll and Seebeck identified the immediate precursor of allicin as alliin [(+)-S-allyl-L-cysteine S-oxide, **3**, $CH_2=CHCH_2S(O)CH_2CH(NH_2)COOH$]. Present evidence indicates that allicin **2** is formed by action of alliinase, a C-S lyase enzyme released by cutting or crushing garlic cloves, on the stable precursor alliin **3**. 2-Propenesulfenic acid (**4**, Scheme 1) is an intermediate in this process. Alliin **3** occurs abundantly in garlic, up to 0.76 ± 0.40% of fresh weight. From onion preparations Virtanen isolated three homo-logues of alliin, namely (+)-S-(prop-1-enyl)-L-cysteine S-oxide [**5**, $CH_3CH=CH-S(O)CH_2CH(NH_2)COOH$], (+)-S-methyl-L-cysteine S-oxide [**6**, $CH_3S(O)CH_2CH-(NH_2)COOH$], and (+)-S-propyl-L-cysteine S-oxide [**7**, $CH_3CH_2CH_2S(O)CH_2CH-(NH_2)COOH$]. The structure of the onion lachrymatory factor (LF, **8**, C_3H_6SO) was independently investigated by Wilkens and Virtanen. 1-Propenesulfenic acid S-H tautomer [$CH_3CH=CHS(O)H$] and propanethial S-oxide [$CH_3CH_2CH=S=O$] were suggested by Virtanen and Wilkens, respectively, as the structure of **8**. More recent work by the author has established that LF (**8**) has the structure (Z)-propanethial S-oxide, and that it most likely originates from rearrangement of 1-propenesulfenic acid O-H tautomer **9** (*5*).

Allicin **2** belongs to a class of unstable and reactive organosulfur com-pounds known as thiosulfinates. It is an antimicrobial agent which affects RNA synthesis in microorganisms and lipid (and cholesterol) biosynthesis in mammals (*6*), yeast and higher plants, inhibiting acetyl-CoA synthetase in the latter (*4*). Thio-sulfinates such as **2** can be synthesized by oxidation of the corresponding disulfide (e.g. **1**) with peracids. Since the above initial discoveries, procedures have been sought to detect and accurately quantitate allicin in fresh and processed garlic and to understand how allicin forms and decomposes. These studies have taken on added importance because of the considerable current interest in biological proper-ties of organosulfur compounds from garlic, especially as related to cancer, and cardiovascular and infectious disease.

After the discovery of alliin **3** and related cysteine S-oxides $RS(O)CH_2CH-(NH_2)COOH$ (**5-7**) in *Allium* species, it was recognized that upon cutting the plants, alliinase-induced cleavage of these precursors could give thiosulfinates $R'S(O)SR''$, where R' and R'' represent methyl, *n*-propyl, 1-propenyl and 2-propenyl groups. These thiosulfinates and derived polysulfides were thought to be the principle source of flavor and aroma in *Allium* species and were viewed as important taxonomic markers. Indirect methods were used to identify and quantify the *Allium*

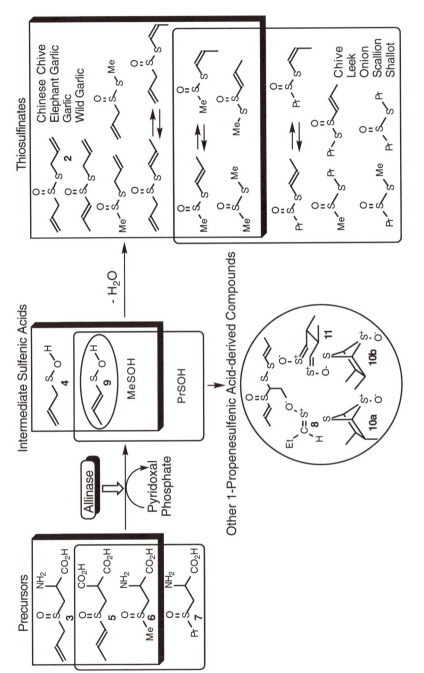

Scheme 1

thiosulfinates, e.g. paper chromatographic separation of cysteine derivatives, $RSSCH_2CH(NH_2)COOH$, of the thiosulfinates (*7a*), or gas chromatographic (GC) analysis of volatile disulfides, $R'SSR''$, that are thought to reflect thiosulfinate composition (*7b*). Paper chromatography suggested the presence of MeS(O)SMe, PrS(O)SPr, AllS(O)SAll, MeS(O)SAll, MeS(O)SPr, and AllS(O)SPr in garlic, onion and Chinese chive extracts, although not all compounds were present in extracts of each plant.

Results

We sought an answer to the question, "What compounds are primarily responsible for the characteristic flavor of freshly cut members of the genus *Allium*?" This information is important in food and flavor chemistry, in chemotaxonomy, in dealing with chemical attraction/repulsion of potential *Allium* insect predators (*8*), and in evaluating possible health benefits of these plants. While GC and GC-MS are of great value in studying compounds of moderate thermal stability, such as those found in distilled oils of garlic and onion, thiosulfinates from *Allium* species are known to decompose on attempted GC analysis (*8*), as shown by the formation of a pair of m/e 144 isomers from allicin **2** (Scheme 2) (*9,10*). The thermal instability of thiosulfinates is associated with the weak S-S bond (bond energy 46 kcal^{-1} mol or less) and the facile pathways available for decomposition (*11*). In view of the complexity of thermal decomposition processes for aliphatic thio-sulfinates, it seemed risky to draw conclusions on *Allium* thiosulfinate composition from analysis of decomposition products found on GC-MS analysis. We therefore undertook a study of the utility of HPLC, "cryogenic" GC-MS, and ^1H NMR methods to directly characterize thiosulfinates in *Allium* extracts (*12,13*). At the same time we sought information on the stability of typical thiosulfinates in various conditions of isolation and chromatography. We were also interested in the chemistry of α,β-unsaturated thiosulfinates of type MeCH=CHS(O)SR, MeCH=CHSS(O)R and MeCH=CHS(O)SCH=CHMe. There are indications that such com-pounds, which have been little studied, show significant biological activity (*14–17*).

Synthesis of Reference Compounds. Reference samples of the various thiosul-finates thought to be present in *Allium* extracts were prepared by oxidation of the corresponding symmetrical or unsymmetrical disulfides $R'SSR''$ or by condensation of sulfinyl chlorides [$R'S(O)Cl$] with thiols [$R''SH$]. Alkyl 1-propenyl or bis(1-propenyl) disulfides were synthesized by stereospecific *cis* or *trans* reduction of 1-propynyl propyl sulfide (conveniently formed by base catalyzed isomerization of 2-propynyl propyl sulfide) followed by lithium-ammonia cleavage of the S-propyl bond, trapping (*E*)- or (*Z*)-1-propenethiolate either with $MeSSO_2Me$ or $PrSSO_2Pr$ forming (*E*)- or (*Z*)-MeCH=CHSSR (R = Me or Pr), or oxidizing the thiolate to bis(1-propenyl) disulfide (Scheme 3). The following thiosulfinates were prepared by these methods: MeS(O)SMe, MeS(O)SPr, MeSS(O)Pr, PrS(O)SPr, MeS(O)-SAll, MeSS(O)All, AllS(O)SAll, (*E/Z*)-MeCH=CHSS(O)Me, (*E/Z*)-MeCH=CHSS-(O)Pr, (*E/Z*)-MeCH=CHSS(O)All, (*E*)-MeCH=CHS(O)SMe, (*E*)-MeCH=CHS(O)-SPr and (*E*)-MeCH=CHS(O)SAll. In the course of these studies it was discovered that (*E*)- or (*Z*)-isomers of MeCH=CHSS(O)R readily interconvert at room temp-erature (see Scheme 4 for proposed mechanism), that monoxidation of (*Z,Z*)- and (*E,Z*)-bis(1-propenyl) disulfide led to *trans*- and *cis*-2,3-dimethyl-5,6-dithiabicyclo-[2.1.1]hexane 5-oxide (**10a,b**, *trans*- and *cis*-zwiebelane) (*18*), respectively, and

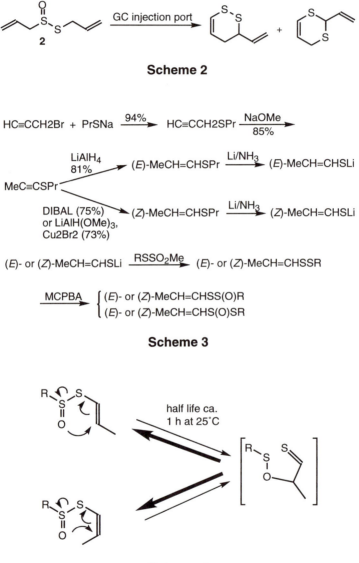

Scheme 2

HC≡CCH2Br + PrSNa $\xrightarrow{94\%}$ HC≡CCH2SPr $\xrightarrow{\text{NaOMe}}_{85\%}$

MeC≡CSPr

$\xrightarrow[81\%]{\text{LiAlH}_4}$ (E)-MeCH=CHSPr $\xrightarrow{\text{Li/NH}_3}$ (E)-MeCH=CHSLi

$\xrightarrow{\substack{\text{DIBAL (75\%)}\\ \text{or LiAlH(OMe)}_3,\\ \text{Cu2Br2 (73\%)}}}$ (Z)-MeCH=CHSPr $\xrightarrow{\text{Li/NH}_3}$ (Z)-MeCH=CHSLi

(E)- or (Z)-MeCH=CHSLi $\xrightarrow{\text{RSSO}_2\text{Me}}$ (E)- or (Z)-MeCH=CHSSR

$\xrightarrow{\text{MCPBA}}$ { (E)- or (Z)-MeCH=CHSS(O)R
(E)- or (Z)-MeCH=CHS(O)SR

Scheme 3

half life ca.
1 h at 25°C

Scheme 4

that bis-oxidation of (*E*,*E*)-bis(1-propenyl) disulfide led to (*Z*,*Z*)-*d*,*l*-2,3-dimethyl-1,4-butanedithial S,S′-dioxide (**11**) (*19*). Compounds **10a,b** and **11** were shown to be present in onion extracts (*18,19*).

Development of Analytical Methods. Both normal phase (Si) and reverse phase (C$_{18}$) HPLC methods were developed to separate isomeric thiosulfinates. The best results were achieved with Si-HPLC with *i*-propanol/hexane gradients, shown in Figure 1 for an onion extract (*12*). We also discovered that separation of most isomeric thiosulfinates could be achieved by GC and GC-MS using wide bore capillary columns (0.53 mm *i.d.*) with cryogenic (0°C) on-column injection and initial column temperature conditions, slow column heating rates (2–5°C/min), and GC-MS transfer line temperatures of 80–100°C, as shown in Figure 2 for the same onion extract as above (*13*).

For HPLC or GC analysis, peeled garlic cloves or onion pieces were homogenized in water (1:10 or 1:1 w/w, respectively) and, after 30 minutes, extracted with methylene chloride and immediately analyzed. A similar procedure was followed with other alliaceous plants [wild garlic (*Allium ursinum*), leek (*Allium porrum* L.), scallion (*Allium fistulosum* L.), shallot (*Allium ascalonicum* auct.), elephant (or great-headed) garlic (*Allium ampeloprasum* L. var. *ampeloprasum* auct.), chive (*Allium schoenoprasum* L.), and Chinese chive (*Allium tuberosum* L.)]. Alternatively, a commercial juicer was used to rapidly extract the juice from onion and other *Allium* species with high water content. The juice was extracted and analyzed as above. Because we sometimes experienced problems in the extraction procedure with severe emulsion formation and found scale-up difficult due to the presence of plant pigments, waxy material and other plant components, we also experimented with vacuum "steam distillation" procedures (*20*). We subjected chopped *Allium* species to high vacuum at room temperature (using an oil bath to prevent the flask contents from freezing), collected the aqueous condensate at liquid nitrogen temperature, and then extracted this distillate with methylene chloride. We found excellent qualitative agreement in thiosulfinate profiles between the samples prepared by distillation and extraction. Furthermore, in the case of the onion distillates, sensory evaluation indicated that these distillates closely reproduce the true flavor of the freshly cut plant. We believe that this method succeeds because of the stabilizing effect of water, through hydrogen bonding (*11*), on the thiosulfinates.

What factors effect the reliability and quantitative reproducibility of *Allium* thiosulfinate analysis by HPLC and "cryogenic" GC-MS methods? Since thiosulfinates are known to be relatively unstable, the effect of varying the elapsed time between homogenization and analysis from one minute to 26 hours was determined for aqueous garlic and onion homogenates maintained at room temperature. In parallel studies using HPLC, "cryogenic" GC-MS, and ^{1}H NMR spectroscopy, it was found that while the total thiosulfinate concentration remained roughly constant during this period, thiosulfinates of type MeCH=CHS(O)SR diminished rapidly and were nearly gone after 6 h, regioisomeric thiosulfinates MeCH=CHSS(O)R were reduced to ca. 50% of their initial value after 26 h, and the initial concentration of MeS(O)SMe was significantly reduced after 6 h. These results explain the failure of previous workers to identify thiosulfinates of type MeCH=CHS(O)SR in onion extracts (*16*). Apart from significant reduction in concentrations of MeCH=CHS(O)SR and a 12% reduction in concentration of MeCH=CHSS(O)R, there was good agreement between the 1 minute and 30 minute

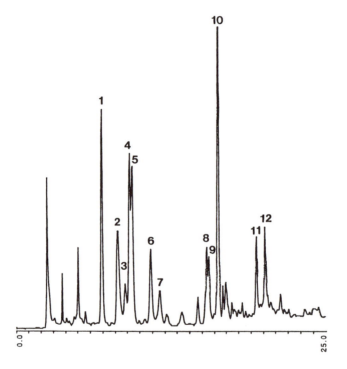

Figure 1. Si-HPLC of onion extract. Peak identification: 1, (*E*)-MeCH=CH-S(O)SPr-*n*; 2, (*Z*)-MeCH=CHSS(O)Pr-*n*; 3, EtCHSO; 4, (*E*)-MeCH=CHSS(O)-Pr-*n*; 5, *n*-PrS(O)SPr-*n*; 6, (*E*)-MeCH=CHS(O)SMe; 7, PhCH₂OH (internal standard); 8, MeSS(O)Pr-*n*; 9, (*Z*)-MeCH=CHSS(O)Me; 10, (*E*)-MeCH=CHS-S(O)Me and MeSS(O)Pr-*n* (overlap); 11, OSCH(CHMe)₂CHSO; 12, MeS(O)S-Me.

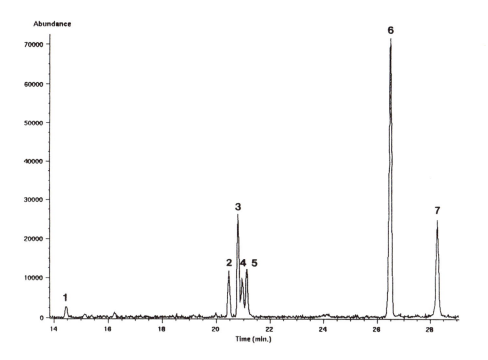

Figure 2. GC-MS total ion chromatograph of same onion extract using a 0.53 mm x 30 m methyl silicone gum capillary column, injector and column programmed from 0°C. Peak identification: 1, MeS(O)SMe; 2, MeS(O)SPr-*n*; 3, (*E,Z*)-MeCH=CHSS(O)Me; 4, MeSS(O)Pr-*n*; 5, (*E*)-MeCH=CHS(O)SMe; 6, (*E,Z*)-MeCH=CHSS(O)Pr-*n*, *trans*-zwiebelane; 7, *cis*-zwiebelane.

samples, supporting our choice of 30 minutes for the analysis protocol. While the "cryogenic" GC-MS methods fail for the garlic thiosulfinates, there was otherwise a good match between the three analytical methods. Other factors affecting our analysis include variation of thiosulfinate content from one part of the plant to another (e.g. leaf, root, rhizome), as others have already noted (*21a*), with plant variety, developmental stage, growing and storage conditions (fertilization, irrigation, humidity, transport). Of course, our analytical methods are of value in determining precisely these differences!

Analytical Results. Table I summarizes the composition of methylene chloride extracts of several common *Allium* species. Some general observations can be made based on this data, together with information from the literature:

1) Gas chromatography and GC-MS, as typically performed with elevated injector, column and GC-MS transfer line temperatures (*12,22*), presents an erroneous picture of the composition of both headspace volatiles and room temperature extracts from *Allium* species. There is no evidence from our HPLC or cryogenic GC-MS analyses of *Allium* extracts for the presence of significant quantities of the polysulfides and thiophenes claimed by prior GC-MS studies, even at concentrations two orders of magnitude lower than that of the thiosulfinates. The answer to the question posed earlier ("What compounds are primarily responsible for the characteristic flavor of freshly cut members of the genus *Allium*?") is "Thiosulfinates and related sulfinyl compounds!"

2) For thiosulfinates RS(O)SCH=CHMe both (*E*)- and (*Z*)-isomers are seen, reflecting the facile room temperature isomerization. However for thiosulfinates RSS(O)CH=CHMe, only the (*E*)-isomer is found in plant extracts. These results are consistent with a [2,3]-sigmatropic rearrangement process possible for thiosulfinates of type RS(O)SCH=CHMe but not those of type RSS(O)CH=CHMe.

3) One way to categorize the different *Allium* spp. is by the types of alkyl groups present in the thiosulfinates. This type of chemotaxonomic information has been used to establish relationships between different members of the *Allium* genus. Only the Chinese chive showed a predominance of *methyl* groups; the methyl group is present in all of the *Allium* species examined. In three of the plants, garlic, elephant garlic, and wild garlic, *allyl* is the major alkyl group. The only other plant containing detectable quantities of allyl is Chinese chive. The *n-propyl* group is the major alkyl group in chive, scallion, shallot, and leek, is present in onion, and is absent in garlic, elephant garlic, wild garlic and Chinese chive. While all of the plants contain 1-propenyl groups, it is the dominant group only in onion. The total percentage of the 1-propenyl group incorporated in onion thiosulfinates, zwiebelanes and bis-sulfine (e.g. 59% for white onions) is somewhat misleading since the majority of the 1-propenyl group generated as 1-propenesulfenic acid winds up as the LF, which is not included in the calculation because most is lost during analysis. For example, in a separate study in which the LF was also isolated it was found that 0.5 mmol of LF was present along with 0.2 mmol total thiosulfinates and zwiebelanes.

4) Garlic grown in warm climates such as Mexico or California shows a typical allyl:methyl:1-propenyl ratio of 80:16:4. Following refrigeration for two months this same garlic shows a ratio of 78:11:11. New York grown garlic (average growing temperature 19–22°C) shows an allyl:methyl:1-propenyl ratio of 95:2:3 while refrigerated New York grown garlic shows a ratio of 89:3:8. Garlic grown in a plains region of India (average growing temperature 30–32°C) shows an allyl:

Table I. Thiosulfinates and Related Compounds from Extracts of *Allium* Species by GC-MS and HPLC (Bracketed) as Mol % of Total

Compound	Garlic Store	Garlic Store (R)	Garlic NY	Elephant Garlic	Wild Garlic	Yellow Onion	Red Onion	Shallot	Scallion	Leek	Chive	Chinese Chive
AllS(O)SAll	[62]	[59]	[89]	[37]	[28]	---	---	---	---	---	---	---
AllS(O)SCH=CHMe-(Z,E)	[5.9]	[18]	[5.3]	[3.7]	[0.9]	---	---	---	---	---	---	---
AllSS(O)CH=CHMe-(E)	[2.1]	[1.6]	[1.6]	[0.6]	---	---	---	---	---	---	---	---
n-PrSS(O)Propenyl-(E)	---	---	---	---	---	(12)[12]	(10)[10]	(14)[14]	(2)[2]	(5)[8]	(3)[3]	---
n-PrS(O)SPropenyl-(Z,E)	---	---	---	---	---	12[10]	29[14]	21[22]	25[17]	5[15]	24[16]	---
n-PrS(O)SPr-n	---	---	---	---	---	9[13]	6[5]	26[27]	35[33]	25[25]	57[58]	---
MeSS(O)Propenyl-(E)	tr	[0.6]	---	tr	---	24[24]	19[26]	8[9]	3[7]	21[12]	1	---
AllS(O)SMe	[8.1]	[7.5]	[1.4]	[17]	[16]	---	---	---	---	---	---	[9]
MeS(O)SPropenyl-(Z,E)	[1.2]	[1.9]	tr	[2.2]	[0.7]	31[25]	27[33]	15[15]	18[22]	29[27]	5[5]	13[5]
MeSS(O)Pr	---	---	---	---	---	1[1]	4[5]	6[2.8]	8[11]	7[5]	4[6]	---
MeS(O)SPr	---	---	---	---	---	1[1]	4[5]	3[1.2]	8[8]	6[5]	7[10]	---
AllSS(O)Me	[18]	[11]	[2.9]	[29]	[34]	---	---	---	---	---	---	[13]
MeS(O)SMe	[2.2]	[0.9]	---	[11]	[20]	10[14]	1[3]	7[9]	1[1]	2[3]	1[2]	74[72]
Zwiebelanes (cis/trans)[a]	---	---	---	---	---	14/6	17/7	17/12	5/3	7.5/3.3	2.3/1.6	---
OSCH(CHMe)2CHSO[a]	---	---	---	---	---	[2]	[11]	[1.8]	---	[3]	---	---
Total % MeS-	16	11	2	35	49	32	21	22	18	30	9	93[d][84]
Total % AllS-	80	78	94	62	50	---	---	---	---	---	---	---
Total % 1-PropenylS-[c]	4	11	3	3	1	50	56	32	30	37	18	7[d][6]
Total % n-PrS-	---	---	---	---	---	18	23	46	52	33	73	---
Total Thiosulfinates[b]	25.6	20.7	14.3	5.2	21	0.35[.4]	0.19[.2]	0.2[.25]	0.1[.08]	0.19[.2]	0.26[.2]	1.7[d][2]

[a] Mol % based upon total thiosulfinate. [b] μmol/g wet weight. [c] For onions, shallot, scallion, leek and chive, based on GC-MS and including zwiebelanes, but not OSCH(CHMe)2CHSO, calcd. as MeCH=CHS(O)SCH=CHMe equiv. [d] GC-MS values exclude allyl methyl thiosulfinates, not detectable by this technique.

methyl:1-propenyl ratio of 74:24:2 while garlic grown in a mountainous region of India (average growing temperature 22–23°C) shows a ratio of 80:18:2. Elephant garlic grown in warm climates, such as Mexico or California, shows an allyl:methyl: 1-propenyl ratio of 61:34:5 while New York grown elephant garlic (average growing temperature 19–22°C) shows a ratio of 67:31:2.

Based on the above data it appears that garlic or elephant garlic grown in a cooler climate (e.g. upstate New York or mountainous regions of India) has a higher allyl:methyl ratio than plants grown in warmer climates. This same trend is seen in the $CH_2=CHCH_2S(O)CH_2CH(NH_2)COOH$ to $CH_3S(O)CH_2CH(NH_2)$-COOH ratio, the former predominating in plants grown in cooler climates. Refrigeration of garlic appears to increase the relative amount of 1-propenyl thiosulfinates, as previously found by Lawson (21b).

Cancer Preventative Properties

Considerable attention has focused on cancer preventative properties of *Allium* species since the 1957 discovery that ethanesulfinothioic acid *S*-ethyl ester, EtS(O)SEt, shows anti-tumor activity (23 24). A fascinating epidemiological study in the People's Republic of China reveals that in a region of China where gastric cancer rates are high, a significant reduction in gastric cancer risk parallels increasing consumption of alliaceous plants. Persons in the highest quartile of intake of these plants experienced only 40% of the risk of those in the lowest (25). Since nitrites, found in processed foods, have been implicated in gastric cancer, it is relevant that gastric juice nitrite concentrations were lower in individuals consuming garlic compared to those who rarely take garlic; the reduction of nitrite concentration in gastric juice after introduction of fresh garlic homogenate was significantly larger than that of controls (26). It is also known that diallyl disulfide inhibits *in vivo* activation of nitrosamines (26). Unsaturated polysulfides found in *Allium* spp. inhibit tumor promotion (24, 27–29) perhaps by enhancing glutathione-dependent detoxification enzymes (30a,b), and also modulate mutagenesis of aflatoxin B1 (30c). Steam distilled onion oil, but not garlic oil, can also function as a weak tumor promoter as well as antipromoter in 7,12-dimethylbenz[*a*]anthracene-initiated mouse skin (31).

Acknowledgments

It is a pleasure to acknowledge the major contributions of my many coworkers and colleagues, whose names are indicated in the references, in our continuing investigation of *Allium* chemistry. I gratefully acknowledge support from the donors of the Petroleum Research Fund, administered by the American Chemical Society, the Herman Frasch Foundation, the National Science Foundation, the NRI Competitive Grants Program/USDA (Award No. 92-37500-8068), Société Nationale Elf Aquitaine, and McCormick & Company Inc. The SUNYA NMR and MS facilities are funded in part by instrument grants from NSF.

Literature Cited

1. Block, E. *Sci. Am.* **1985**, *252*, 114–9.
2. Block, E. In *Folk Medicine: The Art and the Science*; Steiner, R.P., Ed.; American Chemical Society: Washington, DC, 1986; pp 125–37.

3. Virtanen, A. I. *Angew. Chem., Int. Ed. Engl.* **1962**, *1*, 299–306.

4. Block, E. *Angew. Chem., Int. Ed. Engl.* **1992**, *31*, 1135–78.

5. Block, E. *Phosphorus, Sulfur and Silicon* **1991**, *58*, 3–15 and references therein.

6. Sendl, A.; Schliack, M.; Löser, R.; Stanislaus, F.; Wagner, H. *Atherosclerosis* **1992**, *94*, 79–95.

7. a) Fujiwara, M.; Yoshimura, M.; Tsuno, S. *J. Biochem.* **1955**, *42*, 591. b) Saghir, A. R.; Mann, L. K.; Bernhard, R. A.; Jacobsen, J. V. *Proc. Amer. Soc. Hort. Sci.* **1964**, *84*, 386–98.

8. Auger, J.; Lecomte, C.; Thibout, E. *J. Chem. Ecol.* **1989**, *15*, 1847–54; Auger, J.; Lalau-Keraly, F. X.; Belinsky, C. *Chemosphere* **1990**, *21*, 837–43.

9. Brodnitz, M. H.; Pascale, J. V.; Van Derslice, L. *J. Agr. Food Chem.* **1971**, *19*, 273–5.

10. Block, E.; Ahmad, S.; Catalfamo, J.; Jain, M. K.; Apitz-Castro, R. *J. Am. Chem. Soc.*, **1986**, *108*, 7045–55.

11. Block, E.; O'Connor, J. *J. Am. Chem. Soc.* **1974**, *96*, 3921–8, 3929–44.

12. Block, E.; Naganathan, S.; Putman, D.; Zhao, S. H. *J. Agr. Food Chem.* **1992**, *40*, 2418–2430.

13. Block, E.; Putman, D.; Zhao, S. H. *J. Agr. Food Chem.* **1992**, *40*, 2431–8.

14. Block, E.; Zhao, S. H. *J. Org. Chem.* **1992**, *57*, 5815–7.

15. Block, E.; Purcell, P.; Yolen, S. R. *Am. J. Gastroenterol.* **1992**, *87*, 679.

16. Bayer, T.; Breu, W.; Seligmann, O.; Wray, V.; Wagner, H. *Phytochemistry* **1989**, *28*, 2373–7.

17. Wagner, H.; Dorsch, W.; Bayer, T.; Breu, W.; Willer, F. *Prost. Leuk. Essential Fatty Acids* **1990**, *39*, 59–62.

18. Bayer, T.; Wagner, H.; Block, E.; Grisoni, S.; Zhao, S. H.; Neszmelyi, A. *J. Am. Chem. Soc.* **1989**, *111*, 3085–6.

19. Block, E.; Bayer, T. *J. Am. Chem. Soc.* **1990**, *112*, 4584–5.

20. Edwards, N. S. A.; Gillard, R. D.; Groundwater, P. W. *Chem. Ind.* **1991**, 763.

21. a) Mackenzie, I. A.; Ferns, D. A. *Phytochemistry* **1987**, *16*, 763–4. b) Lawson, L. D.; Wood, S. G.; Hughes, B. G. *Planta Med.* **1991**, *57*, 263–70.

22. Sinha, N. K.; Guyer, D. E.; Gage, D. A.; Lira, C. T. *J. Agr. Food Chem.* **1992**, *40*, 842–5.

23. Weisberger, A. S.; Pensky, J. *Science* **1957**, *126*, 1112–5.

24. Lau, B. H. S.; Tadi, P. P.; Tosk, J. M. *Nutrition Research* **1990**, *10*, 937–48.

25. a) You, W. C.; Blot, W. J.; Chang, Y. S.; Ershow, A.; Yang, Z. T.; An, Q.; Henderson, B. E.; Fraumeni, J. F., Jr.; Wang, T. G. *J. Natl. Cancer Inst.* **1989**, *81*, 162–4. b) Blot, W. J. In *Accomplishments in Cancer Research 1987*; Fortner, J. G.; Rhoads, J. E., Eds; J. B. Lippincott: Philadelphia, 1987; pp 231–9. c) Buiatti, E.; Blot, W. *et al. Int. J. Cancer* **1989**, *44*, 611–6.

26. a) Xing, M.; Mei-ling, W.; Hai-xiu, X.; Xi-pu, P.; Chun-yi, G.; Na, H.; Mei-yun, F. *Acta Nutrimenta Sinica* **1982**, *4*, 53–8. b) Yang, C.S. *Cancer Res.* **1991**, *51*, 1509–14.

27. a) Belman, S.; Solomon, J.; Segal, A.; Block, E.; Barany, G. *J. Biochemical Toxicology* **1989**, *4*, 151–60. b) Perchellet, J. P.; Perchellet, E. M.; Belman, S. *Nutr. Cancer* **1990**, *14*, 183–93.

28. a) Wargovich, M. *Carcinogenesis* **1987**, *8*, 487–9. b) Wargovich,M.; Woods, J. C.; Eng, L. V.; Stephens, L. C.; Gray, K. *Cancer Res.* **1988** 6872–5. c) Hayes, M. A.; Goldbert, M. T.; Rushmore, T. H. *Carcinogenesis* **1987**, *8*, 1155–7. d) Brady, J. F.; Ishizaki, H.; Li, D.; Yang, C. S. *Cancer Res.* **1988**, *48*, 5937–40. e) Sumiyoshi, H.; Wargovich, M. *Cancer Res.* **1990**, *50*, 5084–7.

29. Sparnins, V. L.; Barany, G.; Wattenberg, L. W. *Carcinogenesis* **1988**, *9*, 131–4; Wattenberg, L. W. *Proc. Nutr. Soc.* **1990**, *49*, 173–83.
30. a) Gudi, V. A.; Singh, S. V.*Biochem. Pharmacol.* **1991**, *42*, 1261–5; Maurya, A. K.; Singh,S. V.*Cancer Lett.* **1991**, *57*, 121–9. b) Prochaska, H. J.; Santamaria, A. B; Talalay, P. *Proc. Natl. Acad. Sci. USA* **1992**, *89*, 2394–8. c) Tadi, P. P. *Nutr. Cancer* **1991**,*15*, 87–95.
31. Belman, S.; Sellakumar, A.; Bosland, M. C.; Savarese, K.; Estensen, R. D. *Nutr. Cancer* **1990**, *14*, 141–8.

RECEIVED May 17, 1993

Chapter 6

Inhibition of Chemical Toxicity and Carcinogenesis by Diallyl Sulfide and Diallyl Sulfone

J-Y. Hong, M. C. Lin, Zhi Yuan Wang, E-J. Wang, and Chung S. Yang

Laboratory for Cancer Research, College of Pharmacy, Rutgers, The State University of New Jersey, Piscataway, NJ 08855-0789

The effects of diallyl sulfide (DAS) and its metabolite diallyl sulfone (DASO$_2$) on the hepatotoxicity induced by acetaminophen (APAP) as well as on lung tumorigenesis induced by the tobacco-specific carcinogen 4-(methylnitrosamino)-1-(3-pyridyl)-1-butanone (NNK) were studied. APAP at a dose of 0.4 g/kg (for rats) or 0.2 g/kg (for mice) caused a severe hepatotoxicity as manifested by the elevation of serum activities of glutamic-pyruvic transaminase and lactate dehydrogenase, and liver centralobular necrosis. A dose- and time-dependent antidotal effect of oral DASO$_2$ against APAP-induced hepatotoxicity was demonstrated; DAS was slightly less effective. In the carcinogenesis experiments, 100% of female A/J mice treated with a single dose of NNK (100 mg/kg, i.p.) developed lung tumors with an average tumor multiplicity (tumors/mouse) of 7.2. Administration of DAS (200 mg/kg/ day, p.o.) for 3 days prior to NNK treatment decreased the lung tumor incidence to 38% and tumor multiplicity to 0.6. A single dose of DASO$_2$ (100 mg/kg, p.o.) given 2 hr prior to NNK treatment reduced the lung tumor incidence by 50% and tumor multiplicity by 91%. Metabolic activation of NNK was significantly inhibited in the lung and liver microsomes prepared from DAS-treated mice. These results clearly demonstrate that DAS and DASO$_2$ are effective agents against APAP-induced hepatotoxicity and NNK-induced lung tumorigenesis, most probably working by inhibition of the metabolic activation of the related toxicant and carcinogen.

Garlic (*Allium sativum*) has been used widely in culinary practice and as a popular folk medicine for centuries. A major component of fresh garlic is *S*-allylcysteine sulfoxide (alliin), which is odorless and water soluble. The enzyme alliinase converts alliin to allicin, which is unstable and can be further transformed to other garlic compounds including diallyl sulfide (DAS). DAS can also be formed during cooking or after ingestion of garlic. The estimated amount of DAS derived from 1 g

0097–6156/94/0546–0097$06.00/0

of garlic is 30 to 100 µg (*1*). It has been reported that DAS prevents 1,2-dimethyl-hydrazine-induced hepatotoxicity in rats (*2*) and nuclear aberration in mouse colon cells (*3*) as well as inhibits the carcinogenesis induced by 1,2-dimethylhydrazine, benzo[*a*]pyrene, and *N*-nitrosomethylbenzylamine (*4,5*). The mechanisms involved in the chemoprevention of DAS, however, were not clearly understood.

Studies in our laboratory demonstrate that DAS can be metabolized by liver microsomal enzymes to diallyl sulfoxide (DASO) and then to diallyl sulfone (DASO$_2$) (*6*). Our previous work has demonstrated that DAS and its metabolite diallyl sulfone are potent inhibitors of P450 2E1, an enzyme important for the metabolic activation of carbon tetrachloride, *N*-nitrosodimethylamine (NDMA), and acetaminophen (APAP) (*7,8*).

When a single dose of DAS or DASO$_2$ was given to rats, rather selective modulation of hepatic P450 activities was observed. There were no significant changes in the total P450 content and NADPH-P450 reductase activity in the liver microsomes from DAS- or DASO$_2$-treated rats, but the P450 2B1 activity assayed as pentoxyresorufin dealkylation was greatly induced, whereas the P450 2E1 activity assayed as NDMA demethylation was markedly decreased (*7*). The decrease of P450 2E1 activity by DASO$_2$ occurred much more rapidly than by DAS or DASO (*7,8*). Studies *in vitro* with rat liver microsomes revealed that DAS and its metabolites DASO and DASO$_2$ were all competitive inhibitors of P450 2E1-catalyzed reactions. In addition, DASO$_2$ acted as a suicide inhibitor to inactivate P450 2E1 (*6*).

P450 2E1 is a major P450 enzyme constitutively expressed in the liver and is inducible by acetone, ethanol, fasting and diabetes (*9–11*). Many known P450 2E1 substrates are industrial solvents and environmental chemicals, including some toxic and carcinogenic compounds such as CCl$_4$ and NDMA (*12*). P450 2E1-catalyzed metabolic activation is required for CCl$_4$ and NDMA to exert their toxicity and carcinogenicity. Consistent with the inhibitory effects of DAS and DASO$_2$ on P450 2E1, our previous study demonstrated that these two compounds were effective agents against CCl$_4$- or NDMA-induced hepatotoxicity in rats (*8*). In the present communication, we describe the protective effects of DAS and DASO$_2$ against APAP-induced hepatotoxicity and 4-(methylnitrosamino)-1-(3-pyridyl)-1-butanone (NNK)-induced lung tumorigenesis.

Protection Against APAP-induced Hepatotoxicity

APAP is the leading analgesic and antipyretic drug used in the United States. Overdose of APAP is known to cause hepatotoxicity and nephrotoxicity in humans and laboratory animals (*13*). Over 90% of APAP is converted to sulfate and glucuronide conjugates which are subsequently excreted in urine. A small portion of APAP is metabolized by P450 2E1 and 1A2 enzymes to produce *N*-acetyl-*p*-benzoquinone imine (*14,15*), which is either detoxified by the formation of gluta-thione conjugate or arylates the critical cellular proteins to cause toxicity (*16, 17*).

To study the possbile protection against APAP-induced toxicity by DASO$_2$, male Fisher 344 rats (80–90 g) or Swiss Webster mice (8 weeks old) were intragastrically administered APAP suspended in 0.5% tragacanth at a dosage of 0.4 g/kg (for rats) or 0.2 g/kg (for mice). The animals were fasted for 16 hr prior to the APAP administration. DASO$_2$ in distilled water was given orally at different time points after APAP dosing. Twenty-four hr after APAP treatment, the animals were killed and the hepatotoxicity was evaluated by the determination of the serum levels of glutamate-pyruvate transaminase (GPT) and lactate dehydrogenase (LDH) activities as well as the extent of liver necrosis.

APAP caused a significant increase in the levels of serum GPT (5-fold) and LDH (7-fold) activities 24 hr after the treatment. Nearly 75% of the liver was necrotic with a typical centralobular localization.

$DASO_2$ protected the rats against APAP-induced hepatotoxicity in a dose- and time-dependent manner. At a dose of 5 mg/kg given 1 hr after APAP, $DASO_2$ effectively prevented the elevation of serum LDH and GPT activities as well as the development of severe liver necrosis. When rats were given 50 mg/kg $DASO_2$ 1 hr after APAP dosing, a complete protection was observed. The rats had normal serum GPT and LDH levels and the liver morphology was indistinguishable from that of normal rats. When given 3 or 6 hr after APAP, $DASO_2$ exhibited partial protection.

The dose- and time-dependent effects of $DASO_2$ in protection against APAP-induced hepatotoxicity were also demonstrated in mice. When $DASO_2$ (25 mg/kg) was given either immediately or 20 min after APAP dosing, the APAP-caused mouse death, elevation of serum GPT and LDH activities, and liver necrosis were completely prevented. The same dosage of $DASO_2$ showed partial protection when given 1 hr after APAP and was ineffective when given 3 hr after APAP dosing. In both rats and mice, DAS was slightly less effective than $DASO_2$ in protecting against APAP-induced hepatotoxicity under the same dosing conditions.

Inhibition of NNK-induced Lung Tumorigenesis by DAS and $DASO_2$ (*18*)

In addition to the P450 2E1-catalized reactions, our previous work demonstrated that DAS significantly inhibited the metabolism of NNK in rat lung and nasal mucosa (*19,20*). NNK is a tobacco-specific nitrosamine formed by nitrosation of nicotine during the processing of tobacco and cigarette smoking. It is a potent carcinogen in rodents and a likely causative factor in tobacco-related human cancers. P450-catalyzed metabolic activation (α-hydroxylation) is required for NNK to exert its carcinogenic effects. The oxidation of the α-methyl group leads to the formation of formaldehyde and 4-(3-pyridyl)-4-oxo-butyldiazohydroxide, which can alkylate DNA or be converted to 4-hydroxy-1-(3-pyridyl)-1-butanone (keto alcohol). The oxidation of the α-methylene group leads to the formation of 4-oxo-1-(3-pyridyl)-1-butanone (keto aldehyde) and the methylating agent methyldiazohydroxide. In mice, the lung is a major target organ in NNK-induced tumorigenesis (*21*).

To determine the chemopreventive effect of DAS on NNK-induced lung tumorigenesis, female A/J mice were fed AIN-76A diet and were treated at 7 weeks of age with DAS in corn oil for 3 days (200 mg/kg body wt/day, p.o.). Two hours after the final treatment, the mice were either given a single dose of NNK (100 mg/kg body wt, i.p.) and kept for 16 weeks thereafter for the determination of the lung tumor production or killed immediately for the preparation of lung and liver microsomes.

Treatment of female A/J mice in the control group with a single dose of NNK resulted in 100% of mice bearing lung tumors with an average tumor multiplicity of 7.2 ± 1.1. Administration of DAS prior to NNK treatment significantly decreased the lung tumor incidence to 38% and the tumor multiplicity to 0.6 ± 0.2. No difference was observed in animal body weights among the two groups. $DASO_2$, when given at a single dose (100 mg/kg) 2 hr prior to NNK treatment reduced the lung tumor incidence by 50% and tumor multiplicity by 91%. A lower dose of $DASO_2$ (20 mg/kg) caused a 38% reduction in tumor multiplicity but had no effect on the lung tumor incidence.

To study the mechanism by which DAS inhibited the NNK-induced lung tumorigenesis, the effects of DAS on the metabolic activation of NNK in the lung and liver microsomes were examined. Treatment of the female A/J mice with DAS for 3 days (200 mg/kg/day) caused a significant inhibition of NNK metabolism in the lungs and livers. The rates of the formation of the NNK α-hydroxylation products keto aldehyde and keto alcohol were decreased by 70–80% in the lung and liver microsomes. In studies *in vitro*, a dose-dependent inhibition by DAS (0.05–2 mM) on the NNK oxidative metabolism was demonstrated in the mouse lung. This suggests that the decreased metabolic activation of NNK in DAS-treated mice could be partially contributed to the direct inhibition by DAS or its metabolites on the NNK-metabolizing enzyme activity.

Discussion

The present results clearly demonstrate that DAS and its metabolite $DASO_2$ effectively protected the experimental animals from the hepatotoxicity induced by APAP, a known substrate of P450 2E1. DAS and $DASO_2$ were shown to be potent inhibitors of P450 2E1 in our previous studies. Therefore, the protective effect of DAS and $DASO_2$ against the induced hepatotoxicity could be best explained by their ability to inhibit the P450 2E1-catalyzed metabolic activation of APAP. Although the major P450 enzymes responsible for the metabolic activation of NNK remain to be identified, NNK α-hydroxylation was significantly inhibited in the lung and liver microsomes prepared from the DAS-treated A/J mice. This is consistent with the observed remarkable chemopreventive effect of DAS or $DASO_2$ against the NNK-induced lung tumorigenesis.

Mechanisms other than inhibition of the metabolic activation have also been proposed to explain the chemoprotective activity of organosulfur compounds. Trapping of the reactive intermediates could occur by chemical conjugation between the electron-rich sulfur atom and the electrophilic species produced during the metabolic activation (22). Induction of phase II detoxification enzymes such as glutathione *S*-transferase by DAS and other organosulfur compounds have been reported (23). The importance of these mechanisms in protection by DAS or $DASO_2$ against chemical toxicity and carcinogenesis remain to be determined.

A recent epidemiological study indicated that frequent consumption of garlic and other allium vegetables is associated with a lower incidence of stomach cancer (24). The practical application of DAS and other garlic components in the prevention of human cancers needs further study. Individuals exposed to P450 2E1 inducers such as alcohol and isoniazid have elevated P450 2E1 levels and might be more susceptible to the hepatotoxicity induced by protoxicants which are metabolically activated by P450 2E1. $DASO_2$, which is odorless and relatively nontoxic, may be useful in protecting against the induced hepatotoxicity in those individuals.

Acknowledgements

The authors would like to thank Dr. Jinmei Pan and Ms. Dorothy Wong for assistance in the manuscript preparation. This work was supported by NIH Grants CA37037, CA46535, ES03938, ES05022, and Grant 88B18 from the American Institute for Cancer Research.

Literature Cited

1. Yu, T. H.; Wu, C. M.; Liou, Y. C. *J. Agr. Food Chem.* **1989**, *37*, 725–730.
2. Hayes, M. A.; Rushmore, T. H.; Goldberg, M. T. *Carcinogenesis* **1987**, *8*, 1155–1157.
3. Wargovich, M. J.; Goldberg, M. T. *Mutat. Res.* **1985**, *143*, 127–129.
4. Wargovich, M. J. *Carcinogenesis* **1987**, *8*, 487–489.
5. Wargovich, M. J.; Woods, C.; Eng, V. W. S.; Stephens, L. C.; Gray, K. *Cancer Res.* **1988**, *48*, 6872–6875.
6. Brady, J. F.; Ishizaki, H.; Fukuto, J. M.; Lin, M. C.; Fadel, A.; Gapac, J. M.; Yang, C. S. *Chem. Res. Toxicol.* **1991**, *4*, 642–647.
7. Brady, J. F.; Li, D.; Ishizaki, H.; Yang, C. S. *Cancer Res.* **1988**, *48*, 5937–5940.
8. Brady, J. F.; Wang, M.-H.; Hong, J.-Y.; Xiao, F.; Li, Y.; Yoo, J.-S. H.; Ning, S. M.; Fukuto, J. M.; Gapac, J. M.; Yang, C. S. *Toxicol. Appl. Pharmacol.* **1991**, *108*, 342–354.
9. Hong, J.-Y.; Pan, J.; Dong, Z.; Ning, S. M.; Yang, C. S. *Cancer Res.* **1987**, *47*, 5948–5953.
10. Hong, J.-Y.; Pan, J.; Gonzalez, F. J.; Gelboin, H. V.; Yang, C. S. *Biochem. Biophys. Res. Commun.* **1987**, *142*, 1077–1083.
11. Dong, Z.; Hong, J.; Ma, Q.; Li, D.; Bullock, J.; Gonzalez, F. J.; Park, S. S.; Gelboin, H. V.; Yang, C. S. *Arch. Biochem. Biophys.* **1988**, *263*, 29–35.
12. Yang, C. S.; Yoo, J.-S. H.; Ishizaki, H.; Hong, J.-Y. *Drug Metab. Rev.* **1990**, *22*, 147–160.
13. Hinson, J. A. In *Rev. Biochem. Toxicol.* Bend, J. R.; Philpot, R. M. Eds., Elsevier/North-Holland: New York, 1980; pp 103–130.
14. Raucy, J. L.; Lasker, J. M.; Lieber, C. S.; Black, M. *Arch. Biochem. Biophys.* **1989**, *271*, 270–283.
15. Hu, J. J.; Vapawala, M.; Reuhl, K.; Lee, M.-J.; Thomas, P. E.; Yang, C. S. *FASEB J.* **1991**, *5*, A1565.
16. Hinson, J. A.; Monks, T. J.; Hong, M.; Highet, R. J.; Pohl, L. R. *Drug Metab. Dispos.* **1982**, *10*, 47–50.
17. Dahlin, D. C.; Miwa, G. T.; Lu, A. Y. H.; Nelson, S. D. *Proc. Natl. Acad. Sci. USA* **1984**, *81*, 1327–1331.
18. Hong, J.-Y.; Wang, Z.-Y.; Smith, T.; Zhou, S.; Shi, S.; Yang, C. S. *Carcinogenesis* **1992**, *13*, 901–904.
19. Hong, J.-Y.; Smith, T.; Brady, J. F.; Lee, M.; Ma, B.; Li, W.; Ning, S. M.; Yang, C. S. *Proc. Am. Assoc. Cancer Res.* **1990**, *31*, 120.
20. Hong, J.-Y.; Smith, T.; Lee, M.-J.; Li, W.; Ma, B.-L.; Ning, S.-M.; Brady, J. F.; Thomas, P. E.; Yang, C. S. *Cancer Res.* **1991**, *51*, 1509–1514.
21. Hecht, S. S.; Hoffmann, D. *Carcinogenesis* **1988**, *9*, 875–884.
22. Goldberg, M. T. In *Anticarcinogenesis and Radiation Protection* Cerutti, P. A.; Nygaard, O. F.; Simic, M. G. Eds., Plenum: New York, 1987; pp 905–912.
23. Sparnins, V. L.; Barany, G.; Wattenberg, L. W. *Carcinogenesis* **1988**, *9*, 131–134.
24. You, W.-C.; Blot, W. J.; Chang, Y.-S.; Ershow, A.; Yang, Z. T.; An, Q.; Henderson, B. E.; Fraumeni, J. F., Jr.; Wang, T.-G. *J. Natl. Cancer Inst.* **1989**, *81*, 162–164.

RECEIVED September 28, 1993

Chapter 7

Breath Analysis of Garlic-Borne Phytochemicals in Human Subjects

Combined Adsorbent Trapping and Short-Path Thermal Desorption Gas Chromatography–Mass Spectrometry

Reginald Ruiz, Thomas G. Hartman, Karl Karmas, Joseph Lech, and Robert T. Rosen

Center for Advanced Food Technology, Rutgers, The State University of New Jersey, New Brunswick, NJ 08903

Garlic-borne volatile compounds were monitored by breath analysis of human subjects in a time course study following the ingestion of various garlic samples. After specific time intervals, 1 to 1.2 liters of breath were purged and trapped through porous polymer resins and analyzed *via* short path thermal desorption GC-MS methodology. In addition, quantification was achieved by GC with flame ionization detection using an internal standard. Allyl methyl sulfide, diallyl sulfide, diallyl disulfide, *p*-cymene and D-limonene were found consistently in all subjects. Allyl thiol was detected occasionally. The individual appearance and elimination curves for these phytochemicals were found to differ, suggesting that experimental observations related to the pharmacokinetic behavior of the individual compounds. Hydrogen sulfide, a potential breath odor compound, was not efficiently trapped due to its low breakthrough volume in the adsorbent resins. Therefore, this compound was analyzed by direct injection of breath samples using sulfur sensitive flame photometric GC methodology. Preliminary evidence suggests that stomach acid caused increased evolution of this compound during digestion. The procedure described proved to be a useful noninvasive technique for measuring low levels of volatile food-borne phytochemicals on the breath of human subjects.

The history of breath analysis can be traced back as far as 1874 when Anstie discussed alcohol elimination by means of human expired air. The techniques established in the early 1900s for use in respiratory physiology permitted the composition of breath and alveolar air to be determined (*1*). The present use of breath analyses, however, is applied to the medical, food, chemical, and environmental fields. Perhaps its most common application is the breath alcohol test, which is used for medico-legal purposes, as a test for sobriety. In medicine, breath tests are routinely used to measure hydrogen, $^{14}CO_2$, as well as volatile organic compounds to monitor biochemical processes in order to determine intestinal

0097–6156/94/0546–0102$06.00/0

disorders such as malabsorption syndromes and bacterial overgrowth (*2*). In other clinical applications, breath tests are used to monitor malodor and the effectiveness of oral hygiene products (*3*). The analysis of chemical moieties in exhaled human breath as a result of exogenous, environmental, chemical exposure has been studied quite comprehensively. Breath measurements have also been widely employed to evaluate worker exposure to volatile solvents in an occupational setting (*1,4–13*).

Perhaps the common thread through the aforementioned application of breath analysis is the utilization of GC and/or GC-MS methodology. In practice, breath assays fall into two categories: those which require pre-concentration of the sample and those which do not. Breath assays which do not require pre-concentration are generally easier to perform. The former, however, necessitates a greater effort, but offers an unique tool to detect volatile compounds that are too low to be detected in the blood (*2*). The collection, pre-concentration and analysis of breath samples have many advantages. It is a simple, non-invasive method which is a non-traumatic alternative to blood analysis (*14*).

There are a number of sampling devices employed to assay volatile compounds in the breath. Other methods integrate some type of collection device which is integrated into a mass spectrometer (*1,8–10,13,15*). Mass spectrometry is probably the most versatile and useful technique. It overcomes all the major disadvantages associated with collection devices (*14*).

A milestone in the breath analysis was the use of a headspace technique *via* GC-MS. Mackay and Hussen (*16*) wrote a very comprehensive as well as detailed chapter on the headspace techniques in mouth odor analysis. Unfortunately, there is an obvious deficit of published data which addresses the odor causing potential of volatile food-borne chemicals in human breath. There seems to be even fewer sources on the breath analysis for odor components in human breath after the ingestion of garlic (*3, 17–19*).

Haggard and Greenberg (*18*), were among the first pioneers in the breath analysis of alliaceous substances. They concluded that the garlic-borne compounds found on the breath after ingestion did not originate from the blood stream, but instead was an emanation from particles retained by structures in the mouth. This erroneous conclusion could be justified by their analytically inefficient method. Their application of the iodine pentoxide method was less sensitive than the headspace technique employed today. They were obviously not well equipped to observe the excretion kinetics of garlic-borne phytochemicals.

Laasko *et al.* (*20*), were perhaps the first to make the distinction that the composition of sulfur containing volatiles was different when garlic-borne volatiles are generated *in vitro* versus the garlic odor compounds present in garlic breath. By use of headspace technique selective ion monitoring (SIM), they identified two major sulfur components in exhaled air: 2-propene-1-thiol (allyl mercaptan) and diallyl disulfide. Minami *et al.* (*19*), on the other hand, conducted a time course study, to study the effects of fresh garlic on human breath. Their GC with flame photometric detection coupled with GC-MS confirmation methodology, suggests that after garlic ingestion, allyl mercaptan is the major garlic-borne compound, while diallyl disulfide is secondary.

The purpose of this study was to develop a breath analysis protocol for the determination of garlic-borne volatile substances on the breath of human subjects. Because of the numerous garlic products on the market today and advertisement claims of reduction of breath odor, there is need to evaluate the potential of these products to generate flavor and their effect on human breath. Thus, novel methodology is necessary for validating such claims. In addition, non-invasive methods are needed to extrapolate the bloodstream levels of phytochemicals. This study incorporated the analysis of several commercially available garlic products: fresh garlic, dehydrated granular garlic (deodorized and zinc oxide treated), as well as two enteric coated pills and one gelatin encapsulated dehydrated garlic pill.

Materials

Methanol and water solvents were high purity chromatography grade and were obtained from J.T. Baker (Phillipsburgh, NJ). Methyl allyl sulfide, diallyl sulfide, diallyl disulfide, and 2,5-dimethyl thiophene reference standards were purchased from Aldrich (Milwaukee, WI). Tenax and Carbotrap were obtained from Alltech (Deerfield, IL) and Supelco (Bellefonte, PA), respectively. The garlic samples were bought from a local supermarket or donated by a major garlic manufacturing company. The samples consisted of a commercially available deodorized granular garlic, two high flavor generating dehydrated granular garlic powders, fresh garlic, and three garlic pills (one gelatin capsulated and two enteric coated).

Methods

Wet and Dry Garlic Flavor Profiles. Several experiments were conducted to identify garlic-borne volatiles from several commercially available garlic products. By using this approach, the level of pre-existing flavor compounds was established *in vitro*. In addition, the effect of hydration on the alliinase enzyme system to generate garlic-borne flavor compounds was monitored.

In the first assay, one g of a commercially available dry granular garlic was sealed in a Scientific Instrument Services (Ringoes, NJ) solid sample purge and trap apparatus and spiked with 1.0 µl 2,5-dimethyl thiophene internal standard (1.0 µg in 1.0 µl methanol). The purge apparatus was sparged with air at a flow rate of 20 ml per minute at 50°C. The outlet of the purge and trap vessel was fitted with an absorbent trap containing 50 mg Tenax TA (2,6-diphenyl-*P*-phenylene oxide) and 50 mg Carbotrap (activated graphitized carbon). Volatile flavor compounds released into the headspace of the garlic samples were purged with the carrier gas and were quantitatively trapped into an adsorbent tube containing the porous polymer resin. The isolation was carried out for one hour. At the end of the isolation procedure, the adsorbent trap was removed and sealed for further analysis. The apparatus was cleaned and a new trap was fitted to the apparatus along with a new garlic sample and internal standard spike. Three ml water was added to the garlic in the purge and trap cell and mixed into a slurry. The isolation was continued for an additional hour to capture the volatile flavor from the rehydrated garlic. This procedure was repeated for three commercially available granular, deodorized and zinc oxide treated dehydrated garlic samples.

Finally, a third experiment was conducted to measure the rate of total flavor release from the various garlic samples. A 2 g sample of dehydrated granular garlic was added to a 25 ml volume test tube and 20 ml water was introduced. The samples were spiked with 50 ppm 2,5-dimethyl thiophene. The samples were mixed thoroughly and centrifuged at 3000 rpm for ten minutes. An aliquot of supernatant (10 ml) was aspirated and passed through a pre-conditioned solid phase extraction tube packed with 1.5 g Amberlite XAD-2 porous polymer resin. The column was rinsed with 10 ml water to elute sugars and other highly polar species.

After the water rinse, the column was dried for several minutes. Finally, the flavor and aroma compounds were eluted from the column by rinsing with 5 ml methanol. The methanol eluate was concentrated to 1 ml using a gentle stream of nitrogen at room temperature. Samples were ready for analysis.

The adsorbent traps resulting from the isolation procedure were analyzed using a Scientific Instrument Services short path thermal desorption unit which was interfaced into a Varian gas chromatograph with flame ionization detection for quantification [short path thermal desorption gas chromatography with flame ionization detection (SP-TD-GC-FID)]. The tubes were thermally desorbed at 220°C at a rate of 10°C per minute. Flame ionization detection was used and the resulting chromatograms were integrated using a Varian 4290 integrator and a VG Multichrom computer data system. The GC was equipped with a 60 meter x 0.32 μm I.D. J & W Scientific DB-1 capillary column containing a 0.25 μm film thickness. Helium was used as the carrier gas with a flow rate of 1 ml/min which converts to a linear carrier velocity of 20 cm per second. The injector temperature was set at 225°C, and a 10:1 split ratio was used. The program was held at the upper limit for 15 minutes.

Combined SP-TD-GC-mass spectrometry analysis was conducted using a Varian 3400 gas chromatograph directly interfaced to a Finnegan MAT 8230 high resolution magnetic sector mass spectrometer for peak identification and confirmation. All chromatographic conditions were identical to those previously described. The mass spectrometer was operated in an electron ionization mode scanning masses 35–350 continuously at a rate of 1 second per decade. Data was acquired and processed using a Finnegan MAT SS300 data system. Computerized library searches of mass spectra were conducted using the National Bureau of Technology (NIST) mass spectral database. Any mass spectra that were not identified by the computer library were manually searched against the EPA-NIH and Wiley/NBS registries of mass spectra which contain approximately 120,000 entries. Manual interpretation of unknown mass spectra was attempted for those not found in the databases.

Garlic Breath Analysis. A method was developed to effectively apply SP-TD-GC-MS methodology to breath analysis. The first stage in conducting the breath analysis was to designate the target compounds which could be trapped effectively by SP-TD-GC-MS. Upon identification of these phytochemicals, it was necessary to generate a calibration curve for each of the target compounds since internal standard methodology was used. The study was aimed to compare the levels of these specifically targeted garlic-borne phytochemicals from a variety of commercial products: fresh garlic (*Allium sativum*), dehydrated granular garlic products (including deodorized and zinc oxide treatment), as well as a gelatin capsule and two types of enteric coated garlic tablets. Body weights, names, ages and sex of the test subjects were recorded and the doses were calculated.

The dehydrated granular garlic products were evaluated according to the levels of garlic-borne phytochemicals via a time course study. The participant panel was composed of 9 individuals, including men and women whose ages ranged from 21 to 28. These subjects were asked to refrain from eating any garlic products on the testing day. Subjects were fed garlic at a dose of 40 mg per kg of body weight (dry weight basis). The total doses ranged from 40–200 mg per kg of body weight (200 mg/kg for fresh garlic only). On the morning of the test day, subjects were fed the appropriate dose on a plain bagel with cream cheese. Prior to ingesting the garlic however, all subjects were required to give a breath sample which functioned as a control. Breath samples were collected immediately following ingestion and every hour succeeding for a total of 5 sampling times.

The final method of breath collection emerged from 2 stages. Initially, breath samples were collected by instructing the participant to expire through a mouthpiece which was directly attached to a desorption tube. The tube was in turn, connected to a soap bubble meter which indicated when 1 liter of breath had passed through the tube. Unfortunately, this apparatus created a high resistance to expiration. As a result, the breath collection method was modified. Each subject was now instructed to exhale 1.2 liters of breath into a teflon air sampling bag which allowed expired air to be collected without restriction or discomfort. The bag was held in an oven at 50°C for 2 minutes to prevent condensation within the bag. The contents were suctioned through an adsorbent trap containing Tenax TA (100 mg). The trap was purged for 30 minutes with nitrogen gas in order to drive off excess moisture which might have caused freezing and plugging of the GC column. The traps were thermally desorbed directly into the GC and flame ionization detection was utilized for quantification.

GC-MS analyses were performed on several of the breath samples to confirm the structural assignments and GC retention times for the analytes of interest. The instrumentation conditions were identical to those described for the wet and dry flavor profiles. Concentration levels of the individual flavor compounds in all the samples were calculated by dividing the integrated peak area of the internal standard into that obtained for the flavor components using the GC-FID data.

Performance audits were also conducted throughout the whole study to check experimental conditions. Blank GC runs, for example, were implemented periodically in order to insure a clean baseline during each analysis. A test mix which consisted of a known concentration of the target compounds, was also desorbed into the GC in order to reconfirm analyte retention time. In addition, the absorbent traps were also thermally conditioned by passing nitrogen gas through at 40 ml/min in an oven that was temperature programmed from room temperature to 290°C at 6°C/min for four hours.

A second breath analysis was conducted to evaluate three types of garlic pills: two enteric-coated and one gelatin encapsulated dehydrated garlic pill. Garlic-borne volatile compounds were monitored by the breath analysis of five human subjects, in a time course study: 3 males and 2 females. Again, their ages and weights were recorded. During this five week study, each subject was instructed to take each product for a period of five days. On the fifth day, the subject was monitored for garlic-borne phytochemicals via breath analysis. Between testing of each product a period of ten days elapsed in order for the subject to metabolize and excrete any garlic-borne compounds remaining endogenously.

In order to minimize the human variability factor, the subjects were given specific guidelines to follow during each 5 day session of taking each garlic pill by receiving an instruction packet and a supply of pills. The packet provided a calendar which indicated the days they were to report for testing, and which pills they were to be taking. Each subject received a listing of the respective doses. It was important to insure that each subject was ingesting the same allicin production potential on a daily basis for each of the pills tested (enteric coated tablets and the gelcaps). In addition, each participant was required to take their garlic pills after meals to minimize gastric disturbance. The guidelines specifically noted that they should swallow the pills whole and not chew them. This insured that the full dose of garlic was delivered to the gastrointestinal tract to be digested, absorbed and excreted efficiently.

Further, each subject was not permitted to eat any garlic and/or onions for the duration of the study. In fact, a daily diet record was used in order to monitor consumption patterns and the time they took their pills. Each subject was also given a questionnaire in order to evaluate each product. Since this work involved research on human subjects, clearance was obtained from the Rutgers University Review Board (IRB). As a result, each individual was given a consent form to sign.

Each participant was instructed to continue taking their garlic pills after breakfast and lunch of the testing day. These doses maintained the endogenous levels of any garlic-borne phytochemical. In addition, by denoting the times of pill ingestion, the pharmacokinetic behavior of the analytes, with regard to expiration could be monitored over time.

On the testing day, the subjects exhaled 1.2 liters of breath into an air sampling bag, at hourly intervals following ingestion. The garlic-borne compounds were identified and their excretion kinetics were observed. The sample was suctioned through an adsorbent trap containing Tenax TA (100 mg) which had been previously spiked with 250 ng 2,5-dimethylthiophene internal standard for purposes of quantification. The adsorbent traps were thermally desorbed directly into a Varian gas chromatograph with flame ionization detection. SP-TD-GC-MS analysis was also performed on each product and from each subject to confirm structural assignments and GC retention times for the analytes of interest. Again, the instrumentation conditions were identical to those described in the wet and dry flavor profile section.

An additional experiment was conducted in the garlic breath analysis using gas chromatography with flame photometric detection (GC-FPD). Preliminary evidence indicated that H_2S evolution increased during garlic digestion. Because Tenax TA has a relatively low breakthrough volume, it was not an efficient method for H_2S detection. The analysis began by qualitatively generating H_2S gas by reacting Na_2S with HCl in a vial. A 1 ml gas tight syringe was used to inject the headspace gas into the GC, in order to obtain the retention time of the H_2S. The actual breath analysis however, was conducted by giving a 23 year old male 2.27 g dehydrated garlic (40 mg per kg of body weight) on a plain bagel. Again, a teflon air sampling bag was used to capture expired air. Using a Hamilton 1 ml gas tight syringe, direct breath samples were analyzed in a time course study of 5 hours. A Varian 3400 gas chromatograph with a sulfur specific flame photometric detector (GC-FPD) was also utilized in our experiments. The analysis was performed using an 8 inch x 1/8 inch O.D. Chromosil 310 column which was obtained from Supelco (Bellefonte, PA). The helium flow rate was set at 20 ml/min and the column oven was maintained at 50°C. Data was recorded by using a Varian 4400 integrator.

Finally, an experiment was conducted to determine if allicin was detectable via breath analysis. This was necessary to validate that the target compounds were true breath odor components and not thermal degradation artifacts. Again, a subject expired into a teflon air sampling bag after ingestion of garlic. The contents were bubbled through a cryotrap at -80°C which contained 3 ml methanol and diallyl sulfone as an internal standard. The solution was analyzed via HPLC.

Analysis for allicin content were performed using a Varian Model 9010 HPLC with a Varian 9065 diode array UV-Vis detector. Data was acquired and processed using a Varian Star chromatography data system. The HPLC column was a Supelco 25 cm x 4.6 mm I.D. LC-18 with a 5 μm particle size. Injections (25 μl) were made using an isocratic solvent system with a composition of aceto-nitrile:water (30:70). The solvents were ultra high purity HPLC grade certified to have an UV cutoff of 190 nm or below. In addition, the solvents were filtered through a 0.2 μm pore size membrane filter prior to use. The diode array detector was scanned continuously from 190 nm to 310 nm throughout the analysis. The quantitative measurements were made at a wavelength of 195 nm which corresponds to the absorbance maxima for both the allicin and the internal standard. The total run time was approximately 15 minutes. In this assay, the diallyl sulfone internal standard and allicin elute sharply with a relative retention time of 1.39 minutes between peaks.

Results and Discussion

Wet and Dry Garlic Flavor Profiles. The dry flavor profile was designed to evaluate the levels of preformed flavor present in garlic, in various samples prior to hydration. This is a measure of how flavorful the garlic powders are right out of the container. The preformed flavor levels in garlic were compared to those generated upon hydration. The dry sample was a commercially available dehydrated granular garlic. All of the samples showed the same trend. In short, flavor levels increased upon the hydration of dehydrated granular garlic. The dry form contains about 2.2 ppm volatile flavor. Upon hydration, however, the amount of volatile flavor increases approximately to 1100 ppm. This 500 fold increase in volatile flavor production results from the activation of the alliinase enzyme system. In a commercially available dehydrated deodorized garlic, however, we observe only a 100 fold increase in volatile flavor was observed (to approximately 234 ppm). This is due to the fact that the alliinase has been partially deactivated by heat treatment. Although the flavor production is low as compared to fresh or non-heat treated garlic, there is still significant enzyme activity in this product.

It is well known that flavor production in garlic is attributed to allicin (diallyl thiosulfinate), the major thiosulfinate produced by the action of the alliinase system in garlic. The precursor of the allicin is called alliin [(+)-S-2-propenyl-L-cysteine sulfoxide)]. In dehydrated garlic, the enzyme system is stabilized due to the restricted water activity and the abundance of the allicin precursor. Upon hydration (or membrane rupture in fresh garlic), however, the alliinase system is activated and rapidly begins to catalyze the conversion of alliin to allicin. These thiosulfinates are the true primary flavor components of garlic but are chemically unstable and rapidly decompose into a plethora of secondary allylic sulfur containing compounds. The primary compounds detected in the rehydrated garlic include: diallyl disulfide, diallyl sulfide, methyl allyl sulfide, methyl allyl disulfide, dimethyl disulfide, dimethyl trisulfide and methyl allyl trisulfide.

The compounds found in the various samples tested reconfirm the compounds which were cited in the literature. Quantification of total flavor compounds was accomplished using 2,5-dimethyl thiophene as a surrogate internal standard. This compound is not found naturally in fresh garlic flavor but it contains chemical functionality similar to many garlic flavor compounds (sulfur containing heterocycle) so that its extraction efficiency is similar to the naturally occurring garlic flavor compounds. This compound was spiked into the garlic during the flavor analysis at a level of approximately 100 ppm based on the solids content of the various products tested. It was also used to quantify the residual garlic compounds in the breath analysis study. The recovery of the internal standard was determined to be quantitative and highly reproducible throughout all of the experiments, thereby validating the methodology. The data treatment assumes a response factor of 1.0 thus simplifying the calculations and providing for semi-quantitative estimates of concentrations. This is a reasonably accurate method of quantitation because analytical reference standards for all of the hundreds of flavor compounds observed are not readily available for purposes of calculating individual response factors.

The compounds detected in the garlic samples are all characteristic of garlic powders and are natural products (*20*) with the exception of butylated hydroxytoluene (BHT) which was found in products at trace levels. The samples contained primarily allylic sulfur compounds and terpenes. Although not emphasized in current literature, these results indicate that terpenoid compounds were also found to be constituents in garlic flavor. These included *p*-cymene and D-limonene. On the other hand, different levels of allylic sulfur compounds were also found to vary across different families of garlic. Allyl thiol, for example, was sometimes found to be a garlic-borne constituent. The treated samples were all found to contain additional sulfur containing compounds. Many of these components are simple heat abuse by-products of the normal sulfur compounds present in the untreated sample and are commonly found in garlic oils and many other garlic preparations (Tables I and II).

Table I. Compounds Found in a Commercially Available Dehydrated Granular Garlic

Acetic acid	*p*-Cymene
Allyl methyl sulfide	Diallyl disulfide
Allyl methyl trisulfide	Diallyl sulfide
Allyl tetrathiol	Dithiacyclohexane
Butylated hydroxytoluene	2-Fluoronapthalene
Butylene glycol	D-Limonene

Table II. Compounds Found in a Commercially Available Dehydrated Granular Garlic Treated with Zinc Oxide

Acetic acid	Benzothiophene	several Dimethylthiophene
Allyl benzene	Butylated hydroxytoluene	isomers
Allyl methyl disulfide	Butylene glycol	Dimethyl trisulfide
Allyl methyl sulfide	Caryophyllene	Dithiacyclohexane
Allyl methyl trisulfide	α-Copaene	2-Fluoronapthalene
Allyl propyl disulfide	*p*-Cymene	Glycerol diacetate
Allyl tetrathiol	Diallyl disulfide	Glycerol monoacetate
Allyl trithiol	Diallyl sulfide	D-Limonene
Anethole	Diallyl tetrasulfide	D- and 1-Menthone
Benzofuran	2,5-Di-*t*-butyl quinone	Pulegone
	Dimethyl tetrasulfide	1,3,5-Trithiane

Breath Analysis Of Fresh And Dehydrated Granular Garlic. During the design phase of the breath analysis, the results of past experiments indicated the major components shown to be predominant in "garlic breath." Four compounds were found to be present in the breath analysis following ingestion of various garlic products via SP-TD-GC-MS. These compounds include methyl allyl sulfide, diallyl disulfide, diallyl sulfide and allyl thiol. Other sulfur compounds were also sought via GC-MS analysis but were not found in appreciable amounts. Figure 1 represents the mass chromatograms for ions characteristic of garlic-borne phytochemicals on the breath at 3 time intervals after fresh garlic ingestion — 15 min, 1 hr, and 1.5 hours. Although current literature places an emphasis on the sulfur compounds associated with garlic, these results indicate the presence of 2 terpenes, D-limonene and p-cymene.

Figure 1 typifies the results found across all the samples tested. A closer examination of the mass chromatograms clearly illustrates the pharmacokinetic behavior of the sulfur and terpenoid species. As denoted by the relative abundance on the ordinate, the sulfur containing compounds initially appear at high levels and consistently decrease with time. The terpenes, on the other hand, appear at initially low levels and increase with time.

Because surrogate internal standard methodology was used, it was necessary to generate calibration curves in order to calculate the concentration of these phytochemicals. An example is illustrated in Figure 2. Linear calibration curves for each of the target compounds were generated by plotting the ratio of each specific analyte's peak area to the peak area of 2,5-dimethyl thiophene. The line equations derived from these calibration curves were used to calculate the breath concentrations for the individual analytes. The calibration curves are all linear and cover a concentration range that spans three to four orders of magnitude. The correlation coefficients derived from linear regression analysis of the curves were 1.00 in all cases. The dynamic range of these calibration curves were sufficient to encompass all of the data points encountered in the breath analysis experiments.

Quantification of the target compounds was achieved by utilizing these curves. In addition, the elimination curves were generated by plotting these values (Figure 3). The individual elimination curves for the garlic-borne compounds (in all garlic samples) were biphasic in appearance. Levels of methyl allyl sulfide, allyl thiol, diallyl sulfide, and diallyl disulfide were originally high but were found to drop precipitously within a relatively short period of time (1–3 hours). The slope of the elimination curves changes dramatically by leveling off and remaining constant throughout the remainder of the sampling interval. It is believed that the initial high levels and rapid decline of the allylic sulfur compounds are related to their elimination from the oral and pharangeal regions where they are present as residues immediately after the ingestion of garlic. The leveling off portion, however, most likely represents bloodstream elimination of these compounds via gas exchange in the lungs. The slope change of the clearance curve certainly coincides with the time expected for digestion and bloodstream absorption to occur.

As mentioned previously, the terpenoid species were found to increase on the breath with time (Figure 4). The breath concentrations of D-limonene and p-cymene were initially low or nonexistent, but between 1 and 4 hours after ingestion, the levels of the two terpenes rose dramatically. Perhaps this behavior reflects the digestion time necessary to release these species from binding to the food matrix. The increase in slope coincides with the time for absorption and excretion to occur via gas exchange in the lungs.

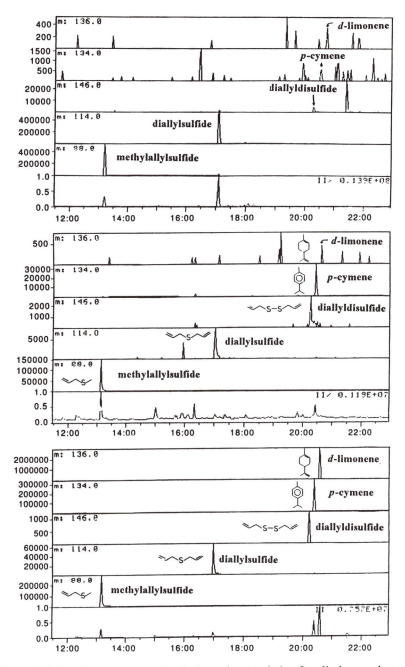

Figure 1. Mass chromatograms for the ions characteristic of garlic-borne phytochemicals on the breath after ingestion of freshly minced raw garlic. Top, 15 min after garlic ingestion; middle, 1 hr after; bottom, 1.5 hr after.

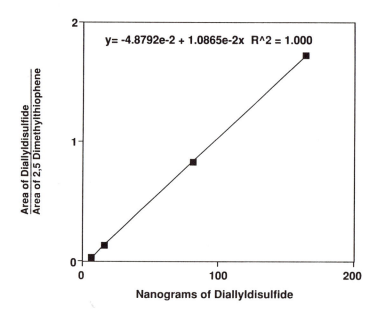

Figure 2. Example of a calibration curve (diallyl sulfide).

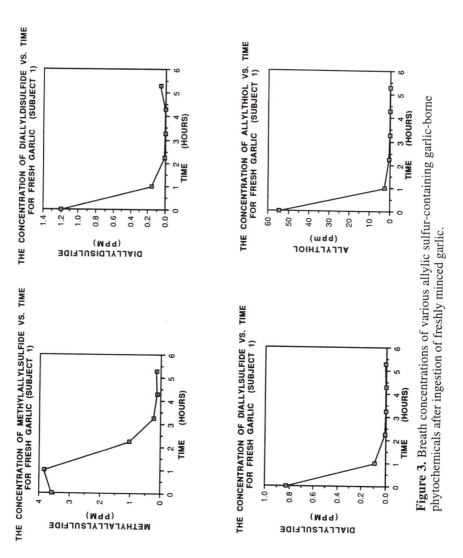

Figure 3. Breath concentrations of various allylic sulfur-containing garlic-borne phytochemicals after ingestion of freshly minced garlic.

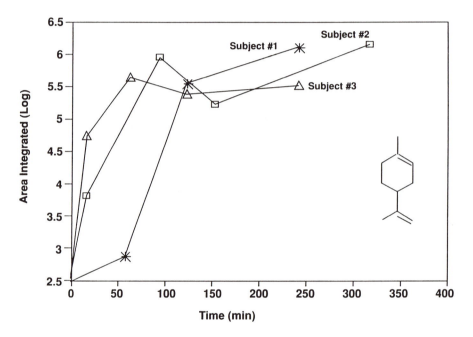

Figure 4. Elimination curve for D-limonene for three subjects.

Although this study incorporated many products and subjects, the results are not surprising. The biphasic character of the elimination curves were found to be consistent for each of the target compounds (Figure 5). The lowest trace was deodorized garlic, while the highest line corresponds to a Chinese garlic with relatively high levels of allicin. The middle range, however, is represented by granular garlic with intermediate levels of allicin.

Lastly, the experiments to detect allicin on the breath via HPLC, revealed that this compound was not present in expired air. Diallyl disulfide, however, was found in appreciable amounts. These results indicate that the compounds detected via GC methodology are true breath odor components and are not thermal degradation artifacts of the primary thiosulfinates. This validates the point that these phytochemicals are actually present on the breath.

Garlic Pill Breath Analysis. In this experiment only methyl allyl sulfide and diallyl disulfide were observed. Detection of methyl allyl sulfide is plotted in figure 6. This figure is representative of the behavior of methyl allyl sulfide that was found across the samples. The concentration of analyte is plotted on the ordinate while the abscissa denotes time. Time zero reflects the time in which the subject ingested the garlic pill(s) after breakfast. Metabolism of the target chemicals were monitored at hourly intervals. In each case, five data points were collected in a one day session. In addition, each line has an arrow to indicate the time in which the subjects took their lunchtime dose. The data was analyzed for the excretion kinetics of these phytochemicals.

In all three products tested, methylallyl sulfide was the only compound to demonstrate a true pattern during the laboratory session. It is believed that the initial increase in slope coincides to garlic-borne chemical excretion via gas exchange in the lungs. The second slope increase may reflect the same response to the lunchtime dose. The granular garlics which were previously tested demonstrated a slightly different pattern. Those curves were biphasic since the slope change for the pills coincided with the time expected for bloodstream absorption and then excretion to occur. The enteric coated B product best demonstrated this pattern. There was no evidence of a high initial level, as seen with the garlic powders, because the pills were swallowed whole. The slight increase in slope may reflect absorption and excretion via gas exchange in the lungs as a result of the lunchtime dose. Both the gelcap garlic and the enteric-coated A product demonstrate similar behavior for methyl allyl sulfide. Diallyl disulfide was detected in such low concentrations for all products that no observable trend could be noted.

In all three garlic pills tested, only trace levels of methyl allyl sulfide and diallyl disulfide were detected. These amounts were negligible. Consequently, these products could be classified as deodorized garlic products because these results were found to be 15 to 20 times lower than the previous breath analyses done with the granular powdered garlic.

A supplementary questionnaire was also distributed to all the subjects in order to evaluate the pills. Most participants experienced "garlic burp." When this occurred, they noted that other individuals noticed their garlic-breath odor. These reactions can simply be explained by the early degradation of the gelcaps in the stomach. Since gastric juice contains gelatinase, the coating is dissolved and garlic powder is free to react with the acidic environment. In addition, gastric disturbance can be expected since volatile gas is generated. Taking these capsules after meals did not seem to totally inhibit the "bounce back" effect of the pill.

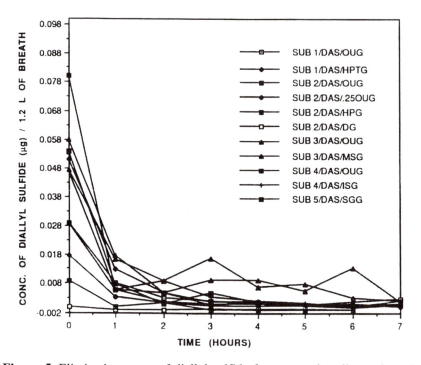

Figure 5. Elimination curve of diallyl sulfide from several garlic products in several subjects. OUG=original untreated garlic, HPTG=high potency treated garlic, HPG=high potency garlic, MSG=medium strength garlic, ISG=intermediate strength garlic, DG=deodorized garlic, SGG=Sunspiced granular garlic.

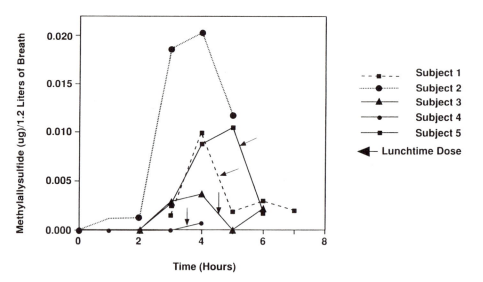

Figure 6. Elimination curve of methylallyl sulfide from an enteric coated garlic tablet.

Although their were some reports of gastric disturbance, they tended to be less critical. It would seem that the enteric coat diminished the incidence of garlic burp because the protective coating was dissolved further down the gastrointestinal tract. The coating was susceptible to the less acidic environment of the duodenum (pH = 6.5). Absorption occurs more readily in the small intestine than the stomach. This may have prevented the "bounce back" effect of any garlic phytochemical. Participants seemed to have preferred a garlic tablet which had a small size, sweet coating and low dose frequency.

Because the protective coating enables both gelcaps and the enteric coated tablets to travel along the gastrointestinal tract, thus minimizing any garlic after-taste, exposure of the dry garlic to the harsh environments may inhibit thiosulfinate production altogether. The water supply in the gastrointestinal tract, is clearly enough to rehydrate the alliinase enzyme system. Due to the presence endogenous metabolites and extreme pH, however, coupled with the protective coating on all of the garlic pills, it is not surprising that low levels of these target compounds were obtained. It can be assumed the presence of these secondary sulfur compounds that were detected on the breath were the result of pre-existing garlic flavor present in the pill prior to ingestion or from a low level production of thiosulfinates *in vivo*.

Thus, these thiosulfinates may not be generated fully as seen *in vitro*. This would obviously challenge any health implications of the product. These allicin degradation compounds could be monitored by dropping these products in an external source of gastric and intestinal fluid to quantify actual flavor generation. This would clearly differentiate between the potential allicin production as opposed to the allicin what is truly formed *in vivo*.

Lastly, in the experiments with hydrogen sulfide detection using GC-FPD, there appears to be a sudden increase then decrease in concentration with the untreated, dehydrated granular garlic (Figure 7). Although hydrogen sulfide is not found in significant concentrations in garlic flavor, it may be produced in the stomach at relatively high levels by the action of gastric acids upon the sulfur compounds in garlic during the digestion process. The zinc oxide ingredient may dissociate at stomach pH and exchange with the chloride ion to form zinc chloride *in vivo*. It is well known that zinc chloride will react with hydrogen sulfide to form zinc sulfide which is a white, nonvolatile precipitate. Hydrogen sulfide has a characteristic rotten egg odor and has a very low sensory threshold. It is likely that its evolution in the stomach will cause gastric disturbance, burping and may be a major contributor to breath odor. In short, the zinc oxide treatment inhibited the release of hydrogen sulfide gas. Tests are currently being conducted to validate this hypothesis.

Conclusion

The study was designed to establish the level of pre-existing flavor imparted by various garlic products when it was dry and rehydrated. In addition, several garlic-borne phytochemicals were identified and found to cause "garlic breath." A further step was taken to observe their pharmacokinetic behavior and excretion kinetics of garlic via breath analysis. This method is quite dynamic because it can be used to evaluate many garlic products on the market in relation to their flavor and breath odor-causing potential.

In short, breath analysis is very useful as a biological monitoring technique to estimate the uptake and excretion kinetics of garlic-borne compounds. This

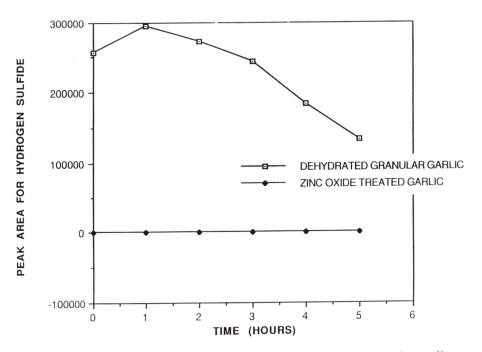

Figure 7. Elimination curve of hydrogen sulfide from two granular garlic products.

breath test is simple, non-traumatic, easily repeated. It is a non-invasive alternative to blood analysis. GC-MS is probably the most versatile and useful technique. Our application of SP-TD-GC-MS has refined the technique of breath analysis.

Acknowledgements

We acknowledge the Center for Advanced Food Technology (CAFT) Mass Spectrometry Lab facility for providing instrumentation support for this project. CAFT is an initiative of the New Jersey Commission of Science and Technology. NJAES publication # F-10569-1-92.

Literature Cited

1. Wallace, L. *Toxicol. Environ.Chem.* **1986**, *12*, 215–236
2. Phillips, M.; Greenberg, J. *J. Chromat.* **1991**, *14*,337–344
3. Tonzetich, J., Ng, S.*Oral Surgery***1976**, *42*,172–181
4. Benoit, F.; Davidson, W.; Lovett, A.; Sabatino, N.; Ngo, A. *International Arch. Occup. and Environ. Health***1985**, *55*,113–20
5. Brugnone, F.; Perbellini, L.; Faccini, G.; Pasini, F.; Bartolucci, G.; DeRosa, E. *International Arch. of Occupa Environ. Health* **1986**, *58*, 105–112
6. Brugnone, F.; DeRosa, E.; Perbellini, L.; Bartolucci, G. *Brit. J. Industrial Med.* **1986**, *43*, 56–61
7. Brugnone, F.; Perbellini, L.; Faccini, G.; Pasini, F.; Danzi, B.; Maranelli, G.; Romeo, L.; Gobbi, M.; Zedde, A. *Amer. J. Industrial Med.* **1989**, *16*, 385–399
8. Gargas, M. *J. Amer. Coll. Toxicol.***1990**, *9*, 447–453
9. Glaser, R.; Arnold, J. *Amer. Industrial Hygien. Assoc. J.* **1989**, *50*, 112–121
10. Gordan, S.; Wallace, L.; Pellizzari, E.; O'Neill, H. *Atmosph. Environ.* **1988**, *22*, 2165–2170
11. Jones, A.; Skagerberg, S.; Yonekura, T.; Sato, A. *Pharmacol. Toxicol.* **1990**, *66*, 62–65
12. Morgan, M.; Phillips, G.; Kirkpatrick, E. In *Biological Monitoring for Pesticide Exposure, Measurement, Estimation and Risk Reduction.* Wang, R.; Franklin, C.; Honeycutt, R.; Reinert, J., Eds.; ACS Symposium Series No. 382; American Chemical Society: Washington, DC, 1988; pp 56–69.
13. Raymer, J.; Thomas, K.; Cooper, D.; Whitaker, D.; Pellizzari, E. *J. Anal. Toxicol.* **1990**, *14*, 337–344
14. Wilson, K. *Scand. J. Work. Environ. Health* **1986**,*12*, 174–192
15. Thomas, K.; Pellizzari, E.; Cooper, S. *J. Anal. Toxicol.* **1991**, *15*, 54–59.
16. Mackay, D.; Hussein, M. In *Analysis of Foods and Beverages, Headspace Techniques.* Charalambous, G., Ed.; Academic Press: New York, 1978; pp 283–357.
17. Hartman, T.; Lech, J.; Rosen, R. *Proceedings of the 39th ASMS Conference on Mass Spectrometry and Allied Topics* **1991**, 641.
18. Haggard, H.; Greenberg, L. *J. Am. Med. Assoc.* **1935**, *104*, 2160–2163
19. Minami, T.; Boku, T.; Katsuhiro, I.; Masanori, M.; Okazaki, Y. *J. Food Science* **1989**, *54*, 763–765
20. Laasko, I.; Seppanen-Laasko, T.; Hiltunen, R.; Muller, B.; Jansen, H.; Knobloch, K. *Planta Medica* **1988** *55*, 257–261.
21. Vermin, G.; Metzger, J.; Fraisse, D.; Scharff, C. *Planta Medica* **1986**, *52*, 96–101.

RECEIVED October 22, 1993

Chapter 8

Sulfur Chemistry of Onions and Inhibitory Factors of the Arachidonic Acid Cascade

Shunro Kawakishi and Y. Morimitsu[1]

Department of Food Science and Technology, Nagoya University, Chikusa, Nagoya 464–01, Japan

Many volatile sulfur compounds and a lachrymatory factor are generated from S-alk(en)yl-L-cysteine sulfoxides by the action of cysteine sulfoxide lyase (CS-lyase) in onions that are cut or crushed. Ten kinds of α-sulfinyl disulfides (AC series) that were human platelet aggregation inhibitors have been isolated and their chemical structures and activities were determined. These products strongly inhibit prostaglandin endoperoxide synthase (PGH synthase) of the arachidonic acid cascade in platelets. Moreover, they have also exhibited a strong inhibitory effect on human 5-lipoxygenase which is concerned with the biosynthesis of leukotriene from arachidonic acid in leukocytes. The relationship between the structures of AC series compounds and their dual activities were studied in detail.

Allium species such as garlic, onion, rakkyo, leek, chive, etc., generate specific sulfur-containing flavors following the breakdown of their plant cells. This flavor generation is due to the formation of volatile sulfur compounds, alk(en)yl disulfides, from S-alk(en)yl-L-cysteine sulfoxides by CS-lyase (EC 4.4.1.4) (*1*). Four kinds of alkyl groups in disulfides specify for their odors in each Allium species. Among these species, garlic (*A. sativum* L.) and onion (*A. cepa* L.) have been widely investigated for their therapeutic effects on vascular diseases such as thrombosis, atherosclerosis, hyperlipidemia and rheumatic arthritis. Furthermore, the extracts and the essential oil of garlic and onion are well known to strongly inhibit human platelet aggregation *in vitro* (*2-4*).

From garlic, (*E,Z*)-ajoene (4,5,9-trithiadodeca-1,6,11-trien 9-*S*-oxide), 2-vinyl-4*H*-1,3-dithiin, diallyltrisulfide (*5,6*) and methyl allyl trisulfide (*7*) have been identified as inhibitors of platelet aggregation. These compounds markedly suppress cyclooxygenase activity of PGH synthase in the arachidonic acid cascade of platelets and 5-lipoxygenase (EC 1.13.11.12) in leukocytes (*8–11*). On the other

---

[1]Current address: School of Food and Nutritional Sciences, University of Shizuoka, Yada 52–1, Shizuoka 422, Japan

0097–6156/94/0546–0120$06.00/0

hand, adenosine (*12*) and alliin (*13*) have been isolated as platelet aggregation inhibitors from onion. Even though the oily extracts of onion also exhibit a strong antiplatelet action, the active compounds were not characterized until now.

Recently, we have isolated and identified several antiplatelet compounds, α-sulfinyl disulfides (named AC series), from the methanol extracts of crushed onion (*14,15*) and their inhibitory action of cyclooxygenase in platelets and human 5-lipoxygenase have been revealed. Moreover, the active AC series were formed through a lachrymatory factor, thiopropanal *S*-oxide, derived from *S*-1-propenyl-L-cysteine sulfoxide by CS-lyase in crushed onion (*16*). In this review paper, we will describe the characterization of AC series antiplatelet factors, their inhibitory activities against platelet aggregation, cyclooxygenase in arachidonic acid cascade of platelets and human 5-lipoxygenase, and their structure-activity relationship.

Formation of Flavors and Lachrymatory Factor in Onion

Four kinds of *S*-alk(en)yl-L-cysteine sulfoxides exist in all *Allium* species and their degradation by CS-lyase generates a characteristic flavor component, dialk(en)yl disulfide (*1*). Their alk(en)yl groups, methyl, *n*-propyl, 1-propenyl and 2-propenyl-(allyl) are present in natural parent compounds. Onion contains three kinds of cysteine sulfoxides having methyl, propyl and 1-propenyl groups. Among them, 1-propenyl is a major group, but the propyl group mainly contributes to the formation of onion flavor. Thus, while the major flavor of onion, dipropyl disulfide, is formed from *S*-*n*-propyl-L-cysteine sulfoxide through *n*-propanesulfenic acid and propyl propanethiosulfinate, the main sulfoxide in onion, *S*-1-propenyl-L-cysteine sulfoxide, is degraded to form a lachrymatory factor, thiopropanal *S*-oxide, by the isomerization of 1-propenesulfenic acid (*17*). As shown in Figure 1, sulfenic acid and thiopropanal *S*-oxide are rapidly decomposed by desulfuring to propanal which is transformed to 2-methyl-2-pentenal by Aldol condensation and dehydration. When onion tissues are homogenized with water, however, the lachrymatory factor is stabilized in tissue suspension (*18*). This phenomenon is closely related to the formation of antiplatelet factors from the lachrymatory compound as described below.

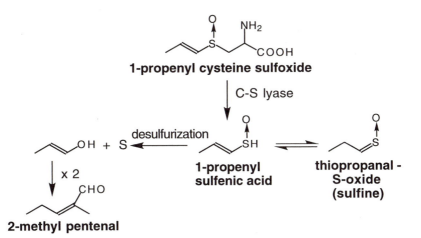

Figure 1. Formation and degradation of lachrymatory factor in onion.

Characterization of Antiplatelet Factors and Their Inhibitory Effects for Human Platelet Aggregation

For the determination of platelet aggregation, platelet-rich plasma (PRP) was prepared from human blood containing citrate by centrifugation at 120 x g for 10 min. Platelet aggregation was measured by turbidimetric method of the reaction mixture of PRP (200 μl) containing 0.2 μg collagen as an aggregation inducer and 1 μl methanolic sample solution using a dual-channel aggregometer. The inhibitory activities were expressed as IC_{50} values, that is, the sample concentration that resulted in 50% inhibition of platelet aggregation (15).

The methanol extract of crushed onion was fractionated by silica gel column chromatography with n-hexane, benzene, chloroform and 20% methanol/ chloroform as eluents. The chloroform fraction showed the strongest inhibitory effects, so this fraction was rechromatographed on a silica gel column with stepwise gradients of chloroform in benzene. The fractions eluted with 50 and 75% chloroform in benzene exhibited the highest activities for platelet aggregation. These eluates were further fractionated by HPLC using a Develosil SI-60-5 column to give 10 active substances named the AC series. Their chemical structures were determined by analyses of their ^1H and ^{13}C-NMR, and EI mass spectra, and all AC series were constituted from a common structural unit of α-sulfinyl disulfide.

The structures and inhibitory activities (IC_{50} values) against platelet aggregation of AC series are shown in Table I. The letters a, b, and c represent the respective diastereoisomers. AC series bore a common structural skeleton with cepaenes (19,20), which were found to be antiasthmatic, antiallergic and antiplatelet compounds in crushed onion. AC series exhibited a strong activity compared to the anti-inflammatory drug aspirin, besides AC-1a and AC-11a, AC-12b was the most effective, as potent as indomethacin.

Chiralcel OB column HPLC was used to separate the diastereoisomer from the respective enantioisomer of AC-1b, AC-11a and AC-11b. Separation was confirmed by the measurement of their CD spectra (data not shown) (21). As shown in Table II, the differences in IC_{50} values between enantioisomers were not so large.

These antiplatelet compounds were probably formed by the interaction of lachrymatory factor, thiopropanal S-oxide and other sulfenic acids. The addition of methyl sulfenic acid to the C_1 position of the lachrymatory, followed by the condensation with another sulfenic acid and dehydration, may transform to α-sulfinyl disulfide (Figure 2). The lachrymatory factor is ordinarily very labile and rapidly decomposed in sliced onion. When onion is rapidly homogenized, it may be stabilized a little to generate α-sulfinyl disulfide.

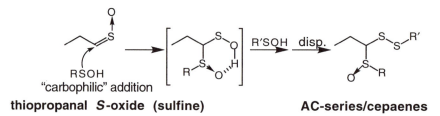

thiopropanal *S*-oxide (sulfine)   **AC-series/cepaenes**

Figure 2. Proposed formation mechanism of α-sulfinyl disulfides (AC series) from lachrymatory factor.

Table I. Inhibitory Activities of AC Series Against Human Platelet Aggregation

	R	IC_{50} (μM)
AC-1a	methyl	67.6
AC-1b	"	18.4
AC-10	propyl	12.8
AC-11a	trans-1-propenyl	48.9
AC-11b	"	11.7
AC-12a	cis-1-propenyl	6.1
AC-12b	"	1.4
AC-13a	(1-methoxy)-propyl	26.4
AC-13b	"	13.1
AC-13c	"	13.1
Indomethacin		2.1
Aspirin		41.2

SOURCE: Reproduced with permission from reference 15. Copyright 1990 Pergamon.

Table II. Inhibitory Activities Of AC Series Enantioisomers Against Human Platelet Aggregation

	IC_{50} (μM)
AC-1b	18.4
AC-1b-(1)	23.2
AC-1b-(2)	16.4
AC-11a	48.9
AC-11a-(1)	49.4
AC-11a-(2)	23.7
AC-11b	11.7
AC-11b-(1)	9.1
AC-11b-(2)	22.9

SOURCE: Reproduced with permission from reference 21. Copyright 1991 Japan Society for Bioscience, Biotechnology, and Biochemistry.

Inhibitory Effects of AC Series on Cyclooxygenase and 5-Lipoxygenase in the Arachidonic Acid Cascade

Inhibition of platelet aggregation by the AC series may be due to the suppression of the enzymes related to the arachidonic acid cascade in platelets, similarly to the inhibitors isolated from garlic. Pharmacological actions of aspirin and indomethacin depend on the inhibition of cyclooxygenase in PGH synthase (Figure 3). The PGH synthase pathway possesses both cyclooxygenase and hydroperoxidase activities, which generate prostaglandins (PGs) and thromboxanes (TXs) from arachidonic acid via PG H_2 and G_2. The enzyme 5-lipoxygenase participates in biosynthesis of leukotriene, a chemical mediator of inflammatory diseases in leukocytes.

We have examined the suppressive effects on prostaglandin biosynthesis of AC series by using rabbit renal microsomes. Furthermore, the inhibitory activities of AC series on 5-lipoxygenase were studied by using an enzyme preparation obtained from gene cloning of human 5-lipoxygenase. The results of our studies on six kinds of AC series — 1a and b, 11a and b, and 12a and b, are shown in Table III. These compounds clearly inhibited PGH synthase and 5-lipoxygenase in the arachidonic acid cascade. Moreover, AC-11a and b were comparable to indomethacin in their inhibition of PGH synthase, and AC-11a and b and 12a and b also inhibited 5-lipoxygenase in a manner similar to AA861, a potent inhibitory drug.

**Table III. Inhibitory Effects of Several α-Sulfinyl Disulfides
(AC Series) on PG Biosynthesis and 5-Lipoxygenase**

	IC_{50} (μM)		
	Platelet aggregation	Prostaglandin biosynthesis	5-Lipoxygenase
AC-1a	67.6	—	27.3
AC-1b	18.4	—	19.6
AC-11a	48.9	1.2	1.4
AC-11b	11.7	1.6	1.5
AC-12a	6.1	ND	2.5
AC-12b	1.4	ND	2.8
Indomethacin	2.1	0.75	—
AA861	—	—	0.3
NDGA	—	—	16.5

ND: not determined

Relationships between Structure and Activity of α-Sulfinyl Disulfide

The α-sulfinyl disulfide structure in AC series is related to ajoene from garlic in possessing sulfinyl and disulfide groups. Many analogues of α-sulfinyl mono-sulfide and disulfide were prepared to study their structure-activity relationship. α-Sulfinyl monosulfide analogues synthesized in this experiment did not inhibit either platelet aggregation or 5-lipoxygenase. α-Sulfinyl disulfide analogues,

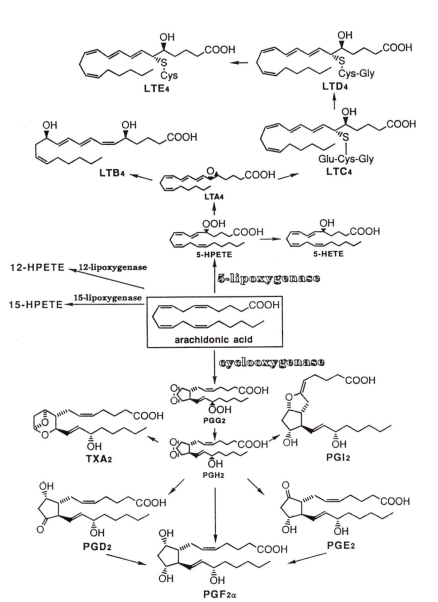

Figure 3. The formation of prostaglandin (PG), thromboxane (TX) and leukotriene (LT) in the arachidonic acid cascade. HETE: Hydroxyeicosatetraenoic acid; HPETE: hydroperoxyeicosatetraenoic acid.

however, showed a clear relationship between their structures and both inhibitory activities as shown in Table IV. α-Alkylthiodisulfide (n=0) did not have inhibitory activity against platelet aggregation or 5-lipoxygenase, but both α-sulfinyl disulfide (n=1) and α-sulfonyl disulfide (n=2) exhibited strong inhibitory activities in both systems. These data suggest that both sulfinyl and disulfide groups in AC series are essential for the development of both activities. Moreover, their activities against platelet aggregation were markedly increased with the chain length of their alkyl groups. In the case of 5-lipoxygenase, no clear relation was observed between inhibitory activity and alkyl chain length.

Table IV. Structure-Activity Relationship of α-Sulfinyl Disulfide Inhibition of Human Platelet Aggregation and Human 5-Lipoxygenase

R	IC_{50} (μM)		
	n = 0	n = 1	n = 2
Human platelet aggregation			
methyl	465	67.6	126.2
		18.4	
ethyl	>1000	26.8	40.6
		12.0	
propyl	>1000	5.1	13.0
		3.0	
Human 5-lipoxygenase			
methyl	ND	27.3	ND
		19.6	
ethyl	>100	ND	50.6
		10.0	
propyl	>100	ND	88.0
		32.1	

Conclusion

Ten kinds of α-sulfinyl disulfides (AC series) were isolated and characterized as active components from onion for the inhibition of platelet aggregation. These active components were derived from the lachrymatory factor, thiopropanal S-oxide, and sulfenic acid, and were structurally similar to ajoene, an antiplatelet factor in garlic, regarding its sulfinyl and disulfide groups. Both sulfur containing groups were essential for development of the activity as antiplatelet factors. The activity of α-sulfinyl disulfides was due to inhibition of cyclooxygenase in PGH synthase of platelets. Moreover, these components were also potent inhibitors of 5-lipoxygenase in leukocytes. Since onion is a widely consumed vegetable, physiological characteristics of onion components as inhibitors of these enzymes have important meaning for human health from the viewpoint of the possible prevention of thrombosis and asthmatic and inflammatory diseases.

Acknowledgments

The authors sincerely thank Drs. T. Matsuzaki and T. Matsumoto of Japan Tobacco Inc. for their kind supply of rabbit renal microsomes and human 5-lipoxygenase.

Literature Cited

1. Whitaker, J. R. In *Adv. Food Res.*; Chichester, C. O., *et al.* Eds.; Academic Press: New York, 1976; Vol. 22, 73–133
2. Baghurst, K. I.; Raj, M. J.; Truswell, A. S. *Lancet* **1977**, *1*, 101.
3. Phillips, C.; Poyser, N. L. *Lancet* **1978**, *1*, 1051.
4. Apitz-Castro, R.; Cabrera, S.; Curz, M. R.; Lederzma, E.; Jain, M. K. *Thrombosis Res.* **1983**, *32*, 155.
5. Block, E.; Ahmad, S.; Jain, M. K.; Crecely, R. W.; Apitz-Castro, R.; Cruz, M. R. *J. Am. Chem. Soc.* **1984**, *106*, 8295.
6. Block, E.; Ahmad, S.; Catafalmo, J.; Jain, M. K.; Apitz-Castro, R. *J. Am. Chem. Soc.* **1986**, *108*, 7045.
7. Ariga, T.; Oshiba, S.; Tamada, T. *Lancet* **1981**, *1*, 150.
8. Makheja, A. N.; Vanderhoek, J. Y.; Bailey, J. M. *Lancet* **1979**, *1*, 781.
9. Makheja, A. N.; Vanderhoek, J. Y.; Bailey, J. M. *Prostaglandins Med.* **1979**, *2*, 413.
10. Srivastava, K. C. *Prostaglandins Leukotrienes Med.* **1984**, *13*, 227.
11. Srivastava, K. C. *Prostaglandins Leukotrienes Med.* **1986**, *24*, 43.
12. Weisenberger, H.; Grube, H.; Koening, E.; Pelzer, H. *FEBS Letters* **1972**, *26*, 105.
13. Liakopulou-Kyriakides, M.; Sinakos, Z.; Kyriakides, D. A. *Phytochem.* **1985**, *24*, 600.
14. Kawakishi, S.; Morimitsu, Y. *Lancet* **1988**, *1*, 330.
15. Morimitsu, Y.; Kawakishi, S. *Phytochem.* **1990**, *29*, 3435.
16. Morimitsu, Y.; Morioka, Y.; Kawakishi, S. *J. Agric. Food Chem.* **1992**, *40*, 368.
17. Brodnitz, M. H.; Pascale, J. V. *J. Agric. Food Chem.* **1971**, *19*, 269.
18. Yagami, M.; Kawakishi, S.; Namiki, M. *Agric. Biol. Chem.* **1980**, *44*, 2533.
19. Bayer, T.; Wagner, H.; Wray, V.; Dorsch, W. *Lancet* **1988**, *1*, 906.
20. Bayer, T.; Breu, W.; Seligmann, O.; Wray, V.; Wagner, H. *Phytochem.* **1989**, *28*, 2373.
21. Morimitsu, Y.; Kawakishi, S. *Agric.Biol.Chem.* **1991**, *55*, 889.

RECEIVED April 14, 1993

Chapter 9

Vinyldithiins in Garlic and Japanese Domestic Allium (*A. victorialis*)

H. Nishimura[1] and T. Ariga[2]

[1]Department of Bioscience and Technology, Hokkaido Tokai University, Sapporo 005, Japan
[2]Department of Nutrition and Physiology, Nihon University School of Agriculture and Veterinary Medicine, 3–34–1 Shimouma, Setagaya, Tokyo 154, Japan

Caucas (*Allium victorialis* L.), a wild plant that grows in northernmost Japan, yielded an essential oil containing more than 80 kinds of volatile components with a composition similar to that of garlic (*A. sativum* L.). The content of the antiaggregation principle allyl methyl trisulfide (MATS) in caucas, however, was 1.5 times higher than that in garlic. Two kinds of vinyldithiins (3,4-dihydro-3-vinyl-1,2-dithiin and 2-vinyl-4H-1,3-dithiin) were detected in the directly extracted caucas oil. Garlic oil and MATS inhibited platelet aggregation induced by most of the known physiological agonists; the aggregation induced by arachidonic acid metabolites, however, was not inhibited by these *Allium* components. Actually, the *Allium* components inhibited production of thromboxane B_2 (TXB$_2$), 12S-hydroxy-5,8,10-heptadecatrienoic acid (HHT), and prostaglandin E_2 (PGE$_2$), as well as 12-hydroxyeicosatetraenoic acid (12-HETE), although the arachidonic acid release reaction was accelerated to some extent, indicating that the antiaggregation is caused by an impairment of arachidonic acid metabolism.

The plants that belong to the *Allium* family have been widely used as nutritious vegetables. The pharmacological effects of garlic, known for many years, have recently been reaffirmed by modern scientific techniques. Among the effects are the well known antithrombotic activities afforded by essential oil of garlic, e.g., fibrinolytic (*1–3*), lipid lowering (*4–6*) and antiaggregation effects (*7–12*). The authors initially found methyl allyl trisulfide (MATS) as an active principle for platelet aggregation inhibition in steam distilled garlic oil (*9*). Since then, several different components have been isolated and identified as principles by many investigators (*10–12*).

One of the reasons why so many active principles have been discovered is the difference in composition of garlic oil preparations (*13–17*). Steam distillation does not give rise to ajoen and vinyldithiins (*10,12,17*), but to MATS (*9,17*), whereas a direct extraction with solvents does not give MATS, but does yield ajoen and vinyldithiins.

0097–6156/94/0546–0128$06.00/0

In the present study, the authors have analyzed caucas, a wild plant growing in Hokkaido, the northernmost island of Japan, that has long been taken by the natives as a pharmacologically active vegetable. Characteristic of the steam-distilled oil components was a high content of MATS. The vinyldithiins 3,4-dihydro-3-vinyl-1,2-dithiin and 2-vinyl-4H-1,3-dithiin were detected as the major components of solvent extracts, and these structures were also found to have anti-aggregation activity. The inhibitory mechanisms of essential oil of garlic and caucas, and MATS, a component of caucas, were extensively studied using assay systems to measure arachidonic acid releasing enzyme action and quantitative analyses of arachidonic acid metabolites, intracellular free Ca^{2+}, and cAMP. In addition, a positive absorption of radiolabeled MATS by platelets was confirmed.

Materials and Methods

Preparation of Essential Oil from Garlic and Caucas. The essential oils were extracted by two different procedures, i.e., direct extraction and steam distillation. For direct extraction, garlic bulbs purchased locally or caucas stalks and leaves freshly harvested in the field were minced into small pieces and then extracted with dichloromethane, in which the minced vegetable (garlic or caucas) was allowed to stand for 2 weeks at room temperature. The essential oil was dehydrated with sodium sulfate and the excess solvent was removed by evaporation at 0°C. For steam distillation, the minced vegetables were homogenized further in water (3 ml/g vegetable) with a Waring blender, and applied to a Linkens-Nickerson steam distillation apparatus. Essential oil components were analyzed by the methods described elsewhere (*14*) using the following equipment: a gas chromatograph (Shimadzu, Tokyo, GC-14A with a 50 m polyethylene glycol 20M-bonded fused silica capillary column), mass spectrometers (JMS-DX300 and JMS-01SG-2, JEOL, Tokyo), and nuclear magnetic resonance (NMR) machines (JNM-GX 270FT, JNM-FX 100, and Burker AM-500 for 1H-NMR; JMN-GX 270FT for 13C-NMR and 2D-NMR, JEOL, Tokyo). High performance liquid chromatography (HPLC) was performed with a JASCO PG-350D equipped with a Unisil Q column and a Shodex RI SE-31 detector (Showadenko Co., Tokyo).

Measurement of Antiaggregation Activity. Platelet-rich plasma (PRP) was prepared from citrated human or rabbit blood. Platelet aggregation was induced by various agonists at concentrations designated in the text, and measured by an aggregation meter (Aggregometer DP-247E, Sienco Inc., U.S.A.). To assay anti-aggregation activity, the *Allium* oil or its component was added to the PRP prior to the addition of an inducer. The oil or its component was diluted with ethanol or acetone and added as a 1 µl aliquot to 300 µl PRP to avoid solvent inhibition. Aggregation curves were analyzed for maximum aggregation and disaggregation. For the former, inhibition percent was obtained as described elsewhere (*11*), and for the latter, disaggregation percent was taken as the negative slope (-%/min) appearing just after the maximum aggregation was obtained.

Radiolabeling of Platelets. Rabbit platelets were labeled with [1-^{14}C]-arachidonic acid (54.5 mCi/mmol, NEN Research Products) according to the method of Bills *et al.* (*18*). To 20 ml PRP 4 µCi of labeled arachidonic acid was added, then, after labeling, free label was removed by washing with HEPES buffer, pH 7.4, containing 1 mM ethylenediaminetetraacetic acid (EDTA).

Thin Layer Chromatography of Platelet Phospholipids. The labeled platelet suspension (445 µl) was treated with 1 µl garlic oil or its components diluted with ethanol. Calcium chloride and bovine thrombin (Mochida Pharmaceuticals, Tokyo; final concentrations 1 mM and 0.1 µl/ml, respectively) were added to suspension and the phospholipids in the platelets were extracted with chloroform and methanol (2:1) containing 0.02% 2,6-di-*t*-butyl-4-methylphenol (butylated hydroxytoluene). The solvent phase, separated by centrifugation at 1500 g for 10 min, was evaporated and the lipid was redissolved in 100 µl chloroform, then applied to a thin layer plate (HPTLC Silica Gel 60 F254, E. Merck, Darmstadt). The plate was developed for about 25 min with a solvent mixture of chloroform, methanol, acetic acid, and water (50:30:4:2), and colored by spraying with 0.02% 2'-7'-dichloro-fluorescein methanol. Each color zone was scraped off and its radioactivity was measured with a liquid scintillation counter (LS 6000TA, Beckman Instruments Inc., CA). Identification of the major platelet phospholipids was made by analyzing authentic phospholipids obtained from Sigma Chemical Co. (St. Louis, MO) with the same TLC plate.

High Performance Liquid Chromatography (HPLC) of Arachidonic Acid Metabolites. [1-^{14}C]-Arachidonic acid-derived radioactive metabolites were separated by an HPLC system (Hitachi 655A, Tokyo) equipped with an Inertsil ODS column, φ4.6 mm x 250 mm (GL Science Co., Tokyo). For elution, a linear gradient formed between 35% acetonitrile, pH 3.5, and 100% acetonitrile was introduced into the column. Since [1-^{14}C]-arachidonic acid left unchanged in the platelet extracts was strongly held by the column, 100% acetonitrile was introduced for about 10 min after the plateau of the gradient had been reached. For identification of the radioactive eluates, authentic labeled materials obtained from Dupont/NEN Research Products were analyzed by the same system. Radioactivity of the eluate collected as 2 ml fractions was measured with a liquid scintillation counter.

Assay of Intracellular Calcium Concentration. Human platelet intracellular free Ca^{2+} concentration ($[Ca^{2+}]i$) was measured by use of Fura-2 acetoxymethyl ester (Dojin Chemical Laboratories, Kumamoto, Japan) and a fluorometer (Hitachi F-2000, Tokyo). Fura-2-loaded human platelets were prepared by the method described elsewhere (*19*), and the concentration was adjusted to 1×10^8 platelets per ml. One µl garlic oil of appropriate dilution was added to a 300 µl aliquot of the platelet suspension, followed by induction of aggregation with 10 µl 0.5 units/ml thrombin. Calculation of $[Ca^{2+}]i$ from the ratio of the fluorescence intensities (emission at 490 nm, excitation at 340 and 380 nm) was made according to the application note supplied by JASCO, Tokyo.

Assay of Cyclic Adenosine Monophosphate (cAMP) in Platelets. cAMP was assayed with a DuPont cAMP[^{125}I] kit. The washed human platelets (albumin gradient and gel filtered platelets, AGFP) prepared by the method of Timmons *et al.* (*20*) were adjusted to a concentration of 1.2×10^8/ml. Garlic oil diluted with ethanol (1 µl) was added to 500 µl AGFP, and the mixture was incubated at 37°C for 30 sec. The conversion of ATP to cAMP was then terminated by the addition of 12% trichloroacetic acid (TCA), after which it was centrifuged at 1500 g for 5 min. The precipitate was washed once with 500 µl 6% TCA. Both supernatants were combined, and washed three times with 4 ml ether. The aqueous phase was lyophilized, redissolved in acetate buffer, pH 5.0, and subjected to the cAMP assay.

Preparation of (^{35}S)-MATS and Its Incorporation. As a starting material, sodium trisulfide was prepared by heating a solid mixture of 5.2 mg anhydrous sodium sulfide and 4.2 mg sulfur containing 1 mCi of ^{35}S, elemental (30 Ci/mg, Amersham Intl. P.L.C., Buckinghamshire, England), at 450°C for 5 h under vacuum (2 mm Hg). The heated mixture was cooled to room temperature and dissolved in 0.5 ml methanol. Then, allyl bromide and methyl bromide, 6 µl each) were added at the same time to the methanolic-sulfide solution kept at -20°C. The (^{35}S)-MATS among the products was isolated by a preparative gas chromatograph equipped with a thermoconductivity detector (GC-9, Shimadzu Co. Tokyo) and a column, 3.2 mm x 2.1 m, packed with Shinchrome E-71 (Shimadzu Co.). Purity of the labeled MATS was estimated to be 99%. The isolated labeled material was dissolved in ethanol and stored at -80°C under a N_2 atmosphere until used.

The uptake of (^{35}S)-MATS by platelets was measured as follows. Washed human platelets (10^6) were suspended in 1 ml HEPES buffer, pH 7.4, containing 5 mM $CaCl_2$, then 3 µl garlic oil containing (^{35}S)-MATS (0.2 µCi) as a tracer was added. After incubation for 5 min at 37°C, the platelets were washed twice with HEPES buffer containing 2 mM EDTA, followed by disruption either by freezing and thawing or with a glass homogenizer. The platelets thus homogenized were centrifuged and their organelles were roughly fractionated. The radioactivity of each fraction was then counted as described above.

Results and Discussion

Essential Oil Components of Caucas. Yurugi *et al.* gave the first report about the ingredients of caucas in their studies on the reaction between thiosulfinates and thiamine (*21*). Nishimura *et al.* reported the identification of the components in essential oil of caucas, and found that 1-propenyl disulfides, 2-methyl-2-pentenal (one of the breakdown products of propanethial S-oxide), and 1-propenyl alk(en)yl components were characteristic in caucas (*22*). In the present study, the essential oils prepared by both direct extraction and steam distillation were analyzed comparatively. As shown in Figure 1, the gas chromatographic patterns were quite different between the two oil preparations. Compared with the steam distillation (Figure 1B), the direct extraction (Figure 1A) gave smaller and fewer peaks at retention times less than 30 min, but extremely high peaks containing vinyldithiins (peak numbers 48 and 53) at around 50 min.

Table I lists the compounds identified. Although quantitative analysis was not performed, caucas was found to have, irrespective of the extraction procedure, methyl allyl disulfide (peak 17), dimethyl disulfide (peak 21), diallyl disulfide (peak 39), and methyl allyl trisulfide (MATS, peak 43). MATS content in caucas has been estimated to be as high as 15.1% (w/w) of the steam distilled oil, about 1.5 times higher than that of garlic (our unpublished data). Several components containing a 1-propenyl radical, which is characteristic of onion [*Allium cepa* L. (*23, 24*)], were also detected — peak numbers 3, 16, 18, 34, 38, 40, 42 and 44. As mentioned above, two kinds of vinyldithiins, 3,4-dihydro-3-vinyl-1,2-dithiin (peak 48) and 2-vinyl-4*H*-1,3-dithiin (peak 53), were detected only in the directly extracted oil. Such a finding is in accordance with that of garlic as reported by Block *et al.* (*13*) and Lawson *et al.* (*17*), and can be explained by the sulfinate decomposition mechanism that is operative during steam distillation (*13*).

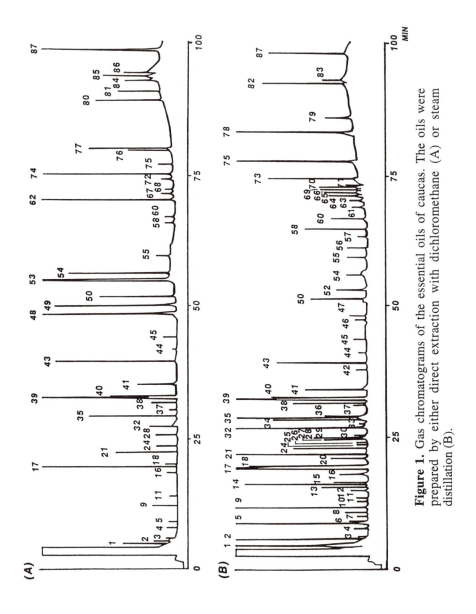

Figure 1. Gas chromatograms of the essential oils of caucas. The oils were prepared by either direct extraction with dichloromethane (A) or steam distillation (B).

Table I. Identification of Essential Oil Components Found in Caucas

Peak	Compound	M^+, m/z	Identification[a]	Extraction[b]
1	$CH_2=CHCH_2SH$	74	MS, Rt, A	S, D
2	$CH_3SCH_2CH=CH_2$	88	MS, Rt, A	S, D
3	$CH_3COCH=CHCH_3$	84	MS, Rt, T	S, D
5	CH_3SSCH_3	94	MS, Rt, A	S, D
7	$CH_2=CHCH_2SCH_2CH_2CH_3$	116	MS, Rt, A	S
8	phenyl–CH_2CH_3	108	MS, T	S
9	$CH_2=CHCH_2SCH_2CH=CH_2$	114	MS, Rt, A	S, D
12	$CH_3CH=CHCH(OH)CH_3$	86	MS, R, T	S
13		106	MS, T	S
14	$CH_3CH_2CH=C(CH_3)CHO$	98	MS, Rt, A	S
15	$CH_3SSCH_2CH_2CH_3$	122	MS, Rt, A	S
16	*cis*-$CH_3SSCH=CHCH_3$	120	MS, Rt, A	S, D
17	$CH_3SSCH_2CH=CH_2$	120	MS, Rt, A	S, D
18	*trans*-$CH_3SSCH=CHCH_3$	120	MS, Rt, A	S, D
20		108	MS, T	S
21	CH_3SSSCH_3	126	MS, Rt, A	S, D
24		134	MS, Rt, A	S
28	$CH_3CH_2CH_2SSCH_2CH_2CH_3$	150	MS, Rt, A	S, D
32	$CH_3SSO_2CH_3$	126	MS, Rt, A	S, D
34	*cis*-$CH_3CH_2CH_2SSCH=CHCH_3$	148	MS, Rt, A	S, D
35	$CH_3CH_2CH_2SSCH_2CH=CH_2$	148	MS, Rt, A	S
38	$CH_3CH=CHSSCH=CHCH_3$[c]	146	MS, Rt, A	S, D
39	$CH_2=CHCH_2SSCH_2CH=CH_2$	146	MS, Rt, A	S, D
40	*trans,trans*-$CH_3CH=CHSSCH=CHCH_3$	146	MS, Rt, A	S, D
41	$CH_3SSSCH_2CH_2CH_3$	154	MS, Rt, A	S, D
42	*cis*-$CH_3SSSCH=CHCH_3$	152	MS, Rt, A	S, D
43	$CH_3SSSCH_2CH=CH_2$	152	MS, Rt, A	S, D
44	*trans*-$CH_3SSSCH=CHCH_3$	152	MS, Rt, A	S, D
45	$CH_2SCH_2SSCH_3$	140	MS, T	S, D
48	$CH_2=CH$–(dithiin ring, S–S)	144	MS, Rt, A	D
49		146	MS, Rt, A	D
50	$CH_2=CHCH_2SSSCH_2CH=CH_2$	178	MS, T	S, D
52		167	MS, T	S
53	$CH_2=CH$–(dithiin ring, S, S)	144	MS, Rt, A	D
54		166	MS, T	S, D
62–67	Polymeric hydrocarbons			

[a] MS = mass spectrometry; Rt = gas chromatographic retention time; T = tentative; A = authentic compound.

[b] S = steam distillation; D = direct.

[c] *cis,cis* or *cis,trans*

Vinyldithiins and Their Antiaggregation Activity. The vinyldithiins isolated from caucas showed antiaggregation activity with strength comparable to that of other known potent sulfide compounds with a straight chain structure, ajoen or MATS (9,10). Figure 2 shows the antiaggregation activity of both vinyldithiins found in caucas. The collagen-induced rabbit platelet aggregation was inhibited by 0.46 mM dithiins, and the inhibition by 3,4-dihydro-3-vinyl-1,2-dithiin was about 5 times stronger than that by 2-vinyl-4H-1,3-dithiin. Vinyldithiin was initially reported to be one of the antiaggregation principles in garlic by Block et al. (10), and the major structures were determined to be 3-vinyl-4H-1,2-dithiin and 2-vinyl-4H-1,3-dithiin (10,12). In this report, the former structure is assigned as 3,4-dihydro-3-vinyl-1,2-dithiin.

Dithiins can be obtained by heat processing of allicin, a sulfinate; therefore, these structures have been thought to be an artifact generated at the moment of GC analysis (25). Evidence that dithiins appear in solvent extracts of garlic or even in oil-macerated garlic, however, may indicate possible generation of dithiins in foods containing cooked garlic or in caucas consumed with some edible oils.

Structure and Activity Relationship. The antiaggregation principles were initially isolated from garlic oil by Ariga et al. (9), and later by Apitz-Castro et al. (10). The inhibitory principles, however, are quite different between the two groups; the former group found methyl allyl trisulfide (MATS) in steam distilled garlic oil, while the latter attributed inhibition of platelet aggregation to (E and Z)-4,5,9-tri-thiadodeca-1,6,11-triene-9-oxide, termed "ajoen," in direct extracts (10–12). We prepared several alk-(en)yl oligosulfides and compared their antiaggregation activity, although none of the synthesized compounds had the same configuration as ajoen (Table II). The ADP-induced human platelet aggregation was strongly inhibited by both dimethyl trisulfide and MATS, but diallyl trisulfide as well as disulfide compounds having methyl, ethyl, propyl, or allyl radicals exhibited only a weak inhibition.

The disaggregation, representing the dissociation of platelet

Table II. Antiaggregation Activity of the Alk(en)ylsulfides

	Inhibitory effect (percent)	
	Inhibition	Disaggregation
M-S-M	2	3
M-S-S-M	3	21
M-S-S-S-M	55	66
E-S-E	2	3
E-S-S-E	8	4
P-S-P	2	3
P-S-S-P	10	3
A-S-A	12	2
A-S-S-A	0	12
A-S-S-S-A	13	4
M-S-A	0	3
M-S-S-S-A	63	34

Each synthesized compound (20 μM) was added to human PRP, followed by stimulation with ADP (20 μM).
M: methyl; S: sulfur; E: ethyl; P: propyl; A: allyl.

aggregates into individual platelets, was also higher for the components with stronger inhibition. Interestingly, the inhibitory strengths of diallyl trisulfide and MATS were quite different, even though they share three sulfur atoms. Therefore, the inhibitory effect of MATS comes from either its methyl radical or from its

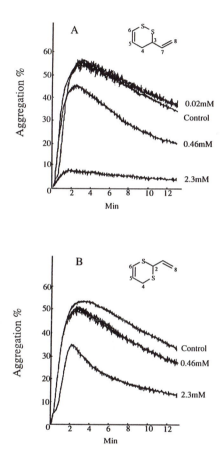

Figure 2. Antiaggregatory effect of vinyldithiins on collagen-induced rabbit platelet aggregation. A: 3,4-dihydro-3-vinyl-1,2-dithiin; B: 2-vinyl-4H-1,3-di-thiin. Collagen: Type I, 32 µg/ml.

unsymmetrical structure. On the other hand, the trisulfide structure in MATS would be marginal in exhibiting platelet aggregation inhibition, since diallyl trisulfide was ineffective. Recently, Lawson *et al.* (*17*) has reported that diallyl trisulfide in steam distilled garlic oil was the most potent inhibitor of collagen-induced platelet aggregation in whole blood. The cause of such discrepancy between the two laboratories is unknown so far.

Antiaggregation Mechanisms of Garlic Oil and MATS

Effects on the Arachidonic Acid-Releasing Mechanism. $[1-^{14}C]$-Arachidonic acid-labeled rabbit platelets were stimulated with thrombin, and the amounts of the label released from various phospholipid moieties of the platelets were measured. First of all, the amount of the label incorporated into each phospholipid was estimated from that of the pre-stimulated platelets (Figure 3, 0 seconds). Of the total radioactivity incorporated into the platelets, 58% was in phosphatidylcholine, 19% in the phosphatidylserine and phosphatidylinositol moieties, and 18% in phosphatidylethanolamine. The labeled platelets could be aggregated by thrombin (0.1 unit/ml) as well as unlabeled platelets, and liberated the label from their phospholipids.

In the presence of garlic oil, which was added just before the stimulation, thrombin-induced platelet aggregation was suppressed, whereas the total amount of the label liberated, which would be free arachidonic acid and/or its metabolites, was markedly increased (Figure 3, panel A.A.). The most prominent phospholipid liberating the label was phosphatidylcholine (PC). The sharp decline in the radioactivity of PC observed by addition of garlic oil indicates that platelets so treated can liberate arachidonic acid from their membranous PC much easier than the untreated control platelets. The phospholipids phosphatidylserine and phosphatidylinositol (PS & PI) also liberated the label, but no significant acceleration by garlic oil was observed. Phosphatidylethanolamine (PE) did not show any significant liberation of the label with or without garlic oil. The antiaggregation principle MATS also enhanced arachidonic acid liberation from PC (data not shown). Thus, garlic oil and its components act to stimulate, not suppress, the arachidonic acid release mechanism, in which phospholipase A_2 and phospholipase C have a central role.

Inhibitory Effect of Garlic Oil and Its Components on the Arachidonic Acid Metabolism in Platelets.

Arachidonic acid, once released from the membrane in response to an agonist, is immediately metabolized to eicosanoids, including prostaglandins (*26*). The essential oil of *Allium* family members is known to act as a blockade to the arachidonic acid metabolism (*8,27*), although there are some conflicting data (*28*). So we performed quantitative analysis of the ^{14}C-arachidonic acid metabolites generated in washed rabbit platelets (AGFP) treated with garlic oil or its component MATS. The platelets prepared could be aggregated fully in response to 50 µM arachidonic acid containing ^{14}C-arachidonic acid (0.15 µCi), and either garlic oil (2 µg/ml) or MATS (0.4 µg/ml, 2.6×10^{-6} M) inhibited the maximum aggregation of the control platelets by 50%; complete inhibition was achieved by 20 µg/ml of either agent. In these aggregation-suppressed platelets, the amount of metabolites derived from ^{14}C-arachidonic acid was markedly decreased as compared with those of the control and larger amounts of ^{14}C-arachidonic acid remained unmetabolized (Figure 4). Unlike acetyl salicylic acid (aspirin), which

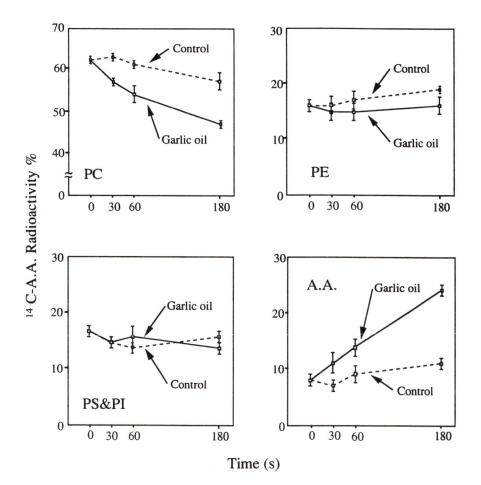

Figure 3. Arachidonic acid liberation from various phospholipids of rabbit platelets in response to thrombin stimulation. PC, phosphatidyl choline; PE, phosphatidylethanolamine; PS&PI, a combined moiety of phosphatidylserine and phosphatidylinositol; A.A., arachidonic acid and/or its metabolites.

inhibits solely cyclooxygenase (*29*), MATS or garlic oil seemed to inhibit both cyclooxygenase and lipoxygenase, since they suppressed production of the metabolites TXB_2 and PGE_2, products of the former enzyme, and 12-HETE, a product of the latter. Cyclooxygenase appeared to be more susceptible to MATS than lipoxygenase, since MATS, at a concentration below that which inhibited 12-HETE production (0.4 μg/ml), caused appreciable inhibition of TXB_2 and PGE_2 formation.

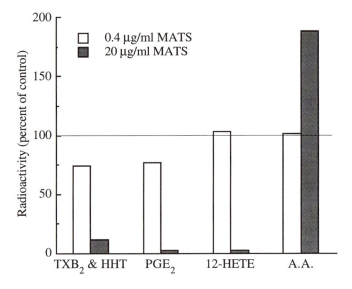

Figure 4. Inhibitory effect of MATS on arachidonic acid metabolism in human platelets. AGFP were stimulated by 50 μM arachidonate containing 0.15 μCi [1-^{14}C]-arachidonate as a tracer.

Garlic Oil and MATS Do Not Inhibit TXA$_2$-Induced Aggregation. Platelet aggregation induced by either TXA_2 or its analogues was not inhibited by garlic oil or MATS. As shown in Figure 5, LASS, a labile substance which stimulates aggregation and is prepared from human platelets by stimulating them with 50 μM arachidonate for 30 s [Willis *et al.* (*30*)], U-46619 (a PGH$_2$ mimic), and STA$_2$ (a TXA$_2$ analog) were able to induce platelet aggregation even in the presence of garlic oil or MATS. These results strongly suggest that both garlic oil and MATS specifically inhibit the metabolic pathways from arachidonic acid to its aggregation-inducing metabolites, PGH_2 or TXA_2.

Effect of Garlic Oil on the Mobilization of Intracellular Ca^{2+}. Changes in the intracellular free calcium concentration ($[Ca^{2+}]i$) of thrombin-stimulated platelets in the presence or absence of garlic oil were measured by use of the Fura-2/AM-loaded rabbit platelets. $[Ca^{2+}]i$ increased in 20–30 s in response to thrombin stimulation, and the increase in Ca^{2+} level was dose dependent. A higher $[Ca^{2+}]i$ led to stronger aggregation: resting platelets registered less than 50 nM $[Ca^{2+}]i$;

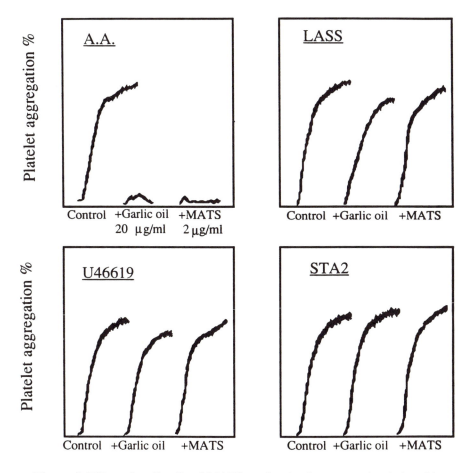

Figure 5. Effect of garlic oil and MATS on the platelet aggregation induced by either arachidonate or its metabolites, LASS or its analogues.

those at 60% aggregation (0.01 units/ml thrombin), 200 nM $[Ca^{2+}]i$; and those at 100% aggregation (0.5 units/ml thrombin), 700 nM $[Ca^{2+}]i$. In the presence of garlic oil (20–200 μg oil/ml of PRP), the platelets lost their capacity for thrombin-induced aggregation by about 60%; the $[Ca^{2+}]i$ in these platelets, however, clearly elevated to a level comparable to that of the control platelets without garlic oil (Figure 6). Elevation of the extracellular fluid Ca^{2+} concentration resulted in an increase in the $[Ca^{2+}]i$ in both garlic oil treated and control platelets (data not presented).

These results indicate that garlic oil affects neither Ca^{2+} mobilization nor Ca^{2+} influx during thrombin-induced aggregation. The observation that platelet aggregation induced by agonists other than arachidonic acid metabolites was inhibited by garlic oil, in spite of the fact that enough Ca^{2+} (more than 500 nM) was mobilized, make it difficult for us to explain the reasons why platelet aggregation capability is lost under these conditions. One of the most plausible explanations is that the platelets, once attacked by garlic oil, may become insensitive to any physiological agonists including arachidonate, but they are still sensitive to the endogenous stimulants, e. g., PGH_2, TXA_2, derived from arachidonic acid. On the other hand, the enhanced release of arachidonic acid from phospholipids by garlic might give rise to more PI metabolites, e. g., diacylglycerol or inositol triphosphate, and lead to a high $[Ca^{2+}]i$ resulting from Ca^{2+} released either from membrane-bound calcium or from granules, mostly located in the dense tubular system. Perhaps the platelets do not aggregate in response to raised $[Ca^{2+}]i$ only, and for aggregation, certain numbers of agonist molecules must bind to the platelets. Of course, impairment of some effectors responding to Ca^{2+} should also be taken into account. Possible effectors may be plasma membrane proteins necessary for platelet-platelet interaction, as has been suggested by Apitz-Castro et al. (28).

Effect of Garlic Oil on the Intracellular cAMP Concentration. It is well known that cAMP affects several events required for platelet aggregation, and its increased level leads to platelet inactivation via protein kinase activation or Ca^{2+} lowering effect. Actually, the antiaggregation factor PGI_2 elevated the intracellular cAMP concentration more than 30-fold as compared with the control (Table III). Since garlic oil did not inhibit aggregation induced by arachidonate metabolites as described above, it is predictable that garlic oil would not have any elevating effect on cAMP. As expected, garlic oil did not show any effect on the cAMP level (Table III). Thus, the ineffectiveness of garlic oil on platelet calcium mobilization was revealed by two means, $[Ca^{2+}]i$ and cAMP assays.

Table III. Effect of Garlic Oil on the cAMP Concentration of Human Platelets

	cAMP pmol per 6×10^7 cells[a]
Control	1.8 ± 1.0
PGI_2 (0.1 mM)	67.5 ± 15.2
Garlic oil (10 μg)[b]	2.5 ± 1.5
(100 μg)	2.2 ± 1.1

[a]Mean ± S.D. (n=4) [b]μg/ml AGFP

MATS-Uptake by Platelets. Ingestion of garlic, its essential oil, or even an active principle, MATS, gives rise to decreased capacity of platelets to aggregate. No evidence has been obtained, however, to indicate that platelets absorb such garlic

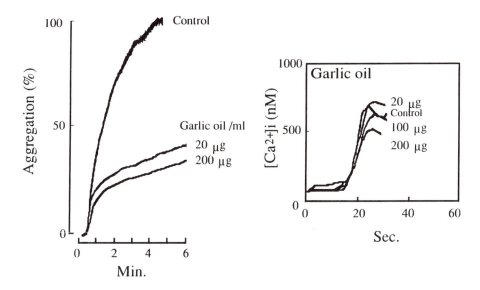

Figure 6. Effect of garlic oil on the aggregation and the intracellular Ca^{2+} concentration of human platelets stimulated with thrombin. The left curves show platelet aggregation, and the right, $[Ca^{2+}]i$ increase. The amount of garlic oil represents $\mu g/ml$ of washed platelets.

components. In advance of studying *in vivo* turnover of the components, we examined the *in vitro* uptake of MATS using synthesized (^{35}S)-MATS and washed human platelets. As shown in Figure 7, the majority of (^{35}S)-MATS was absorbed by platelets. Distribution to the membrane reached 92% of the total label absorbed, and only 8% was shared by mitochondria, microsomes and cytosol. This suggests that while the antiaggregation principle in garlic may be absorbed by the platelets *in vitro*, what amount of ingested principle is absorbed by platelets in blood is yet unknown.

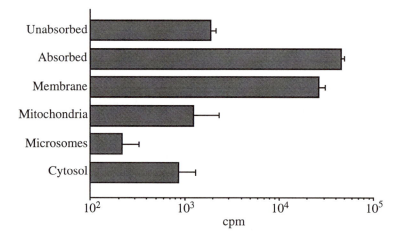

Figure 7. (^{35}S)-MATS uptake by human platelets.

Acknowledgments

We thank Dr. Hanny C. Wijaya and Hideka Suzuki for preparation and analysis of essential oils. This work was supported by the Ministry of Education of Japan (Grant 622-5002) and Nihon University (The Interdisciplinary General Joint Research Grant in 1989).

Literature Cited

1. Bordia, A.; Joshi, H. K.; Sanadhya, Y. K.; Bhu, N. *Atherosclerosis* **1977**, *28*, 155–159.
2. Chutani, S. K.; Bordia, A. *Atherosclerosis* **1981**, *38*, 417–421.
3. Jain, R. C.; Andleigh, H. S. *Br. Med. J.* **1969**, *1*, 514.
4. Augusti, K. T.; Mathew, P. T. *Experientia* **1974**, *15*, 468–469.
5. Bordia, A.; Arora, S. K.; Kothari, K.; Jain, D. C.; Rathore, B. S.; Rathore, A. S.; Dobe, M. K.; Bhu, N. *Atherosclerosis* **1975**, *22*, 103–109.
6. Cang, M. L. W.; Johnson, M. A. *J. Nutr.* **1980**, *110*, 931–936.
7. Bordia, A. *Atherosclerosis* **1978**, *30*, 355–360.
8. Makheja, A. N.; Vanderhoeck, J. Y.; Bailey, J. M. *Lancet* **1979**, *I*, 781.
9. Ariga, T.; Oshiba, S.; Tamada, T. *Lancet* **1981**, *1*, 150–151.

10. Block, E.; Ahmad, S.; Jain, M. K.; Crecely, R. W.; Apitz-Castro, R.; Cruz, M. R. *J. Am. Chem. Soc.* **1984**, *106*, 8295–8296.

11. Apitz-Castro, R.; Escalante, J.; Vargas, R.; Jain, M. K. *Thromb. Res.* **1986**, *42*, 303–311.

12. Block, E.; Ahmad. S.; Catalfamo, L.; Jain, M. K.; Apitz-Castro, R. *J. Am. Chem. Soc.* **1986**, *108*, 7045–7055.

13. Block, E. *Scientific American* **1985**, *March*, 94–99.

14. Yu, T.; Wu, C.; Liou, Y. *J. Agric. Food Chem.* **1989**, *37*, 725–730.

15. Yu, T.; Wu, C.; Chen, S. *J. Agric. Food Chem.* **1989**, *37*, 730–734.

16. Yu, T.; Wu, C.; Liou, Y. *J. Food Sci.* **1989**, *54*, 632–635.

17. Lawson, L. D.; Ranson, D. K.; Hughes, B. G. *Thromb. Res.* **1992**, *65*, 141–156.

18. Bills, T. K.; Smith, J. B.; Silver, M. J. *Biochim. Biophys. Acta* **1976**, *424*, 303–314.

19. Smith, J. B.; Dangelmaier, C.; Selak, M. A.; Ashby, B.; Daniel, J. *Biochem. J.* **1992**, *283*, 889–892.

20. Timmons, S.; Hawiger, J. *Thromb. Res.* **1978**, *12*, 297–306.

21. Yurugi, S.; Matsuoka, T.; Togashi, M. *J. Pharm. Soc. Jpn.* **1954**, *74*, 1017.

22. Nishimura, H.; Wijaya, C. H.; Mizutani, J. *J. Agric. Food. Chem.* **1988**, *36*, 563–566.

23. Freeman, G. G.; Whenham, R. J. *J. Sci. Food Agr.* **1975**, *26*, 1869–1886.

24. Freeman, G. G.; Whenham, R. J. *J. Sci. Food Agr.* **1975**, *26*, 1333–1346.

25. Borndnitz, M.H.; Pascale, J.V.; Van Derslice, L. *J. Agric. Food Chem.* **1971**, *19*, 273–275.

26. Bills, T. K.; Smith, J. B.; Silver, M. J. *J. Clin. Invest.* **1977**, *60*, 1–6.

27. Ali, M.; Thomsom, M.; Alnaqeeb, M. A.; Al-Hassan, J. M.; Khater, S. H.; Gomes, S. A. *Prostaglandins Leukotrienes Essential Fatty Acids* **1990**, *41*, 95–99.

28. Apitz-Castro, R.; Cabrera, S.; Cruz, M. R.; Ledezma, E.; Jain, M. K. *Thromb. Res.* **1983**, *32*, 155–169.

29. Roth, G. J.; Stanford, N.; Majerus, P. W. *Proc. Natl. Acad. Sci. USA* **1975**, *72*, 3073–3076.

30. Willis, A. L.; Vane, F. M.; Kuhn, D. C.; Scott, C. G.; Pettrin, M. *Prostaglandins* **1974**, *8*, 453–507.

RECEIVED April 28, 1993

Chapter 10

Thermal Decomposition of Alliin, the Major Flavor Component of Garlic, in an Aqueous Solution

Tung-Hsi Yu[1], Chi-Kuen Shu[2], and Chi-Tang Ho[1]

[1]Department of Food Science, Cook College, Rutgers, The State University of New Jersey, New Brunswick, NJ 08903
[2]Bowman Gray Technical Center, R.J. Reynolds Tobacco Company, Winston-Salem, NC 27102

Aqueous solutions of alliin, the major flavor precursor of garlic, were heated in a closed system at 180°C under different pH conditions. The volatile flavor compounds generated were isolated by Likens-Nickerson simultaneous steam distillation/solvent extraction and quantified and identified by GC and GC-MS. A total of 49 volatile compounds were identified in this study. Except allyl alcohol and acetaldehyde, the majority of compounds identified were sulfur-containing compounds. A mechanism for the initial transformation of alliin to allyl alcohol and cysteine and further decomposition of cysteine was proposed which could explain the formation of many sulfur-containing compounds such as methyl sulfides, thiazoles, trithiolanes and dithiazines in this study.

Intact garlic cloves contain alliin (S-allylcysteine S-oxide), a colorless and odorless compound. It is also well-known that the enzyme allinase, which is activated when the cellular tissue of garlic is disrupted, converts alliin to allicin (1). Boiling the garlic bulb or homogenizing the garlic bulb with alcohol containing limited quantities of water deactivates the enzyme; alliin is not converted to allicin and no pungent odor can be detected from the garlic samples (1,2). Allicin is very unstable and can be converted to allyl sulfides and dithiins, which contribute to the flavor of garlic products (1,3-6). pH and/or thermal effects on the volatile decomposition compounds of allicin, garlic homogenate, garlic oil and diallyl disulfide have been studied (5,7-11), but the thermal effects on the volatiles formed from the decomposition of alliin have not been studied.

Several papers discuss the content of alliin in garlic products. Stoll and Seebeck (12) reported that intact garlic contains 0.24% (w/w) alliin. Iberl et al. (13) found that the alliin content in fresh garlic bulbs from different origins was approximately 0.9%. Ziegler and Sticher (14) reported that the alliin content in various garlic samples, including fresh garlic, dried extracts and garlic preparations ranges from <0.1 to 1.15%.

0097–6156/94/0546–0144$06.00/0

There are also several different reports about the stability of alliin. Stoll and Seebeck (*1*) determined that alliin was stable in an aqueous solution even at high temperature, but if it was heated in a diluted methanol or ethanol to 100°C, the solution immediately became a dark red color and ammonia and carbon dioxide were given off (*1*). Sreenivasamurthy *et al.* (*3*) found that alliin remained stable during long term storage, in both an aqueous extract and dehydrated garlic powder.

Garlic is used worldwide as one of the most common spices in food preparation. The contributions of allicin, the primary flavor compound of garlic, and allyl sulfide compounds (garlic oil), the secondary flavor compounds of garlic, have been well studied. On the other hand, the contribution of alliin through thermal decomposition to the flavor of food products has not yet been studied. In this paper, we report on our studies of the volatile compounds from thermal decomposition of alliin in aqueous solutions of different pHs.

Experimental

Synthesis of Alliin. Alliin was synthesized from L-deoxyalliin according to the method of Iberl *et al.* (*13*) with slight modification as shown below.

Synthesis and Purification of L-Deoxyalliin (*S*-Allyl-L-cysteine). L-Cysteine (1 mol) was suspended in 3 l absolute ethanol and kept in an ice bath. Allyl bromide (1.1 mol, 99%, Aldrich Chemical Co., Milwaukee, WI) was added to the stirred suspension followed by the addition of 3.5 mol NaOH as a 20 M aqueous solution. After stirring for 1 hr, a solution of crude deoxyalliin formed and was acidified to pH 5.3 by glacial acetic acid at 30°C. After crystallization at 4°C, the white needle-like crystals were filtered, washed twice with absolute ethanol and dried at 50°C. They were recrystallized at 4°C by dissolving the substance in a small portion of boiling water containing 1% glacial acetic acid. This solution was poured into a 15-fold amount of boiling ethanol. The solvent immediately turned turbid with the appearance of small plates of deoxyalliin. After cooling down to room temperature, the solution was left to stand at 4°C. Filtration was followed by washing with ethanol and drying at 50°C. This product was purified L-deoxyalliin.

Synthesis and Purification of L-Alliin (*S*-allyl-L-cysteine sulfoxide). L-Deoxyalliin (0.3 mol) was dissolved in 500 ml distilled water, then 0.6 mol hydrogen peroxide (30% w/w, Aldrich) was added slowly to the solution, and stirring was continued for 24 hours at room temperature. The solvent was evaporated at 60°C under vacuum. The dry, white residue was dissolved in a 500 ml mixture of acetone:water:glacial acetic acid (65:34:1) at boiling temperature. While cooling down to room temperature, long white needles started to grow. To complete precipitation of the oxidized amino acid, the solution was kept at 4°C overnight. After filtration of the precipitate, it was washed with acetone:water: glacial acetic acid (65:34:1), followed by washing with absolute ethanol. The precipitate was then dried at 50°C. The product was pure L-alliin.

Thermal Decomposition of L-Alliin. L-Alliin (0.005 mol) was dissolved in 100 ml distilled water. The solution was adjusted to pH 3, 5, 7 or 9, then added to a 0.3 liter Hoke SS-DOT sample cylinder (Hoke Inc., Clifton, NJ) and sealed. The solution was heated at 180°C in an oven for 1 hr. After being cooled down to room temperature, a decomposed L-alliin mass was obtained.

Isolation of the Volatile Compounds. The total reaction mass was simultaneously distilled and extracted into diethyl ether using a Likens-Nickerson apparatus. After distillation, 5 ml heptadecane stock solution (0.0770 g in 200 ml diethyl ether) was added to the isolate as an internal standard. After being dried over anhydrous sodium sulfate and filtered, the distillate was concentrated to about 5 ml using a Kurdena-Danish apparatus fitted with a Vigreaux distillation column, then slowly concentrated further under a stream of nitrogen in small sample vial to a final volume of 0.2 ml.

Gas Chromatographic Analysis. A Varian 3400 gas chromatograph equipped with a fused silica capillary column (60 m x 0.25 mm i.d.; 1 µm thickness, DB-1, J & W Inc.) and a flame ionization detector was used to analyze the volatile compounds. The operating conditions were as follows: injector temperature, 270°C; detector temperature, 300°C; helium carrier flow rate, 1 ml/min; temperature program, 40°C (5 min), 2°C/min, 260°C (60 min). A split ratio of 50:1 was used.

GC-MS Analysis. The concentrated isolate was analyzed by gas chromatography-mass spectrometry (GC-MS) using Hewlett-Packard 5840A gas chromatograph coupled to a Hewlett-Packard 5985B mass spectrometer equipped with a direct split interface and the same column used for the gas chromatography. The operating conditions were the same as described above. Mass spectra were obtained by electron ionization at 70 eV and an ion source temperature of 250°C.

Identification of the Volatile Compounds. Identification of the volatile compounds in the isolate was mostly based on GC-MS, and information from the GC retention index (I_k) using a C_5-C_{25} mixture as a reference standard. The structural assignment of volatile compounds was accomplished by comparing the mass spectral data with those of authentic compounds available from the Browser-Wiley computer library, NBS computer library or previously published literature (15,16). The retention indices were used for the confirmation of structural assignments.

Results and Discussion

The pH changes, final appearance and flavor description of thermal decomposition solutions of alliin are listed in Table I. After thermal decomposition, the pH of the solution changed, the color developed and various flavors were also generated.

Table I. Initial pH, Final pH, Final Appearance and Flavor Description of Thermally Decomposed Alliin in Aqueous Solution

Initial pH	Final pH	Final Appearance	Flavor Description
3.0	4.0	earthy-yellow	sour sulfuryl with black mushroom and slight roast meaty
5.0	4.5	slight dark-green	ethereal, sour, sulfuryl, popcorn-like, roast meaty
7.0	6.3	earthy, dark-yellow	sulfuryl, roast nutty-meaty
9.0	8.0	slight brown-yellow	sulfuryl, roast nutty-meaty

The volatile compounds identified after thermal decomposition of alliin in aqueous solutions of different pHs are listed in Table II. Allyl alcohol and acetaldehyde were the dominant volatile compounds found in all samples. The formation of allyl alcohol from alliin could be explained by the mechanism shown in Figure 1. [2,3]-Sigmatropic rearrangement of alliin may lead to intermediate sulfenate (**I**). The reduction of sulfenate (**I**) will yield allyl alcohol and cysteine. A similar mechanism has been proposed by Park *et al.* (*17*) to explain the formation of acrolein from S-(3-chloro-2-propenyl)cysteine.

Table II. Volatile Compounds Identified from the Thermal Decomposition of Alliin in Aqueous Solutions of Different pH at 180°C

Compound	mg/mol			
	pH 3	pH 5	pH 7	pH 9
1-propene	1.1	39.9	n.d.*	1.3
sulfur dioxide	34.1	258.7	0.7	9.4
acetaldehyde	132.1	1199.6	364.0	131.3
allyl alcohol	1452.9	93.3	1149.5	929.2
ethyl acetate	83.5	232.3	0.8	33.4
acetic acid	1.4	72.7	n.d.	n.d.
2-butenal	n.d.	n.d.	n.d.	1.5
4-pentenal	n.d.	n.d.	n.d.	10.1
2-methyl-1,3-dioxane	n.d.	18.2	n.d.	n.d.
4-methyl-1,3-dioxane	n.d.	10.4	n.d.	n.d.
thiazole	9.8	9.9	25.6	1.8
acetal	35.5	26.4	1.6	trace**
dimethyl disulfide	n.d.	11.1	n.d.	n.d.
2-methylthiazole	n.d.	7.5	4.5	1.2
2-methylpyridine	13.4	n.d.	n.d.	n.d.
ethyl methanesulfinate	n.d.	5.1	n.d.	n.d.
methyl ethyldisulfide	n.d.	62.0	n.d.	n.d.
2-ethyl-1,3-dioxolane	n.d.	n.d.	0.9	trace
methyl propyldisulfide	n.d.	7.5	n.d.	n.d.
methyl allyldisulfide	n.d.	7.7	n.d.	n.d.
dimethyl trisulfide	n.d.	11.5	n.d.	n.d.
1,2-dithiacyclopentane	trace	1.7	n.d.	n.d.
2-formylthiophene	4.8	n.d.	1.5	13.3
3-formylthiophene	4.2	3.5	trace	5.6
2-acetylthiazole	3.2	359.5	16.7	0.7
2,5-dithiahexane	n.d.	n.d.	n.d.	0.5
3-methyl-1,2-dithiolane-4-one	n.d.	n.d.	n.d.	1.9
2-ethylthiazole	n.d.	n.d.	n.d.	3.0
methyl propyltrisulfide	n.d.	3.6	n.d.	n.d.
1,3-dithiane	2.6	n.d.	n.d.	n.d.
2-formyl-3-methylthiophene	trace	n.d.	n.d.	n.d.
3,5-dimethyl-1,2,4-trithiolane	n.d.	n.d.	19.1	3.8

Continued on next page

Table II. Continued

Compound	mg/mol			
	pH 3	pH 5	pH 7	pH 9
2-methyl-1,3-dithiacyclohexane-5-one	n.d.	n.d.	4.5	0.6
methyl-1,2,3-trithiacyclopentane	22.3	188.0	n.d.	5.3
2,4,6-trimethylperhydro-1,3,5-dithiazine	n.d.	n.d.	70.0	5.1
1,2,3-trithiacyclohexane	4.4	n.d.	n.d.	0.5
dimethyl tetrasulfide	n.d.	5.3	n.d.	n.d.
dipropyl trisulfide	trace	n.d.	n.d.	n.d.
3-ethyl-5-methyl-1,2,4-trithiolane	n.d.	n.d.	10.9	n.d.
3-ethyl-1,2,4-trithiacyclo-hexane	n.d.	n.d.	13.4	n.d.
methyl-1,2,3,4-tetrathiacyclo-hexane	n.d.	57.5	n.d.	n.d.
1,2,3,4-tetrathiacycloheptane	214.8	9.7	n.d.	0.8
3,6-dimethyl-1,2,4,5-tetra-thiacyclohexane	n.d.	n.d.	6.9	trace
4,6-dimethyl-1,2,3,5-tetra-thiacyclohexane	n.d.	20.9	n.d.	3.5
4-ethyl-6-methyl-1,2,3,5-tetrathiacyclohexane	trace	trace	14.9	13.7
total	2020.1	2723.5	1705.5	1180.7

*n.d.: not detected
**trace: <0.5 mg/mol

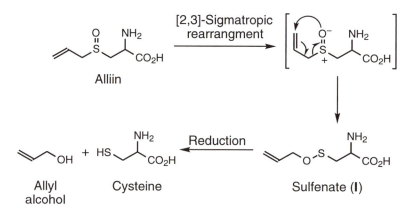

Figure 1. Proposed mechanism for the formation of allyl alcohol and cysteine from alliin.

Among the volatiles identified in this study, acetaldehyde, ethyl acetate, thiazole, 2-methylthiazole, 2-methylpyridine, 3-formylthiophene, 2-acetylthiazole 3-methyl-1,2-dithiolane-4-one, 3,5-dimethyl-1,2,4-trithiolane, 2,4,6-trimethyl-perhydro-1,3,5-dithiazine, 3,6-dimethyl-1,2,4,5-tetrathiacyclohexane and 4,6-dimethyl-1,2,3,5-tetrathiacyclohexane have been identified from the thermal decomposition products of cysteine (*18,19*). It is reasonable to assume that these compounds were also generated from the decomposition of cysteine, which is formed as the result of thermal decomposition of alliin in this study.

Obata and Tanaka (*20,21*) studied the photolysis of cysteine and cystine and observed that hydrogen sulfide, ammonia, carbon dioxide and acetaldehyde were produced. Fujimaki *et al.* (*22*) pyrolyzed cysteine and cystine separately at 270–300°C under reduced pressure and nitrogen. They identified several highly volatile compounds including hydrogen sulfide, ammonia, acetaldehyde. Boelens *et al.* (*23*) studied the degradation of cysteine/cystine in terms of the primary products and secondary products. They postulated a mechanism for the interaction of acetaldehyde and hydrogen sulfide in which both came from the decomposition of the amino acids. The interaction of acetaldehyde and hydrogen sulfide formed bis(1-mercaptoethyl) sulfide which was oxidized to 2,4,6-trimethyl-1,3,5-trithiane or to 2,4,6-trimethylperhydro-1,3,5-dithiazine in the presence of ammonia. In addition, bis(1-mercaptoethyl) sulfide was disproportioned into different sulfides including ethyl disulfide and ethyl trisulfide. Instead of ethyl sulfides, methyl sulfides, methyl disulfide, methyl trisulfide, and methyl tetrasulfide were identified in this study. After the formation of cysteine from the decomposition of alliin, cysteine would further decompose into small volatile compounds, such as acetaldehyde, hydrogen sulfide and ammonia. These small volatile compounds would then react with each other to form heterocyclic sulfur-containing compounds. The proposed mechanisms for the formation of some of the heterocyclic sulfur-containing compounds are shown in Figure 2.

2-Acetylthiazole and 2-methylthiazole were also generated from the decomposition of cysteine. Their probable formation mechanisms are shown in Figure 3. Some of the oxygen-containing compounds, such as acetic acid, ethyl acetate, acetal and 2-butenal, are postulated to be generated from acetaldehyde. Their possible formation mechanisms are shown in Figure 4.

Similarly to the degradation of cysteine (*18*), the volatile profiles of the degradation of alliin were found to be highly pH dependent. At pH 5, a degradation occurred leading predominately to dioxanes, methyl sulfides, 2-acetylthiazole, methyl-1,2,3-trithiacyclopentane and methyl-1,2,3,4-tetrathiacyclohexane. At lower pH (pH 3), the formation of acetal, 2-methylpyridine and 1,2,3,4-tetrathiacycloheptane was favored. On the other hand, the formation of 3,5-dimethyl-1,2,4-trithiolane, 2,4,6-trimethylperhydro-1,3,5-dithiazine and 4-ethyl-6-methyl-1,2,3,5-tetrathiacyclohexane was favored under neutral or alkaline pH (pH 7 or 9) conditions.

Conclusions

During thermal treatment in aqueous solution, alliin is postulated to decompose into allyl alcohol and cysteine. Cysteine then decomposes further into acetaldehyde, hydrogen sulfide and ammonia. These small volatile compounds then react with each other to form methyl sulfides, thiazoles, thrithiolanes and cyclic sulfur-containing volatile compounds. The pH condition of the aqueous solution affects

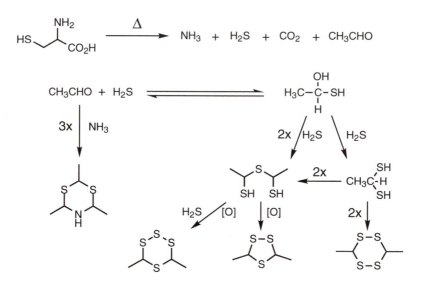

Figure 2. Proposed mechanisms for the formation of several volatile compounds from cysteine.

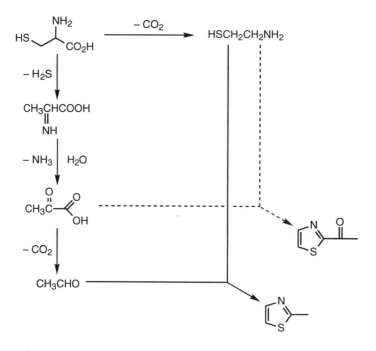

Figure 3. Proposed mechanism for the formation of 2-acetylthiazole and 2-methylthiazole from cysteine.

the pattern of thermal decomposition of alliin to form different kinds of volatile compounds, which contribute different flavor sensations to the decomposed products.

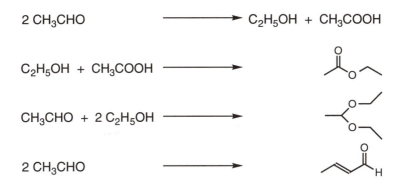

Figure 4. Proposed mechanisms for the formation of some of the oxygen-containing volatile compounds from acetaldehyde.

Acknowledgements

We thank Mrs. Joan Shumsky for her secretarial aid. New Jersey Agricultural Experiment Station Publication No. D-10205-9-92 supported by State Funds.

Literature Cited

1. Stoll, A.; Seebeck, E. *Adv. Enzymol.* **1951**, *11*, 377–400.
2. Ueda, Y.; Sakaguchi, M.; Hirayama, K.; Miyajima, R.; Kimizuka, A. *Agric. Biol. Chem.* **1990**, *54*, 163–169.
3. Sreenivasamurthy, V.; Sreekantiah, K. R.; Johar, D. S. *J. Sci. Ind. Res.* **1961**, *20C*, 292–295.
4. Block, E. *Sci. Amer.* **1985**, March, 94–99.
5. Yu, T. H.; Wu, C. M. *J. Food Sci.* **1989**, *54*, 977–981.
6. Yu, T. H.; Wu, C. M.; Liou, Y. C. *J. Agric. Food Chem.* **1989**, *37*, 725–730.
7. Yu, T. H.; Wu, C. M. *J. Chromatogr.* **1989**, *462*, 137–145.
8. Yu, T. H.; Wu, C. M.; Chen, S.Y. *J. Agric. Food Chem.* **1989**, *37*, 730–734.
9. Yu, T. H.; Wu, C. M.; Liou, Y.C. *J. Food Sci.* **1989**, *54*, 632–635.
10. Yu, T. H.; Wu, C. M. *Food Sci.(ROC)* **1988**, *15*, 385–393.
11. Block, E.; Iyer, R.; Grisoni, S.; Saha, C.; Belman, S.; Lossing, F. P. *J. Amer. Chem. Soc.* **1988**, *110*, 7813–7827.
12. Stoll, A.; Seebeck, E. *Helv. Chim. Acta.* **1948**, *31*, 189–210.
13. Iberl, B.; Winkler, G.; Muller, B.; Knobloch, K. *Planta Med.* **1990**, *56*, 320–325.
14. Ziegler, S. J.; Sticher, O. *Planta Med.* **1989**, *55*, 372–378.
15. Kawai, T. *Thiadiazine and dithiazines in volatiles from heated seafoods, and mechanisms of their formation*; Shiono Koryo Kaisha, Ltd.: Yodogawaku, Osaka, Japan, 1991; pp 14–88.

16. Zhang, Y. *Thermal generation of food flavors from cysteine and glutathione*, Ph.D. Thesis, Rutgers University, New Brunswick, New Jersey, 1991.
17. Park, S. B.; Osterloh, J. D.; Vamvakas, S.; Hashmi, M.; Andres, M. W.; Cashman, J. P. *Chem. Res. Toxicol.* **1992**, *5*, 193–201.
18. Shu, C. K.; Hagedorn, M. L.; Mookherjee, B. D.; Ho, C.-T. *J. Agric. Food Chem.* **1985**, *33*, 442–446.
19. Zhang, Y.; Chien, M.; Ho, C.-T. *J. Agric. Food Chem.* **1988**, *36*, 992–996.
20. Obata, Y.; Tanaka, H. *Agric. Biol. Chem.* **1965**, *29*, 191–195.
21. Obata, Y.; Tanaka, H. *Agric. Biol. Chem.* **1965**, *29*, 196–199.
22. Fujimaki, M.; Kato, S.; Kurata, T. *Agric. Biol. Chem.* **1969**, *33*, 1144–1151.
23. Boelens, M.; van der Linde, L. M.; de Valois, P. J.; van Dort, H. M.; Takken, H. J. *J. Agric. Food Chem.* **1974**, *22*, 1071–1076.

RECEIVED July 6, 1993

OTHER SULFUR-CONTAINING PHYTOCHEMICALS

Chapter 11

Chemoprotection by 1,2-Dithiole-3-thiones

T. W. Kensler[1,2], J. D. Groopman[1], B. D. Roebuck[3], and T. J. Curphey[4]

Departments of [1]Environmental Health Sciences and [2]Pharmacology and
Molecular Sciences, Johns Hopkins Medical Institutions,
Baltimore, MD 21205
Departments of [3]Pharmacology and Toxicology and [4]Pathology, Dartmouth
Medical School, Hanover, NH 03755

1,2-Dithiole-3-thiones are five-membered cyclic sulfur-containing
compounds with antioxidant, chemotherapeutic, radioprotective as
well as cancer chemoprotective properties. One substituted dithiolethione, oltipraz [5-(2-pyrazinyl)-4-methyl-1,2-dithiole-3-thione],
which was originally developed as an antischistosomal agent, has
been shown to protect against cancers induced in rodents by several
different classes of chemical carcinogens in multiple tissues, including lung, trachea, forestomach, colon, breast, skin, liver and
urinary bladder. Induction of electrophile detoxication enzymes,
resulting in diminished carcinogen-DNA adduct formation and
reduced cytotoxicity, appears to be an important component of the
anticarcinogenic action of oltipraz and other 1,2-dithiole-3-thiones.
Phase I clinical trials of this drug are presently underway in the
United States.

Several substituted 1,2-dithiole-3-thiones exhibit chemotherapeutic and radioprotective, as well as chemoprotective, properties (1–4). One of these agents,
oltipraz [5-(2-pyrazinyl)-4-methyl-1,2-dithiole-3-thione], has shown significant antischistosomal activity in experimental animals and in humans. Cure rates of up to 90
percent have been achieved with single doses of oltipraz in field trials (4). Another
substituted 1,2-dithiole-3-thione in clinical use is anethole dithiolethione [5-(p-
methoxyphenyl)-1,2-dithiole-3-thione]. Anethole dithiolethione is used in Canada
and Europe as a choleretic and a stimulant of salivary secretion to counteract the
dryness of the mouth caused by psychotropic drugs (5). In general, clinical studies
have indicated minimal side effects with these dithiolethione drugs. The structures
of oltipraz, anethole dithiolethione, and unsubstituted 1,2-dithiole-3-thione are
shown in Figure 1.

During the course of studies on the mechanisms of antischistosomal activity
of 1,2-dithiole-3-thiones, Bueding et al. (6) initially noted that administration of
oltipraz to mice infected with *Schistosoma mansoni* caused a reduction in the glutathione stores of the parasites, but increased levels of glutathione in many tissues of

0097–6156/94/0546–0154$06.00/0

the host (7). Subsequent studies demonstrated that oltipraz, anethole dithiolethione, and related 1,2-dithiole-3-thiones were potent inducers of enzymes concerned with the maintenance of reduced glutathione pools as well as enzymes important to electrophile detoxication. Notably, elevated NAD(P)H:quinone reductase, epoxide hydrolase, glutathione S-transferase (GST) and UDP-glucuronosyl transferase (UGT) activities have been observed in many organs of rats and mice treated with oltipraz. The elevation of electrophile detoxication enzymes has been recognized as characteristic of the action of many chemoprotective agents as exemplified by the antioxidants BHA, BHT and ethoxyquin (8). Like these compounds, oltipraz has subsequently been shown to be an effective chemoprotective agent in nearly a dozen different models of experimental carcinogenesis.

A broad range of anticarcinogenic activity coupled with its apparently low mammalian toxicity has prompted the continued development of oltipraz as a potential human chemoprotective agent. It is currently undergoing Phase I trials in the United States to determine its pharmacokinetics and dose-limiting side effects during low dose, chronic administration to humans (9).

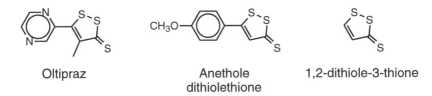

| Oltipraz | Anethole dithiolethione | 1,2-dithiole-3-thione |

Figure 1. Structures of oltipraz, anethole dithiolethione, and 1,2-dithiole-3-thione.

Chemoprotection in Experimental Models

The biochemical manifestations of oltipraz in schistosome-infected mice prompted Bueding to predict that this drug might have cancer chemoprotective properties. The initial confirmation that 1,2-dithiole-3-thiones may exert chemoprotective effects *in vivo* came from the demonstration that oltipraz and anethole dithiolethione protected against the hepatotoxicity of carbon tetrachloride and acetaminophen in mice (10). Subsequent studies have demonstrated protection by oltipraz against the acute hepatotoxicities of allyl alcohol and acetaminophen in the hamster (11,12) and aflatoxin B_1 in the rat (13). Toxin-induced elevations in liver function tests were blunted in all cases. Pretreatment with oltipraz also substantially reduced the mortality produced by either single or chronic exposure to aflatoxin B_1 (13).

To directly test the cancer chemoprotective activity of oltipraz, Wattenberg and Bueding (14) examined the capacity of oltipraz to inhibit carcinogen-induced neoplasia in mice. Oltipraz was administered either 24 or 48 hours before treatment with each of three chemically diverse carcinogens: diethylnitrosamine, uracil mustard, and benzo[a]pyrene. This sequence of oltipraz and carcinogen administration was repeated once a week for 4 to 5 weeks. As depicted in Figure 2, oltipraz reduced by nearly 70 percent the number of both pulmonary adenomas and tumors of the forestomach induced by benzo[a]pyrene. Pulmonary adenoma formation induced by uracil mustard or diethylnitrosamine was also significantly reduced by oltipraz pretreatment, but to a lesser extent. Oltipraz has subsequently been shown to have chemoprotective activity against different classes of carcinogens targeting the breast (*15,16*), colon (*17*), liver (*18*), skin (*19*), trachea (*15*) and bladder (*15*). Unfortunately, the full experimental details of only a few of these studies have been published, so that it is difficult at this time to compare the potency and efficacy of oltipraz as an anticarcinogen against other chemoprotective agents affecting the same target sites. Nonetheless, the potency of oltipraz is highlighted by the observations in F344 rats that dietary concentrations of 200 and 400 ppm oltipraz significantly reduced tumor incidence and multiplicity in azoxymethane-induced intestinal carcinogenesis (*17*) while 750 ppm oltipraz in the diet afforded complete protection against aflatoxin B_1-induced hepatocarcinogenesis (*18*). Moreover, dietary concentrations as low as 100 ppm engendered >90 percent reductions in the hepatic burden of presumptive preneoplastic lesions in the aflatoxin model (*20*). To date, one negative study with oltipraz has been reported. Pepin *et al.* (*21*) have observed that feeding mice 250 ppm oltipraz had no effect on pulmonary tumorigenesis induced by the tobacco specific nitrosamine 4-(methylnitrosamino)-1-(3-pyridyl)-1-butanone. Whether effects of tobacco specific carcinogens can be attenuated in other target organs such as pancreas and bladder remains to be established. Moreover, the effects of oltipraz on other tobacco-related carcinogens such as the aromatic amines *o*-toluidine and 4-aminobiphenyl have not as yet been investigated.

The role of oltipraz in combination chemoprotection is being studied. When fed at 40 and 80 percent of the maximum tolerated dose to mice, oltipraz had no effect in preventing urinary bladder carcinogenesis induced by *N*-butyl-*N*(4-hydroxybutyl)-nitrosamine. When combined with α-difluoromethylornithine, however, significant, dose-dependent inhibition was observed (*22*). Oltipraz in combination with either the retinoid *N*-(4-hydroxyphenyl)retinamide or β-carotene was also very effective against diethylnitrosamine-induced respiratory carcinogenesis (*23*). Thus, the use of oltipraz in combination with agents exhibiting different mechanisms of action appears promising.

Mechanisms of Chemoprotection by 1,2-Dithiole-3-thiones

Although oltipraz exerts chemoprotective activity in a variety of animal models, very few studies have addressed possible anticarcinogenic mechanisms in these systems. Almost all of the chemoprotection protocols tested to date have involved concomitant exposure to both carcinogen and oltipraz. It seems likely, therefore, that oltipraz is affecting the metabolism and/or disposition of carcinogens (Figure 3). Thus, possible mechanisms to explain the observed protective effects of oltipraz might include [1] inhibition of phase I enzymes to retard metabolic activation; [2] induction of phase I enzymes (i.e., cytochrome P-450s) to enhance carcinogen detoxication; [3] induction of phase II xenobiotic-metabolizing enzymes (e.g., GSTs, UGTs) to enhance carcinogen detoxication and elimination; [4] nucleophilic trapping of reactive intermediates; and [5] enhancement of DNA repair processes. Post-initiation effects [6] of oltipraz may also occur but have not been extensively examined in anticarcinogenesis bioassays.

Oltipraz inhibits aflatoxin B_1-induced hepatocarcinogenesis by modifying the metabolism and disposition of the carcinogen. In particular, alterations in the balance of competing pathways of the ultimate carcinogen, aflatoxin-8,9-oxide, directly modulate the availability of this epoxide for binding to DNA. Anticarcino-

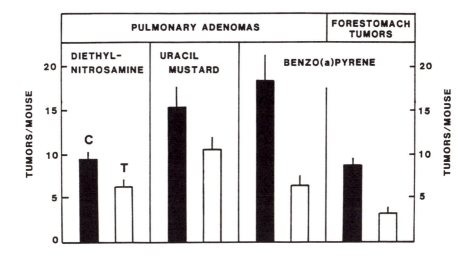

Figure 2. Effect of pretreatment with oltipraz on carcinogen–induced neoplasia in female mice. C, control; T, pretreated with oltipraz. (Adapted from ref. *17*).

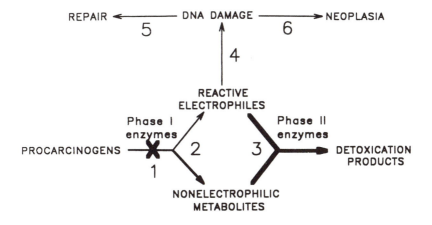

Figure 3. General effects of oltipraz on carcinogen metabolism and disposition. Numbers indicate possible mechanisms of action; see text for details.

genic concentrations of oltipraz in the diet markedly induce the activities of GSTs in tissues of the rat to facilitate the conjugation of glutathione to the aflatoxin-8,9-oxide, thereby enhancing its elimination and coordinately diminishing DNA adduct formation (20,24). Feeding 750 ppm oltipraz for 1 week before exposure to aflatoxin B_1 increases the initial rate of the biliary elimination of aflatoxin-glutathione conjugate by nearly three-fold (25). Concordantly, feeding oltipraz led to three to four-fold increases in the specific activity of rat liver GST (20). In vitro studies have shown that purified rat GSTs of the alpha class will conjugate aflatoxin-8,9-oxide to glutathione (26). Three-fold elevations in hepatic levels of the alpha GST subunit Ya have been measured 3 to 7 days after inclusion of oltipraz in the diet (27). Slot blot analysis using a full-length cDNA probe for the rat GST Ya gene showed that the steady state levels of hepatic GST Ya mRNA were elevated in response to oltipraz. Nuclear runoff experiments suggested that the initial increases in hepatic GST Ya mRNA and protein levels in response to oltipraz were mediated through transcriptional activation of the GST Ya gene (27). Several regulatory elements controlling the expression and inducibility of the Ya subunit of rat GST have been identified (28). A 41 bp element in the 5'-flanking region of the rat GST Ya gene, termed the "Antioxidant Response Element," appears likely to mediate induction by oltipraz, but the molecular details of the signaling pathway for the inductions of GST Ya and other phase II enzymes by oltipraz or other 1,2-dithiole-3-thiones remain to be fully established. While the role of GSTs as determinants of sensitivity of aflatoxin B_1 hepatocarcinogenesis appears reasonable, other phase II enzymes, notably UGTs and NAD(P)H:quinone reductase, are likely to exert important effects on the detoxication of other carcinogens and need to be explored in greater detail. A significant attribute of oltipraz is the responsiveness of many tissues (e.g., liver, lung, intestinal mucosa, kidney) to the enzyme inductive actions of this drug.

Oltipraz is a very weak inducer of some cytochrome P-450 enzymes. Incubation with aflatoxin B_1 of microsomes prepared from rats fed oltipraz produces moderately larger amounts of some oxidative metabolites than do control microsomes (20), although no increases in aflatoxin-8,9-oxide have been observed in hepatic microsomes from oltipraz treated rats (29). An evaluation by Putt et al. (29) of the effect of feeding oltipraz to rats on the induction of different classes of hepatic cytochrome P-450s, as determined by Western blotting, indicates no effects on 1A1, 2B1, 2C12, 3A1/2 and 3A4 levels while small increases are seen with 2C11 and 2E1. Direct addition of oltipraz to untreated microsomes also affects the metabolism of aflatoxin B_1. Addition of 100 μM oltipraz to microsomal incubations results in substantial inhibition of the formation of aflatoxin-8,9-oxide (29). The nature of this inhibition has not been characterized as yet, nor is it known whether pharmacologically achievable concentrations in the liver affect cytochrome P-450 activities in vivo. Although oltipraz reduces the levels of aflatoxin-8,9-oxide in vitro, oltipraz does not appear to function as a direct nucleophilic trap for aflatoxin-8,9-oxide. This conclusion is supported by studies indicating no difference in the levels of aflatoxin-N^7-guanine as measured by HPLC when aflatoxin-8,9-oxide is added directly into a solution containing calf thymus DNA in the absence or presence of excess amounts of oltipraz. Moreover, chromatographic analysis provided no evidence for the formation of aflatoxin-oltipraz adducts (25).

Oltipraz and other inducers of electrophile detoxication enzymes clearly protect against the DNA damaging actions of procarcinogens. It is less clear, however, how such mechanisms can explain the anticarcinogenic effects in models where direct-acting carcinogens are used (e.g., *N*-nitroso-*N*-methylurea (*16*)). Perhaps elevation of cellular thiol levels leads to the trapping of reactive intermediates. Moreover, it is very apparent from molecular dosimetry studies in the aflatoxin model that reduction in target organ DNA adduct burden substantially underestimates the ultimate degree of chemoprotection against disease endpoints (*18,30,31*). It is important to consider, therefore, additional actions by these agents. Cohen and Ellwein (*32*) have recently highlighted the importance of cell proliferation in the neoplastic process. With genotoxic chemicals they have noted a modest dose-dependent increase in tumor prevalence at low doses of carcinogens as a consequence of DNA interactions whereas the much greater increase at higher doses reflects the synergistic influence of increased cell proliferation. The cytotoxic actions of these carcinogens produces increased compensatory proliferation of surviving cells. The powerful protective effect of oltipraz against the cytotoxic actions of hepatotoxins suggests that this agent may act indirectly to reduce cell proliferation, thereby enhancing its anticarcinogenic activity.

Structure-Activity Relationships in Chemoprotection by 1,2-Dithiole-3-thiones

A large number of compounds been synthesized and subsequently employed in structure-activity analysis of the antischistosomal (*6*), enzyme inductive (*7,20,33, 34*), radioprotective (*2*), and chemoprotective (*16,18,35*) actions of 1,2-dithiole-3-thiones. Unlike the situation with the antischistosomal activity of oltipraz, it is quite evident that the cancer chemoprotective activity of this drug is exclusively embodied in the dithiolethione nucleus of the molecule. For example, anethole dithiolethione and unsubstituted 1,2-dithiole-3-thione, which are devoid of antischistosomal activity, show anticarcinogenic activity in animal bioassays. Moreover, unsubstituted 1,2-dithiole-3-thione is a very potent and effective inducer of phase II enzymes such as GSTs and NAD(P)H:quinone oxidoreductase in cell culture (*33*) and in rat liver (*20*). 1,2-Dithiole-3-thione is the most effective of a series of analogues of oltipraz in reducing aflatoxin-DNA adduct formation *in vivo* in rat liver. Substitutions at the 4 or 5 carbons of the dithiolethione nucleus typically reduced protective activity. Conversion of the thione at the 3 position to a ketone also diminished activity (*20*). Dietary concentrations as low as 10 ppm of 1,2-dithiole-3-thione engender substantial protection against aflatoxin B_1-induced tumorigenesis (*35*).

1,2-Dithiole-3-thione possesses two characteristics that make it an attractive chemoprotective compound. First, as shown in Figure 4, 1,2-dithiole-3-thione is a monofunctional inducer (*20,33,36*). Bifunctional inducers, typified by TCDD, phenobarbital, ethoxyquin, and butylated hydroxytoluene, elevate the activities of both phase I enzymes such as cytochrome P450 and phase II enzymes such as the GSTs. In contrast, monofunctional inducers elevate the activities of phase II enzymes selectively (*37*). The monofunctional nature of the induction of phase II enzymes by 1,2-dithiole-3-thione is elegantly demonstrated by De Long *et al.* (*33*) through the use of mutant (Ah receptor function defective) Hepa 1c1c7 cells in which phase I enzymes cannot be induced but phase II enzymes are. This monofunctional characteristic minimizes the potential complications of enhanced procarcinogen activation through the ancillary induction of certain cytochrome P-450s. Second, like oltipraz, 1,2-dithiole-3-thione induces phase II enzymes in multiple tissues following dietary administration, thereby enhancing the potential for protection in a variety of organs.

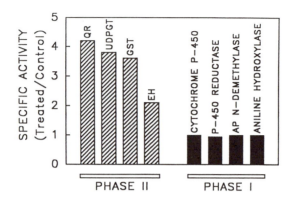

Figure 4. Induction patterns of phase I and phase II enzymes in the livers of rats fed 0.03% 1,2-dithiole-3-thione. (Reproduced with permission from reference 35. Copyright 1992 IRL Press.)

The Role of 1,2-Dithiole-3-thiones in Chemoprotection by Various Plant Species

In 1958, Jirousek and Stárka (*38*) reported, without experimental details, the isolation of two 1,2-dithiole-3-thiones from cabbage and Brussels sprouts. On the basis of UV spectrum, polarographic and chromatographic behavior, and several simple chemical and color tests, these workers tentatively identified one of the dithiolethiones as the parent 1,2-dithiole-3-thione itself. The second dithiolethione was suggested to have an oxygenated aromatic ring at position 5. Further work on the structure of these two substances have not appeared in the literature. Recently, Marks and coworkers (*39*), using current separation and detection techniques, reinvestigated the question of whether the parent dithiolethione occurs naturally in cabbage. These workers were able to show that 1,2-dithiole-3-thione was not present in their cabbage sample at concentrations greater than 1 ppb. Marks and coworkers also examined an extract of Brussels sprouts for the presence of the parent as well as other dithiolethiones. The extract was separated by HPLC and fractions showing strong absorbance at 415 nm were examined by GC-MS for mass fragmentation patterns characteristic of dithiolethiones. No such patterns were detected, suggesting that the parent or other dithiolethiones were not present in the extract from Brussels sprouts. In view of the many factors that might influence the levels of dithiolethiones in plants, the insensitivity of UV detection, the likely complexity of plant extracts, and the inability of GC-MS to detect highly polar molecules (for example, glycoside conjugates), the question of whether dithiole-thiones occur naturally in plants of the *Brassica* family must still be regarded as unresolved. The work of Marks and coworkers, however, makes it less likely that structurally simple dithiolethiones are components of these plant species.

A possibility that seems not to have been considered is that chemoprotective agents related to the dithiolethiones may arise *in vivo* by biotransformation of naturally-occurring precursors. For example, 3*H*-1,2-dithiole has been detected in cooked white asparagus (*40*), in garlic (*41–43*), and in the Amazonian garlic bush (*44*). Oxidation of this compound at C-3 to produce 1,2-dithiole-3-one, a known chemoprotective agent (*20*), might be expected to be facile. Similarly, the methyl-ated derivatives of 3*H*-1,2-dithiole found in garlic oil (*43*) might be expected to give rise to methylated 1,2-dithiole-3-ones *in vivo*. The great potency of the 1,2-dithiole-3-thiones as chemoprotective agents and the widespread occurrence of complex organosulfur compounds in plants whose consumption is associated with chemoprotective activity, suggest that an open mind should be kept concerning the possibility that 1,2-dithiole-3-thiones or the related 1,2-dithiole-3-ones play a role in chemoprotection by these plants.

Conclusions

1,2-Dithiole-3-thiones are a large class of five membered cyclic sulfur-containing compounds with antioxidant, chemotherapeutic, radioprotective, and chemo-protective properties. 1,2-Dithiole-3-thiones are unique among classes of cancer chemoprotective compounds in that they were first identified on the basis of their biochemical properties, which include induction of carcinogen detoxication enzymes, and subsequently evaluated to confirm their anticarcinogenic activities. Two dithiolethiones, oltipraz and anethole dithiolethione, have been extensively used as medicines in humans. The most comprehensively studied 1,2-dithiole-3-thione, oltipraz, has recently been demonstrated to have chemoprotective activity against several classes of carcinogens in many target organs of rodents, including liver, breast, colon, pancreas, lung, trachea, forestomach, bladder, and skin. While many different classes of both natural and synthetic experimental chemoprotectors (i.e., phenolic antioxidants, isothiocyanates, flavonoids, indoles, cinnamates, coumarins, terpenes and others) induce electrophile detoxication enzymes, oltipraz may offer the earliest and easiest prospect for examining the role of enzyme induction as a protective strategy in humans. Unlike the situation with many of the phytochemicals, substantial preclinical research has already been conducted with oltipraz to establish its safety and efficacy in animals. Moreover, Phase I clinical trials of oltipraz are presently underway. Subsequent trials with this agent might be most appropriately targeted towards individuals at high risk for occupational or environmental exposures to genotoxic carcinogens. Conversely, the apparently limited presence of 1,2-dithiole-3-thiones in natural products suggests that they are not likely to contribute substantively to nutritionally-based interventions in the general population.

Acknowledgments

We gratefully acknowledge support for our work in protection against cancer from the National Cancer Institute (CA39416 and CA44530) and the National Institute of Environmental Health Sciences (Center Grant ES03819). J.D.G. is recipient of Research Career Development Award CA01517.

Literature Cited

1. Lozac'h. N.; Vialle, J. In *The Chemistry of Organic Sulfur Compounds*, Kharasch, N.; Meyers, C. Y., Eds; Pergamon Press: Oxford, 1966; Vol. 2, pp 257–285.
2. Teicher, B. A.; Stemwedel, J.; Herman, T. S.; Ghoshal, P. K.; Rosowsky, A. *Br. J. Cancer* **1990**, *62*, 17–22.
3. Kensler, T. W.; Groopman, J. D.; Roebuck, B. D.; In *Cancer Chemoprevention*, Wattenberg, L.; Lipkin, M.; Boone, C. W.; Kelloff, G. J. Eds.; CRC Press: Boca Raton, FL, 1992; pp 205–226.
4. Archer, S. *Ann. Rev. Pharmacol.* **1985**, *25*, 485–508.
5. Häusler, R.; Ritschard, J. *Rev. Suisse Praxis Med.* *1979, 68*, 1063–1068.
6. Bueding, E.; Dolan, P.; Leroy, J-P. *Res. Commun. Chem. Pathol. Pharmacol.* **1982**, *37*, 293–303.
7. Ansher, S. S.; Dolan, P.; Bueding, E.; *Food Chem. Toxicol.* *1986, 24*, 405–415.
8. Prochaska, H. J.; De Long, M. J.; Talalay, P. *Proc. Natl. Acad. Sci. USA* **1985**, *82*, 8232–8236.
9. Benson, A. B. III; Mobarhan, S.; Ratain, M.; Sheehan, T.; Berezin, F.; Giovanazzi–Bannon, S.; Ford, C.; Rademaker, A. *Proc. Amer. Assoc. Clin. Oncol.* **1992**, *11*, 145.
10. Ansher, S. S.; Dolan, P.; Bueding, E. *Hepatology* **1983**, *3*, 932–935.
11. Davies, M. H.; Schamber, G. J.; Schnell, R. C. *The Toxicologist* **1987**, *7*, 219.
12. Davies, M. H.; Schamber, G. J.; Schnell, R. C. *Toxicol. Appl. Pharmacol.* **1991**, *109*, 17–28.
13. Liu, L-Y.; Roebuck, B. D.; Yager, J. D.; Groopman, J. D.; Kensler, T. W. *Toxicol. Appl. Pharmacol.* **1988**, *93*, 442–451.
14. Wattenberg, L. W.; Bueding, E. *Carcinogenesis* **1986**, *7*, 1379–1381.
15. Boone, C. W.; Kelloff, G. J.; Malone, W. E. *Cancer Res.* **1990**, *50*, 2–9.
16. Steele, V. E.; Sigman, C. C.; Boone, C. W.; Kelloff, G. J. *Proc. Amer. Assoc. Cancer Res.* **1992**, *33*, 159.
17. Rao, C. V.; Tokomo, K.; Kelloff, G.; Reddy, B. S. *Carcinogenesis* **1991**, *12*, 1051–1055.
18. Roebuck, B. D.; Liu, Y-L.; Rogers, A. R.; Groopman, J. D.; Kensler, T. W. *Cancer Res.* **1991**, *51*, 5501–5506.
19. Helmes, C. T.; Becker, R. A.; Seidenberg, J. M.; Schindler, J. E.; Kelloff, G. *Proc. Amer. Assoc. Cancer Res.* **1989**, *30*, 177.
20. Kensler, T. W.; Egner, P. A.; Dolan, P. M.; Groopman, J. D.; Roebuck, B. D.; *Cancer Res.* **1987**, *47*, 4271–4277.
21. Pepin, P.; Bouchard, L.; Nicole, P.; Castonguay, A. *Carcinogenesis* **1992**, *13*, 341–348.
22. Thomas, C. F.; Detrisac, C. J.; Moon, R. C.; Kelloff, G. J. *Proc. Amer. Assoc. Cancer Res.* **1992**, *33*, 169.
23. Moon, R. C.; Rao, K. V. N.; Detrisac, C. J.; Kelloff, G. J. In *Cancer Chemoprevention*, Wattenberg, L. W.; Boone, C. W.; Kelloff, G. J.; Lipkin, M., Eds; CRC Press: Boca Raton, FL, 1992, pp 83–94.
24. Kensler, T. W.; Egner, P. A.; Trush, M. A.; Bueding, E.; Groopman, J. D. *Carcinogenesis* **1985**, *6*, 759–763.

25. Kensler, T. W.; Davidson, N. E.; Egner, P. A.; Guyton, K. Z.; Groopman, J. D.; Curphey, T. J.; Liu, Y.-L.; Roebuck, B. D. In *Mycotoxins, Cancer and Health*, Bray, G. A.; Ryan, D. H., Eds; LSU Press: Baton Rouge, LA, 1991; pp 359–371.

26. Coles, B.; Meyer, D. J.; Ketterer, B.; Stanton, C. A.; Garner, R. C. *Carcinogenesis* **1985**, *6*, 693–697.

27. Davidson, N. E.; Egner, P. A.; Kensler, T. W. *Cancer Res.* **1990**, *50*, 2251-2255.

28. Rushmore, T. H.; King, R. G.; Paulson, K. E.; Pickett, C. B. *Proc. Natl. Acad. Sci. USA* **1990**, *87*, 3826–3830.

29. Putt, D. A.; Kensler, T. W.; Hollenberg, P. F. *FASEB J.* **1991**, *5*, A1517.

30. Kensler, T. W.; Egner, P. A.; Davidson, N. E.; Roebuck, B. D.; Pikul, A.; Groopman, J. D. *Cancer Res.* **1986**, *46*, 3924–3931.

31. Groopman, J. D.; DeMatos, P.; Egner, P. A.; Love-Hunt, A.; Kensler, T. W. *Carcinogenesis* **1992**, *13*, 101–106.

32. Cohen, S. M.; Ellwein, L. B. *Science* **1990**, *249*, 1007–1011.

33. De Long, M. J., Dolan, P., Santamaria, A. B.; Bueding, E. *Carcinogenesis* **1986**, *7*, 977–980.

34. Davies, M. H.; Blacker, A. M.; Schnell, R. C. *Biochem. Pharmacol.* **1987**, *36*, 568–570.

35. Kensler, T. W.; Groopman, J. D.; Eaton, D. L.; Curphey, T. J.; Roebuck, B. D. *Carcinogenesis* **1992**, *13*, 95–100.

36. Talalay, P.; De Long, M. J.; Prochaska, H. J. *Proc. Natl. Acad. Sci. USA* **1988**, *85*, 8261–8265.

37. Prochaska, H. J.; Talalay, P. *Cancer Res. 1988, 48*, 4776–4782.

38. Jirousek, L.; Stárka, J. *Naturwissenschaften* **1958**, *45*, 386–387.

39. Marks, H. S.; Leichtweis, H. C.; Stoeswand, G. S. *J. Agric. Food Chem. 1991, 39*, 893–895.

40. Tressl, R.; Bahri, D.; Holzer, M.; Kossa, T. *J. Agric. Food Chem.* **1977**, *25*, 459–463.

41. Guo, H.; Cui, L. *Fenxi Ceshi Tongbao* **1990**, *9*, 11–16.

42. Chang, K. W.; Hwang, J. Y.; Woo, I. S. *Han'guk T'oyang Piryo Hakhoechi* **1988**, *21*, 183–193.

43. Ding, Z.; Ding, J.; Yang, C.; Saruwatari, Y. *Yunnan Zhiwu Yanjiu 1988*, *10*, 223–226.

44. Zoghbi, M. D. G. B.; Ramos, L. S.; Maia, J.; Guilherme, S.; Da Silva, M. L.; Luz, A. I. R. *J. Agric. Food Chem. 1984, 32*, 1009–1010.

RECEIVED October 4, 1993

Chapter 12

Chemoprevention of Colon Cancer by Thiol and Other Organosulfur Compounds

Bandaru S. Reddy and Chinthalapally V. Rao

Division of Nutritional Carcinogenesis, American Health Foundation, Valhalla, NY 10595

Epidemiological studies suggest that consumption of garlic and cruciferous vegetables rich in organosulfur compounds is associated with a reduced risk for cancer development including cancer of the colon in man. We used laboratory animal models to test isolated organosulfur compounds and their substituted analogues as chemopreventive agents. The effect of two dose levels of dietary oltipraz [5-(2-pyrazinyl)-4-methyl-1,2-dithiole-3-thione], a substituted dithiolethione, anethole trithione (ATT), and diallyl disulfide (DDS) on azoxymethane (AOM)-induced colon carcinogenesis was studied in male F344 rats. The maximum tolerated doses (MTD) of oltipraz, ATT or DDS added to a semipurified diet were found to be 500, 250 or 250 ppm, respectively. Dietary oltipraz, ATT, and DDS at levels of 40% MTD and 80% MTD were tested as inhibitors of AOM-induced colon cancer. The results demonstrated that these dietary organosulfur compounds significantly inhibited colon carcinogenesis in a dose-dependent manner.

A role for nutrition in the etiology of cancer is increasingly apparent. Cancers of several major sites, including colon, stomach, breast, endometrium and prostate, are directly or indirectly associated with dietary factors. In the United States and other Western countries, colorectal cancer is one of the leading causes of cancer deaths (1). Although several epidemiological and experimental studies have demonstrated that diets high in total fat and low in fiber are generally associated with increased risk for colon cancer (2–6), the etiology of colon cancer is multifactorial and complex in that it may arise from the combined actions of environmental factors — such as low levels of a number of, as yet unidentified, genotoxic agents — and endogenous formation of tumorigenic substances. Strategies for cancer prevention involving reduction or elimination of human exposure to these environmental factors may not always be possible. An alternative approach with a potential for more immediate impact is to identify safe and effective agents which prevent the

0097–6156/94/0546–0164$06.00/0

endogenous formation or enhance detoxification of carcinogens and/or inhibit tumor promotion and progression (*7–9*).

Epidemiological studies also suggest that generous consumption of fruits and vegetables is associated with a reduced risk for development of several types of cancer including cancer of the colon in humans (*2,10,11*). Although the nature of constituents of these fruits and vegetables responsible for reduced risk has not been fully elucidated, it is apparent that the foodstuffs are a source of naturally occurring anticarcinogenic agents that hinder the formation of carcinogens from precursors in the body or that act protectively to lessen or eliminate the effects of carcinogens and tumor promoters (*8*).

Green and yellow vegetables, including cabbage, brussels sprouts and other cruciferous vegetables contain several organosulfur compounds including isothiocyanates and dithiolethiones (*12,13*). Among dithiolethiones, oltipraz, a substituted form of dithiolthione, [5-(2-pyrazinyl)-4-methyl-1,2-dithiole-3-thione], has been used in man as an antischistosomal drug (*14*). Recent work has also focused on organosulfur compounds found in garlic (*Allium* species) namely diallyl sulfide, diallyl disulfide and diallyl trisulfide (*8*). In animal models, Belman (*15*) demonstrated that garlic oil inhibited mouse skin tumorigenesis. Several of these agents have been tested in short term screening and long term efficacy studies in laboratory animals for their chemopreventive properties (*9,16*).

The purpose of this chapter is to provide an overview of the inhibitory effect of dietary organosulfur compounds — namely oltipraz, anethole thrithione, and diallyl disulfide — in colon carcinogenesis. The focus will be on the results thus far generated in our laboratory. The structures of oltipraz anethole thrithione and diallyl disulfide are shown in Figure 1.

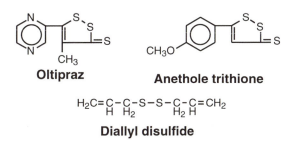

Oltipraz   **Anethole trithione**

$H_2C=\underset{H}{C}-\underset{H_2}{C}-S-S-\underset{H_2}{C}-\underset{H}{C}=CH_2$

Diallyl disulfide

Figure 1. Chemical structures of organosulfur compounds

Animal Models for Colon Cancer

A variety of compounds, namely, 1,2-dimethylhydrazine (DMH), azoxymethane (AOM), methylazoxymethanol acetate (MAM), 3,2′-dimethyl-4-aminobiphenyl (DMAB), methylnitrosourea (MNU), and *N*-methyl-*N*′-nitro-*N*-nitrosoguanidine (MNNG), that are carcinogenic to the colon have been used in a number of animal models to study the effect of dietary constituents on tumorigenesis at this site (*17, 18*). These carcinogens produce both benign adenomas and adenocarcinomas similar to those observed in humans. There is a consensus in the literature that, both morphologically and clinically, the colon tumors induced in rats by these agents resemble relevant neoplasms in man. The spreading of rat colon adenocarcinomas via lymphogenic region to regional and distal lymph nodes by implantation in peritoneum and hematogenous dissemination in lungs and other viscera also bear much resemblance to human phenomenon. Additionally, these rodent models have been used for over two decades as unique tools for systematic studies of risk factors for colon cancer observed in the human setting and for determining whether or not suspected etiologic factors can be reproduced under controlled laboratory conditions. Thus, as is described here, a number of major elements observed in humans could not have been established without careful, deliberate investigations carried out in laboratory animals.

In all our studies presented here, colon cancer was initiated with AOM in male F344 rats to evaluate potential chemopreventive properties of organosulfur compounds. Beginning at 5 weeks of age, groups of male F344 rats were fed a modified AIN-76A (control) diet and experimental diets containing 40 and 80% of the maximum tolerated doses (MTD) of organosulfur compounds (Table I). After two weeks on the diets, groups of animals intended for carcinogen treatment received 15 mg/kg body weight AOM subcutaneously once per week for 2 weeks or one injection of 30 mg/kg body weight. The experiments were terminated 52 weeks after the AOM treatment. Colon tumors were subjected to histopathologic evaluation by routine procedures (18).

Table I. Percentage Composition of Experimental Semi-purified Diets

Ingredients	Control diet[a]	Experimental diet[b]
Casein	20.0	20.0
DL-Methionine	0.3	0.3
Corn starch	52.0	52.0
Dextrose	13.0	13.0
Corn oil	5.0	5.0
Alphacel	5.0	5.0
Mineral mix, AIN	3.5	3.5
Vitamin mix, AIN revised	1.0	1.0
Choline bitartrate	0.2	0.2
Chemopreventive agents	0	40 and 80% MTD

[a]Adapted from American Institute of Nutrition Reference Diet (AIN-76A) with modification of the source of carbohydrate.
[b]Chemopreventive agents were added to the diets at the expense of corn starch.

Maximum Tolerated Doses of Organosulfur Compounds

Prior to efficacy studies of organosulfur compounds, the maximum tolerated dose (MTD) of each test agent added to AIN-76A semipurified diet was determined in male F344 rats. At 5 weeks of age, groups of male F344 rats were fed the AIN-76A diet (control) and experimental diets containing 5 concentrations of each test agent. Body weights were recorded twice weekly for 10 weeks. At the end of 10 weeks, all animals were sacrificed and the organs were examined grossly for any abnormalities. Based on body weights and clinical signs of toxicity, the MTDs of test agents were calculated. The MTD is defined as the highest dose of the agent in the diet that causes no more than a 10% body weight decrement as compared to the appropriate control diet group and does not produce mortality or any clinical signs of toxicity that would be predicted to shorten the natural life span of the animal. The MTDs of organosulfur compounds were: anethole trithione, 250 ppm; oltipraz, 500 ppm; and diallyl disulfide, 250 ppm.

Inhibitory Effect of Anethole trithione and Oltipraz on Colon Carcinogenesis

Oltipraz and other dithiolethiones have been shown to have a broad spectrum of anticarcinogenic actions against several types of carcinogens in a variety of organs (*19*). A pioneering study by Wattenberg and Bueding (*20*) indicated that oral administration of oltipraz inhibited benzo[a]pyrene (B[a]P)-, diethylnitrosamine (DEN)-, and uracil mustard-induced pulmonary and/or forestomach tumors in mice. Oltipraz has also been shown to protect against hepatotoxicity induced by acetaminophen, carbon tetrachloride (*21*), and aflatoxin B_1 (*22*). Roebuck *et al.* (*23*) have shown that oltipraz inhibits aflatoxin-induced liver cancer in F344 rats.

Because of the low toxicity and inhibitory effect of oltipraz against diverse carcinogens, we evaluated the efficacy of this compound and another substituted dithiole thione, namely anethole trithione, in colon carcinogenesis. In these studies, experimental diets containing 40 and 80% MTD levels of oltipraz or anethole trithione were fed 2 weeks before and during carcinogen treatment and until termination of study at 52 weeks. The body weights of animals fed the control diet and experimental diets containing oltipraz and anethole trithione were comparable. The results of our study, which are summarized in Table II, demonstrate that feeding 200 (40% MTD) and 400 ppm (80% MTD) oltipraz significantly inhibited the incidence of AOM-induced colon adenocarcinomas. Anethole trithione at 100 (40% MTD) and 200 ppm (80% MTD) in the diet also suppressed the incidence and multiplicity of colon adenocarcinomas in a dose-dependent manner (Table II).

Table II. Effect of Organosulfur Compounds on AOM-induced Colon Tumor Incidence in Male F344 Rats

Organosulfur compound tested	Colon adenocarcinoma incidence (%)
Experiment 1[a]	
Control diet	72
Oltipraz (200 ppm)	47*
Experiment 2[b]	
Control diet	33
Anethole trithione (100 ppm)	11*
" (200 ppm)	8*
Diallyl disulfide (100 ppm)	11*
" (200 ppm)	6*

[a] Invasive and noninvasive colon adenocarcinomas.
[b] Invasive colon adenocarcinomas.
*Significantly different from its respective control diet group, $p < 0.05$.

We also measured serum oltipraz levels at different time points during the period of the study (Table III). Serum oltipraz levels were maintained throughout the study. The animals fed 80% MTD (400 ppm) of oltipraz showed increased levels of serum oltipraz as compared to those fed 40% MTD (200 ppm) of oltipraz.

In order to understand the mechanism by which dietary oltipraz exerts its inhibitory action on AOM-induced colon carcinogenesis, we assessed the effect of this agent on hepatic and colonic mucosal activities of glutathione S-transferase (GST) and tyrosine protein kinase (TPK) and on the hepatic and colonic DNA methylation in rats treated with AOM (24). The results show that dietary oltipraz significantly enhanced hepatic and colonic GST activity which is accompanied by a decrease in DNA methylation (7-methylguanine and O^6-methylguanine) in the colon and liver (Table IV).

Table III. Effect of dietary oltipraz on serum oltipraz levels in male F344 rats

Weeks on oltipraz diet	Amount of oltipraz in diet	
	200 ppm	400 ppm
1	277 ± 43[a]	426 ± 55
8	331 ± 33	505 ± 47
16	304 ± 58	536 ± 71
24	352 ± 43	584 ± 43

[a]Nanograms oltipraz per ml serum (mean value $\pm$ SD; n=9).

Although the significance of 7-methylguanine in tumor formation has been questioned, the formation and persistence of O^6-methylguanine in the DNA of target tissue is closely correlated with carcinogenicity (25,26). Many chemo-preventive agents including several phenolic antioxidants have been shown to inhibit carcinogenesis by modulating the activities of detoxifying enzymes that are involved in the metabolism of carcinogens (27,28). It has also been reported that oltipraz elevates several enzymes such as epoxide hydrolase and NADP(H)-quinone reductase (27) that are involved in the inactivation of electrophilic species.

There is a possibility that oltipraz alters the metabolism of carcinogen in the target tissue. AOM is first metabolized to methylazoxymethanol (MAM), a prox-imate carcinogen, in the liver by the microsomal mixed function oxidases, and MAM can be further metabolized in the liver to a DNA-alkylating species, presumably the methyldiazonium ion, which is considered to be an ultimate carcinogen (29,30). Additional studies have shown that the MAM that is not metabolized in the liver is carried to colon by circulation and activated to methyldiazonium ion (31) which can methylate colonic cellular nucleophiles, including DNA. For example, several colon tumor inhibitors have been shown to interact with AOM metabolism and to inhibit DNA alkylation and carcinogenicity (30, 32). Therefore, the possibility exists that dietary oltipraz alters the metabolism of AOM in the liver and colon mucosa thereby inhibiting colon tumor formation. It is also possible that increased GST activity in the liver and colon of animals fed the oltipraz decreases the level of DNA adduct formation by modulating AOM metabolism.

Dietary oltipraz significantly inhibited colonic mucosal and liver tyrosine protein kinase (TPK) activity (Table IV). Evidence from various studies indicates that the phosphorylation of proteins at tyrosine residues plays an important role in the regulation of cellular growth and differentiation (33). TPK activity is associated with cellular receptors for epidermal growth factor, platelet-derived growth factors, insulin receptor (34) and several oncogenes (33). Increased expression of TPK activity may be responsible for unlimited growth (35). Increased levels of TPK

Table IV. Effect of Dietary Oltipraz on AOM-induced Hepatic and Colonic Mucosal Enzymes and DNA Methylation in Male F344 Rats

Experimental group	Glutathione S-transferase (μmol/mg protein/min)	7-Methylguanine (μmol/mol guanine)	O^6-Methylguanine (μmol/mol guanine)	Tyrosine protein kinase (pmol ^{32}P/mg protein/min)
Liver				
Control	4.9 ± 0.17	6343 ± 667	728 ± 63	333 ± 15
Oltipraz	22.7 ± 1.70***	2185 ± 228***	306 ± 32***	146 ± 14***
Colon				
Control	1.1 ± 0.08	1567 ± 26	232 ± 24	220 ± 13
Oltipraz	7.1 ± 0.20***	284 ± 23***	157 ± 11*	158 ± 8**

All values are the mean ± SE from 6 rats. Asterisks indicate significant difference from respective control, *: $p < 0.05$; **: $p < 0.01$; ***: $p < 0.001$.)

activity have been found in neoplastic tissue (*33,34*). In the present study, AOM-induced TPK activity in the cytosol and membrane fractions of liver and colonic mucosa was significantly inhibited by oltipraz. The mechanism of TPK inhibition by oltipraz in unclear. There is, however, a possibility that the inhibition by oltipraz might be similar to that of Genistein- and Erbstatin-like specific TPK inhibitors. These agents inhibit TPK activity by altering DNA topoisomerase II activity and by acting at the level of the ATP site (*36,37*).

Inhibitory Effect of Diallyl Disulfide on Colon Carcinogenesis

Investigations were carried out to determine the chemopreventive properties of diallyl disulfide on AOM-induced colon carcinogenesis in male F344 rats. The experimental protocols were as described above. All animals were fed 40 (100 ppm) and 80% (200 ppm) MTD of this agent in AIN-76A diet starting 2 weeks before and during carcinogen administration and until termination of the study at 52 weeks after carcinogen treatment. The body weights of animals fed the experimental diets containing 100 and 200 ppm of diallyl disulfide were comparable to those fed the control diet. Diallyl disulfide at 100 and 200 ppm in the diet significantly inhibited the incidence and multiplicity of colon adenocarcinomas (unpublished observations; Table II). Experiments are in progress to determine the mechanism(s) by which diallyl disulfide inhibits colon carcinogenesis.

Summary and Conclusions

In conclusion, the results of our studies demonstrate that several organosulfur compounds — namely oltipraz, anethole trithione and diallyl disulfide — inhibit colon carcinogenesis in a laboratory animal model. The combined observations from several laboratories indicate a broad spectrum of chemopreventive properties of oltipraz and diallyl disulfide in several cancer models. The inhibition of colon carcinogenesis by dietary oltipraz is associated with an increase in liver and colonic GST activity and reduced formation of DNA adducts. In addition, dietary oltipraz modulates liver and colonic TPK activity that has been shown to play a role in the regulation of cellular growth and differentiation. We believe that the organosulfur compounds have great potential as chemopreventive agents for colon cancer.

Acknowledgments

We thank Donna Virgil for preparation of the manuscript. This work was supported by USPHS Grant CA-17613 and Contracts NO1-CN-85095-01 and NO1-CN-85095-05 from the National Cancer Institute.

Literature Cited

1. Boring, C. C.; Squires, T. S.; Tong, T. *CA–A Cancer J. Clinicians*, **1992**, *42*, 19–38.
2. Committee on Diet, Nutrition, and Cancer, National Research Council *Diet, Nutrition and Cancer*, National Academy Press: Washington, DC, 1982; pp 358–370.

3. Reddy, B. S. In *Diet and Colon Cancer: Evidence from Human and Animal Model Studies*; Reddy, B. S., Cohen, L. A., Eds.; CRC Press: Boca Raton, FL, 1986; pp 47–65.
4. Wynder, E. L.; Kajitani, T.; Ishikawa, S.; Dodo, H.; Takano, A. *Cancer* **1969**, *23*, 1210–1220.
5. Willet, W. C.; Stampfer, M. J.; Colditz, G. A.; Rosner, B. A.; Speizer, F. E. *N. Engl. J. Med.* **1990**, *323*, 1664–1669.
6. Lanza, E.; Shankar, S.; Trock, B. In *Macronutrients*; Micozzi, M. S., Moon, T. E., Eds.; Marcel Dekker, Inc: New York, 1992; pp 293–319.
7. Boone, C. W.; Kelloff, G.; Malone, W. E. *Cancer Res.* **1990**, *50*, 2–9.
8. Wattenberg, L. W. In *Chemoprevention of Cancer by Naturally Occurring and Synthetic Compounds*; Wattenberg, L. W.; Lipkin, M.; Boone, C. W.; Kelloff, G., Eds.; CRC Press: Boca Raton, FL, 1992; pp 19–39.
9. Kelloff, G. J.; Boone, C. W.; Malone, W. F.; Steele, V. E. In *Cancer Chemoprevention*; Wattenberg, L. W.; Lipkin, M.; Boone, C. W.; Kelloff, G. J., Eds.; CRC Press: Boca Raton, FL, 1992; pp 41–56.
10. Colditz, G. A.; Branch, L. G.; Lipnick, R. J.; Willett, W. C.; Rosner, B.; Posner, B. M.; Hennekens, C. H. *Am. J. Clin. Nutr.* **1985**, *41*, 32–36.
11. Boyd, J. N.; Babish, J. G.; Stoewsand, G. S. *Fd. Chem. Toxicol.* **1982**, *20*, 47–52.
12. Jirousek, L. *Collect Czech. Chem. Commun.* **1957**, *22*, 1494–1502.
13. Jirousek, L.; Starka, L. *Nature 45*, 386–387.
14. Bueding, E.; Dolan, P.; Leroy, J. P. *Res. Commun. Chem. Pathol. Pharmacol.* **1982**, *37*, 297–303.
15. Belman, S. *Carcinogenesis* **1983**, 1063–1065.
16. Wargovich, M. J. In *Cancer Chemoprevention*; Wattenberg, L. W.; Lipkin, M.; Boone, C. W.; Kelloff, G. J., Eds.; CRC Press: Boca Raton, FL, 1992; pp 195–203.
17. Reddy, B. S. In *Animal Experimental Evidence on Macronutrients and Cancer*; Micozzi, M. S., Moon, T. E., Eds.; Marcel Dekker, Inc: New York, 1992; pp 33–54.
18. Reddy, B. S.; Maruyama, H.; Kelloff, G. *Cancer Res.* **1987**, *47*, 5340–5346.
19. Kensler, T. W.; Groopman, J. D.; Roebuck, B. D. In *Chemoprevention by Oltipraz and Other Dithiolethiones*; Wattenberg, L. W.; Lipkin, M.; Boone, C. W.; Kelloff, G. J., Eds.; CRC Press: Boca Raton, FL, 1992; p 205.
20. Wattenberg, L. W.; Bueding, E. *Carcinogenesis* **1986**, *7*, 1379–1381.
21. Ansher, S. S.; Dolan, P.; Bueding, E. *Hepatology* **1983**, *3*, 932–935.
22. Liu, T.-Y.; Roebuck, B. D.; Yager, J. D.; Groopman, J. D.; Kensler, T. W. *Toxicol. Appl. Pharmacol.* **1988**, *93*, 442–451.
23. Roebuck, B. D.; Liu, T.-Y.; Rogers, A. R.; Groopman, J. D.; Kensler, T. W.; *Cancer Res.* **1991**, *51*, 5501.
24. Rao, C. V.; Nayini, J.; Reddy, B. S. *Proc. Soc. Exp. Biol. Med.* **1991**, *197*, 77–84.
25. Rogers, K. J.; Pegg, A. E. *Cancer Res.* **1977**, *37*, 4082–4087.
26. Bull, A. W.; Burd, A. D.; Nigro, N. D. *Cancer Res.* **1981**, *41*, 4938–4941.
27. Kensler, T. W.; Egner, P. A.; Trush, M. A.; Bueding, E.; Groopman, J. D. *Carcinogenesis* **1985**, *6*, 759–763.
28. De Long, M. J.; Prochaska, H. J.; Talalay, P. *Cancer Res.* **1985**, *45*, 546–551.
29. Fiala, E. S. *Cancer* **1977**, *40*, 2436–2445.
30. Zeduck, M. S. *Prev. Med.* **1980**, *9*, 346–351.

31. Fiala, E. S.; Sohn, O. S.; Hamilton, S. R. *Cancer Res.* **1987**, *47*, 5939–5943.
32. Fiala, E. S. In *Inhibition of Carcinogen Metabolism and Action by Disulfiram, Pyrazole, and Related Compounds*; Plenum Publishing: New York, 1981; pp 23–69.
33. Hunter, T.; Cooper, J. A. *Annu. Rev. Biochem.* **1985**, *54*, 897–930.
34. Yarden, Y. *Annu. Rev. Biochem.* **1988**, *57*, 443–478.
35. Cooper, J. A.; Sefton, B. M.; Hunter, T. *Methods Enzymol.* **1983**, *99*, 387–402.
36. Markovits, J.; Linassier, C.; Fosse, P.; Couprie, J.; Pierre, J.; Sablon, A. J.; Saucier, J. M.; Pecq, J. L.; Larsen, A. K. *Cancer Res.* **1989**, *49*, 5111–5117.
37. Akiyama, T.; Ishida, J.; Nakagawa, S.; Ogawara, H.; Watanabe, S.; Itoh, N.; Shibuya, M.; Fukami, Y. *J. Biol. Chem.* **1987**, *262*, 5592–5595.

RECEIVED October 5, 1993

Chapter 13

Inhibition of Esophageal Tumorigenesis by Phenethyl Isothiocyanate

G. D. Stoner[1], A. J. Galati, C. J. Schmidt, and M. A. Morse[1]

Department of Pathology, Medical College of Ohio, Toledo, OH 43614

Phenethyl isothiocyanate, a naturally occurring constituent of cruciferous vegetables, is a potent inhibitor of nitrosamine-induced esophageal cancer. F-344 rats fed diets containing phenethyl isothiocyanate at 1.5, 3 and 6 mmol/kg diet, before and during treatment with the carcinogen N-nitrosobenzylmethylamine, developed 89–100% fewer esophageal tumors than carcinogen-treated control rats. Phenethyl isothiocyanate exhibited inhibitory effects against both preneoplastic lesions and neoplastic lesions. The effects of phenethyl isothiocyanate (10, 25, 50 and 100 μM) on DNA methylation by N-nitrosobenzylmethylamine in cultured explants of rat esophagus were also investigated. Phenethyl isothiocyanate produced a dose-dependent inhibition in the levels of DNA methylation at the N^7 (20–89%) and O^6 (55–93%) positions of guanine. Therefore, a strong correlation was observed between the inhibitory effects of phenethyl isothiocyanate *in vivo* and *in vitro*.

Esophageal cancer in humans occurs worldwide and ranks seventh among cancers in order of frequency of occurrence in both sexes combined (*1*). Epidemiological studies indicate that nutritional and environmental factors play a major role in the etiology of esophageal cancer. An increased risk for developing esophageal cancer is correlated with tobacco smoking (*2,3*), consumption of alcoholic beverages (*4,5*) and of salt-pickled and moldy foods (*6*). The molds which can contaminate foods include members of the *Fusarium* species (*7*), which produce several toxins, and *Geotrichum candidum* (*8*), which promotes the formation of nitrosamine carcinogens. Research in the Transkei region of South Africa and in China suggests that N-nitroso compounds and their precursors are probable etiological factors in esophageal cancer in these high-incidence areas (*6,8*). Most esophageal tumors in these high-incidence areas are classified histologically as squamous cell carcinomas.

[1]Current address: Department of Preventive Medicine, The Ohio State University, 300 West 10th Avenue, Columbus, OH 43210

0097–6156/94/0546–0173$06.00/0

Animal model studies have shown that many nitrosamines are esophageal carcinogens in the rat (*9,10*). Early studies by Druckrey, *et al.* (*9*) indicated that asymmetrical nitrosamines could readily induce squamous cell carcinomas in the rat esophagus, the most potent being *N*-nitrosobenzylmethylamine (NBMA, Figure 1) (*11*). In metabolism studies of several nitrosamines in explant cultures of rat esophagus, the highest level of metabolite binding to DNA was observed with NBMA (*12*); therefore, a strong correlation exists between the carcinogenic potency of NBMA *in vivo* and its binding to esophageal DNA *in vitro*.

N-Nitrosobenzylmethylamine   **Phenethyl isothiocyanate**
(NBMA)   **(PEITC)**

Figure 1. Structures of NBMA and PEITC.

In vitro studies and investigations in experimental animals have shown that foods contain a number of compounds with the ability to inhibit chemically-induced cancer (*13*). Among these are the isothiocyanates, including phenethyl isothiocyanate (PEITC, Figure 1) (*14–18*). PEITC is a primary product of thio-glucosidase-catalyzed hydrolysis of gluconasturtiin, a glucosinolate compound found in certain cruciferous vegetables (*19*). Pretreatment of rats with PEITC inhibited the induction of mammary tumors by 7,12-dimethylbenz[*a*]anthracene (DMBA) (*14*). Addition of PEITC to the diet inhibited DMBA and benzo[*a*]pyrene (BP)-induced lung and forestomach tumors in mice (*14*). Recent studies have demonstrated the ability of PEITC to inhibit the metabolism and DNA methylation of a series of nitrosamines both *in vivo* and *in vitro* (*20–23*). Moreover, PEITC was shown to inhibit lung tumor induction in rats (*15*) and in mice (*17*) by the tobacco-specific nitrosamine, 4-(methylnitrosamino)-1-(3-pyridyl)-1-butanone (NNK). Finally, when provided in the diet at concentrations of 3 and 6 mmol/kg, PEITC inhibited the induction of esophageal tumors in rats by NBMA (*18*).

In this paper, we report the results of testing PEITC at five concentrations (0.33, 0.75, 1.5, 3 and 6 mmol/kg) in the diet for its ability to inhibit NBMA tumorigenesis in the esophagus of F-344 rats. Results from the *in vivo* bioassay are compared with the inhibitory effects of PEITC on the metabolism and DNA-damaging effects of NMBA in cultured rat esophageal tissues.

Materials and Methods

Chemicals. [methyl-^{3}H]NBMA (5 Ci/mmol; purity, >97%) was purchased from Moravek Biochemicals, Inc., Brea, California. Unlabeled NBMA (purity, >98%) was obtained from the National Cancer Institute Chemical Carcinogen Repository at Midwest Research Institute, Kansas City, Missouri. Standards for N^7-methyl-guanine (N^7-MeGua) and O^6-methylguanine (O^6-MeGua) (purity, 98%) were purchased from Chemsyn Science Laboratories, Lenexa, Kansas. PEITC (purity, >99%) was obtained from Aldrich Chemical Company, Milwaukee, Wisconsin.

The purity of all chemicals was checked by high performance liquid chromatography (HPLC).

DNA Binding and Adduct Studies. PEITC was evaluated for its ability to inhibit the interaction of NBMA metabolites to the DNA of rat esophagus. The protocol for these studies has been described in detail (*18*), and is summarized briefly as follows:

Explant Culture. Esophageal explants were prepared and cultured in chemically defined CMRL-1066 medium as described (*24*). The explants were cultured for 6 hours and then exposed for 12 hours to 1 μM [methyl-^{3}H]NBMA concurrently with PEITC at concentrations of 10, 25, 50 and 100 μM. Control explant cultures were incubated for 12 hours in medium containing 1 μM [methyl-^{3}H]NBMA and 1% dimethyl sulfoxide, the solvent for NBMA. After incubation, the tissues from each series were pooled and stored at -85°C for subsequent analyses.

Isolation of Radiolabeled Carcinogen Bound to DNA. Explant DNA was isolated by a modification of the method of Gupta (*25*). The purified DNA was dissolved in 200 μl of 0.01 X standard saline-citrate and quantitated spectrophotometrically. The radioactivity associated with the DNA was determined in a liquid scintillation counter.

Analysis of DNA Adducts. N^7-MeGua and O^6-MeGua adducts in rat esophageal DNA were analyzed as described previously (*18*). Briefly, DNA isolated as described above was hydrolyzed with 0.1 M HCl at 70°C for 30 minutes. The hydrolysates were analyzed by HPLC using a Whatman Partisil strong cation exchange column (SCX-10; 4.5 x 250 mm) and a Radiomatic Flo-One/Beta detector. Samples were eluted isocratically with a buffer solution containing 75 mM ammonium phosphate:methanol (85:15), pH 2.5, at a flow rate of 1.0 ml/minute, and the column effluent was monitored at 254 nm. Standards of nonradioactive N^7-MeGua and O^6-MeGua were added to all the samples to ascertain retention times of the major NBMA-DNA adducts.

Bioassay. Male F-344 rats, 6–8 weeks old, were purchased from Harlan Sprague-Dawley, Indianapolis, Indiana. After two weeks of acclimatization to the animal facility, the rats were randomized into 10 groups, each consisting of 15 animals. The treatments administered to the groups are summarized in Table I. Groups 1 and 5 were given AIN-76A diet, while groups 2–4 and 6–10 received AIN-76A diet containing PEITC at concentrations ranging from 0.33 to 6 mmol/kg diet. PEITC was mixed in the diet weekly, using a Hobart mixer, and stored at 4°C before use. Previous studies have shown that PEITC is stable in the diet for at least 10 days (*15*). The concentration of PEITC in the diet was monitored by random sampling of portions of the mixed diet and was found to be homogeneous throughout. After 2 weeks of feeding the respective diets, NBMA (0.5 mg/kg body weight) was administered by s.c. injection once per week for 15 weeks. The total dose of NBMA administered per rat was 7.5 mg (0.050 mmol)/kg body weight. Vehicle-control rats were given s.c. injections of dimethyl sulfoxide in distilled water (1:4, v:v) for the same period. After completion of the treatment with either NBMA or vehicle, all groups were maintained on their respective diets (with or without

Table I. Incidence and Multiplicity of Esophageal Tumors After Treatment with Vehicle, PEITC, NBMA and NBMA plus PEITC

Group[a]	Treatment		Tumor Incidence		Tumor Multiplicity	
	NBMA[b] (mg/kg body wt.)	PEITC[c] (mmol/kg diet)	Percent of rats with tumors	Inhibition (%)	Tumors/rat (mean ± S.D.)	Inhibition (%)
1	0	0	0	—	0	—
2	0	0.75	0	—	0	—
3	0	3.00	0	—	0	—
4	0	6.00	0	—	0	—
5	0.5	0	100	—	11.5 ± 4.5	—
6	0.5	0.33	100	0	10.7 ± 1.1	7
7	0.5	0.75	100	0	5.7 ± 1.2	50
8	0.5	1.50	60	40	0.9 ± 0.2[d]	93
9	0.5	3.00	13	87	0.1 ± 0.3[d]	99
10	0.5	6.00	0	100	0.0[d]	100

[a] Each group contained 15 animals.
[b] Administered s.c. once per week for 15 weeks.
[c] PEITC was fed in the diet 2 weeks before, during and for 8 weeks after administration of NBMA.
[d] Significantly different from Group 5 ($p < 0.05$).

PEITC) for 8 additional weeks. Food consumption was measured weekly, and body weights were recorded every 2 weeks during the 25-week bioassay.

At the termination of the bioassay, a complete gross autopsy was performed on each animal. The esophagus was excised *in toto*, opened longitudinally, placed flat on white index cards with the epithelium uppermost, and fixed in 10% buffered formalin. Quantification of esophageal tumors (≥0.5 mm in diameter) was performed as described previously (*18,26*). Sections of esophagi from each group were scored for the presence of preneoplastic lesions (i.e., acanthosis and hyperkeratosis, leukoplakia, and leukokeratosis) and neoplastic lesions (i.e., papilloma and carcinoma) (*27*).

Statistical Analysis

All data were analyzed using SAS on a personal computer (*18*). The data on DNA binding of NBMA were analyzed utilizing a test of linear contrast. These data, as well as the N^7 and O^6 adduct formation data, were expressed in terms of percentage of inhibition. Ninety-five percent confidence intervals were calculated. Food consumption and body weight data were analyzed using analysis of variance. The food intake over the study period was evaluated on a per-cage basis. Weight gain was evaluated over the study period as the difference in weight between week 25 and week 1. Four predefined 1 d.f., contrasts of mean weight gain were calculated: NBMA-treated groups versus groups not treated with NBMA; PEITC-treated groups versus groups not treated with PEITC; the interaction between treatment with PEITC and treatment with NBMA; and low-dose PEITC groups versus high-dose PEITC groups. Tumor incidence data were analyzed using χ^2 tests. The data on frequency of tumors per rat were evaluated using a Wilcoxon rank sum test.

Results

In Vitro Studies. DNA binding studies showed that PEITC elicited a dose-dependent inhibition ($p<0.01$) of NBMA metabolite binding to the DNA of rat esophageal explants (Table II). Ten, 25, 50 and 100 μM PEITC inhibited the binding of NBMA metabolites to DNA by 54, 77, 87 and 96%, respectively.

The effect of PEITC (10 and 100 μM) on the formation of N^7-MeGua and O^6-MeGua adducts by NBMA in esophageal explant DNA is also shown in Table II. At 10 and 100 μM, PEITC inhibited the formation of N^7-MeGua by 20 and 89%, and of O^6-MeGua by 55 and 93%, respectively, in esophageal DNA. The inhibition in the formation of both adducts was significant ($p<0.05$) only in explants treated with 100 μM PEITC.

In Vivo **Studies.** At the beginning of the bioassay, the average weight of animals was the same in all ten groups (data not shown), but weight gain throughout the study was not equivalent for all groups. There was a significant reduction ($p<0.05$) in weight gain (10% or less) in all groups receiving PEITC at 6 and 3 mmol/kg diet compared to those not receiving PEITC in the diet (data not shown). Furthermore, a significant ($p<0.05$) difference in weight gain (approximately 5%) was observed between groups fed the high dose of PEITC (groups 4 and 10 in Table I) and those given the lower doses of PEITC (groups 2, 3, 6, 7, 8 and 9 in Table I) (data not shown). No other differences between the groups were observed. The dietary intake was the same for all groups during the 25 week bioassay (data not shown).

Table II. Effect of PEITC on DNA Binding and DNA Adduct Formation in Cultured Rat Esophagus

PEITC Concentration	DNA binding[a] (pmol/mg DNA)	Adduct Formation (pmol/mg DNA)[a]	
		N^7-MeGua	O^6-MeGua
0	58.3 ± 10.7[b]	8.81	0.40
10	27.1 ± 6.5 (54)	7.06 (20)	0.18 (55)
25	13.7 ± 3.3 (77)	N.D.[c]	N.D.
50	7.5 ± 1.0 (87)	N.D.	N.D.
100	2.3 ± 0.1 (96)	0.94 (89)	0.03 (93)

[a] Data from four experiments; numbers in parenthesis are percentage of inhibition.
[b] Mean $\pm$ S.D.
[c] N.D.: not determined

At necropsy, the esophageal mucosa of the control animals was normal. No tumors were observed in vehicle-treated rats (group 1, Table I) or in rats that received PEITC only (groups 2–4, Table I). Tumors were observed only in the esophagi of NBMA-treated animals. No metastases or invasion of tumor cells to adjacent tissues was found. Most tumors were exophytic and were distributed along the entire length of the esophagus.

The tumor incidence in NBMA-treated rats (group 5) was 100% (Table I). PEITC, at concentrations of 1.5, 3 and 6 mmol/kg diet, decreased the tumor incidence to 60, 13 and 0%, respectively. The observed average number of 11.5 esophageal tumors per rat in NBMA-treated animals (Table I) was similar to that observed in a previous study in which a comparable dose of PEITC was administered (26). PEITC, at concentrations of 1.5, 3 and 6 mmol/kg diet, produced a significant inhibition (p<0.01) in the multiplicity of esophageal tumors. At 0.75 mmol/kg diet, PEITC reduced the multiplicity of esophageal tumors by 50%, but this reduction was not significant (p>0.05).

Histopathological analysis of the esophageal lesions induced by NBMA indicated that PEITC, at 1.5, 3 and 6 mmol/kg diet, caused a significant inhibition in the occurrence of all preneoplastic (hyperkeratosis and acanthosis, leukoplakia and leukokeratosis) and neoplastic (papilloma and carcinoma) lesions (data not shown).

Discussion

The present study extends previous observations in which PEITC was found to be a potent inhibitor of NBMA-induced esophageal carcinogenesis in male F-344 rats (18). When administered in the diet at concentrations of 1.5, 3 and 6 mmol/kg, PEITC caused a 93, 99 and 100% inhibition in the multiplicity of NBMA-induced esophageal tumors, respectively (Table I). The development of both preneoplastic and neoplastic lesions was inhibited by PEITC. On a molar dose basis, PEITC is approximately 10-fold more active than ellagic acid as an inhibitor of NBMA tumorigenesis in the rat esophagus (18,26). The differences in inhibitory activity between PEITC and ellagic acid could be due to differences in absorption rate and/or in mechanism(s) of action. Diallyl sulfide, a thioether found in garlic, was also found to be a potent inhibitor of NBMA-tumorigenesis in the rat esophagus

(*28*). Due to different protocols for the administration of diallyl sulfide and PEITC, however, a comparison of the relative chemopreventive efficacy of the two compounds cannot be made.

In vitro studies using explant cultures of rat esophagus indicated that PEITC inhibits the binding of NBMA metabolites to rat esophageal DNA, and the formation of N^7-MeGua and O^6-MeGua adducts in esophageal DNA. When added to the medium at a concentration of 100 μM, PEITC inhibited the binding of NBMA metabolites to DNA by approximately 95% and the formation of DNA adducts by approximately 90%. Thus, as expected, there was a strong correlation between the extents of inhibition of DNA binding and the formation of methylated adducts. Ellagic acid, at 100 μM, produced only a 45% inhibition in the formation of N^7-MeGua and O^6-MeGua adducts in the DNA of rat esophageal explants treated with 1 μM [methyl-^{3}H]NBMA (*29*). These data are in agreement with the observation that PEITC is a more potent inhibitor of NBMA-induced tumors in the rat esophagus than ellagic acid. Moreover, they indicate that DNA methylation *in vitro* may be an effective marker for screening potential inhibitors of NBMA carcinogenicity (*30*).

The precise mechanism(s) by which PEITC inhibits NBMA tumorigenesis in the rat esophagus is unknown. It is probable, however, that PEITC inhibits the metabolism of NBMA. PEITC could influence the detoxification of NBMA and/or elicit specific effects on the cytochrome P-450 enzymes involved in NBMA activation. Labuc *et al.* (*31*) have shown that microsomes prepared from rat esophagus metabolize NBMA at a high rate to yield benzaldehyde and a methylating agent in a cytochrome P-450-dependent reaction (NBMA debenzylase). NBMA is also metabolized at a high rate in rat liver, but it is not toxic or carcinogenic in the liver (*32*). Based upon its structure, PEITC and its metabolites are unlikely to act as scavengers of the reactive intermediates produced in the metabolic activation of NBMA. It is more likely that PEITC inhibits the oxidative metabolism of NBMA. The binding of isothiocyanates to protein through sulfhydryl or amino groups has been demonstrated (*33*). As indicated in a previous report (*15*), this type of conjugation could occur between PEITC and specific esophageal cytochrome P-450 isozymes responsible for NBMA activation. Clearly, additional studies are needed to elucidate the specific biochemical mechanisms for the inhibition of NBMA-induced esophageal tumorigenesis by PEITC.

Acknowledgments

We thank Dr. Diana Morrissey and Mr. Steven Wagner for assisting in the development of a portion of the data in this report and Diana Carter for typing the manuscript.

Literature Cited

1. Parkin, D. M.; Stjernsward, J.; Muir, C. S. *Bull. WHO* **1984,** *62*, 163–182.
2. Wynder, E. L.; Bross, I. J. *Cancer* **1961,** *14*, 389–413.
3. Warwick, G. P.; Harington, J. S. *Adv. Cancer Res.* **1973,** *17*, 81–229.
4. Tuyns, A. J.; Pequignot, G.; Abbatucci, J.S. *Int. J. Cancer* **1979,** *23*, 443–447.
5. Walker, E. A.; Castegnario, M.; Garren, L.; Toussaint, G; Kowalski, B. *J. Natl. Cancer Inst.* **1979,** *63*, 947–951.
6. Yang, C. S. *Cancer Res.* **1980,** *40*, 2633–2644.

7. Hsia, C. C.; Tsain, B-L.; Harris, C. C. *Carcinogenesis* **1983,** *4,* 1101–1107.
8. Warwick, G. P.; Harington, J. S. *Adv. Cancer Res.* **1973,** *17,* 81–229.
9. Druckrey, H. In *Topics in Chemical Carcinogenesis;* Nakahara, W.; Takayama, S.; Sugimura, T; Odashima, S., Eds.; University Park Press: Baltimore, MD, 1972; pp 73–101.
10. Bulay, O.; Mirvish, S. S. *Cancer Res.* **1979,** *39,* 3644–3646.
11. Lijinsky, W.; Saavedra, J. E.; Rueber, M. D.; Singer, S. S. *J. Natl. Cancer Inst.* **1982,** *68,* 681–684.
12. Autrup, H; Stoner, G. D. *Cancer Res.* **1982,** *42,* 1307–1311.
13. Wattenberg, L. W. *Cancer Res.* **1985,** *45,* 1–8.
14. Wattenberg, L. W. *J. Natl. Cancer Inst.* **1977,** *58,* 295–298.
15. Morse, M. A.; Wang, C-X.; Stoner, G. D.; Mandal, S.; Conran, P. B.; Amin, S. G.; Hecht, S. S.; Chung, F-L. *Cancer Res.* **1989,** *49,* 549–553.
16. Morse, M. A.; Amin, S. G.; Hecht, S. S.; Chung, F-L. *Cancer Res.* **1989,** *49,* 2894–2897.
17. Morse, M. A.; Elkind, K. I.; Amin, S. G.; Hecht, S. S.; Chung, F-L. *Carcinogenesis* **1989,** *10,* 1757–1759.
18. Stoner, G. D.; Morrissey, D. T.; Heur, Y-H.; Daniel, E. M.; Galati, A. J.; Wagner, S. A. *Cancer Res.* **1991,** *51,* 2063–2068.
19. Van Etten, C. H.; Daxenbichler, M. E.; Williams, P. H.; Kwolek, W. F. *J. Agric. Food Chem.* **1976,** *24,* 452–455.
20. Chung, F-L.; Juchatz, A.; Vitarius, J.; Hecht, S. S. *Cancer Res.* **1984,** *44,* 2924–2928.
21. Chung, F-L.; Wang, M.; Hecht, S. S. *Carcinogenesis* **1985,** *6,* 539–543.
22. Doerr-O'Rourke, K.; Trushin, N.; Hecht, S. S.; Stoner, G. D. *Carcinogenesis* **1991,** *12,* 1029–1034.
23. Murphy, S. E.; Heiblum, R.; King, P. G.; Bowman, D.; Davis, W. J.; Stoner, G. D. *Carcinogenesis* **1991,** *12,* 957–961.
24. Stoner, G. D.; Pettis, W.; Haugen, A.; Jackson, F.; Harris, C. C. *In Vitro* **1981,** *17,* 681–688.
25. Gupta, R. C. *Proc. Natl. Acad. Sci. USA* **1984,** *81,* 6943–6947.
26. Mandal, S.; Stoner, G. D. *Carcinogenesis* **1990,** *11,* 55–61.
27. Pozharisski, K. M. In *Pathology of Tumors in Laboratory Animals*; Turusov, V. S., Ed.; Part 1, IARC Scientific Publication No. 5; International Agency for Research on Cancer: Lyon, France, 1973; Vol. 1, pp 87–100.
28. Wargovich, M. J.; Woods, C.; Eng, V. W. S.; Stephens, L. C.; Gray, K. *Cancer Res.* **1988,** *48,* 6872–6875.
29. Mandal, S.; Shivapurkar, N. M.; Galati, A. J.; Stoner, G. D. *Carcinogenesis* **1988,** *9,* 1313–1316.
30. Stoner, G. D. In *Cellular and Molecular Targets for Chemoprevention*; Steele, V. E.; Stoner, G. D.; Boone, C. W.; Kelloff, G. J., Eds.; CRC Press, Boca Raton, Florida, 1992; pp 247–256.
31. Labuc, G. E.; Archer, M. C. *Cancer Res.* **1982,** *42,* 3181–3186.
32. Mehta, M.; Labuc, G. E.; Urbanski, S. J.; Archer, M. C. *Cancer Res.* **1984,** *44,* 4017–4022.
33. Drobnica, L.; Gemeiner, P. In *Protein Structure and Evolution*; Fox, J. L.; Deyl, Z.; Blazej, A., Eds.; Marcel Dekker, Inc., New York, 1976; pp 105–115.

RECEIVED July 28, 1993

Chapter 14

High-Performance Liquid Chromatographic Determination of Glucosinolates in *Brassica* Vegetables

J. M. Betz and W. D. Fox

Center for Food Safety and Applied Nutrition, U.S. Food and Drug Administration, Washington, DC 20204

Glucosinolates are naturally occurring constituents of *Brassica* vegetables. The term refers to a class of more than 100 sulfur-containing glycosides that yield thiocyanate, nitrile, or isothiocyanate derivatives upon enzymatic hydrolysis. These compounds are important because of their potential toxicity and because epidemiologic and other evidence indicates that some of them may inhibit some carcinogenic processes when consumed as part of the normal diet. Accordingly, quantitation of these compounds in processed and unprocessed foods has become important. Existing analytical methods are time consuming and labor intensive. Solid phase extraction (SPE) of broccoli extracts, followed by reverse phase ion pair high performance liquid chromatography (HPLC) of intact, nonderivatized glucosinolates, provides a rapid and simple method for evaluation of glucosinolate loss during food processing.

The varieties of the genus *Brassica* that are closely related to cabbage and are consumed as greens or vegetables were all derived from sea cabbage (*Brassica oleracea* var. *sylvestris* L.), a coastal species found wild in much of western Europe (*1*). According to Nieuwhof (*2*), cruciferous plants have been used medicinally and as food for millennia. Gout, diarrhea, coeliac trouble, stomach trouble, deafness, headache, wound healing, and mushroom poisoning are some of the conditions for which they were used (*2*). As food these plants were eaten raw, cooked, or pickled (*3–5*).

Chemistry

The major bioactive constituents of the Brassicaceae (Cruciferae) are a class of sulfur-containing glycosides called glucosinolates and their stable thiocyanate, nitrile, and isothiocyanate hydrolysis products. The glucosinolates generally occur in the form of a potassium salt, whose general structure is shown in Figure 1 (*6*). The R groups (aglycones) are unstable alkyl or aryl side chains which, upon being released by enzymatic hydrolysis, are responsible for the characteristic flavors and odors of the *Brassicas* (*7–12*). The glycone portion of nearly all glucosinolates consists of a single β-D-glucose moiety, although at least two exceptions have been noted (*13,14*).

More than 100 glucosinolates have been reported (*15*). These compounds are distributed over eleven families of dicotyledonous angiosperms (*16*). The glucosinolates previously identified in broccoli are shown in Table I.

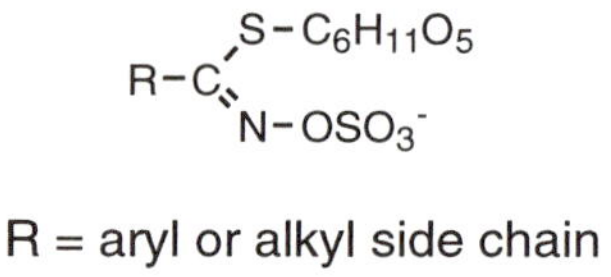

$$R-C\overset{S-C_6H_{11}O_5}{\underset{N-OSO_3^-}{}}$$

R = aryl or alkyl side chain

Figure 1. General glucosinolate structure.

Table I. Glucosinolates in Broccoli

R group	Trivial name
3-Methylsulfinylpropyl-	Glucoiberin[a]
Allyl-	Sinigrin[a]
2(*R*)-Hydroxy-3-butenyl-	Progoitrin[a,b]
2(*S*)-Hydroxy-3-butenyl-	*epi*-Progoitrin[a,b]
4-Methylthiobutyl-	Glucoerucin
4-Methylsulfinylbutyl-	Glucoraphanin
2-Phenylethyl-	Gluconasturtiin[a]
3-Indolylmethyl-	Glucobrassicin[a]
3(*N*-Methoxy)indolylmethyl-	Neoglucobrassicin

[a]Standards available for this study.
[b]These isothiocyanates cyclize to form oxazolidinethiones.

The current systematic approach to glucosinolate nomenclature was proposed by Ettlinger *et al.* (*17*). In this system, the name of the side chain was prefixed to the term glucosinolate. Readers desiring more information on this subject should consult the works of VanEtten and Tookey (*18*), Challenger (*19*), Kjaer (*20*), and Fenwick *et al.* (*21*). Despite the number of glucosinolates which have been described, individual species generally produce only a few types in abundance (*21*). More important are the quantity and pattern of distribution of individual glucosinolates within species, which depend on factors such as plant part examined (*22*), stage of development, horticultural variety, cultivation conditions, climate, and agricultural practices (*23–27*).

The volatile constituents of *Brassica* vegetables are generated by the hydrolysis of glucosinolates by myrosinase (thioglucoside glucohydrolase, EC 3:2:3:1) (*19,28*). This enzyme is located in specialized myrosin cells in Cruciferae (*29*). Details of the complex interrelationships between the glucosinolates, myrosinase and its isoenzymes, cofactors, and the individual hydrolysis products have emerged slowly (*29–31*). When cells of *Brassica* are ruptured in the presence of water, endogenous myrosinase catalyzes the hydrolysis of glucosinolate into glucose and an unstable product which undergoes a spontaneous Lossen-type rearrangement. The rearrangement yields bisulfate ion and the characteristic thiocyanate, nitrile, or isothiocyanate product, the exact nature of which depends on several additional factors, although the nitrile is usually favored (*32–34*). Glucosinolates treated with exogenous myrosinase at neutral pH yield aglycones which undergo rearrangement to produce the volatile isothiocyanates (*35*). Glucosinolates treated with the enzyme under acidic conditions (pH 3) yield nitriles rather than isothiocyanates, as do those hydrolyzed in the presence of Fe^{++} and those which possess a β- or γ-hydroxyl group in the parent side chain (*15,21,33,36*). These latter products (hydroxyisothiocyanates) are unstable and spontaneously cyclize to form the oxazolidine-2-thiones (OZTs).

   Indole glucosinolates yield either the 3-carbinol (I3C) (via an unstable isothiocyanate), upon hydrolysis at pH 7, or the 3-nitrile [indole-3-acetonitrile (I3A)], and subsequently indole-3-acetic acid, under acidic conditions (*36–38*). The 3-carbinol may condense to the diindolyl methane or react with ascorbic acid to form ascorbigenin (*39*).

   Epithionitrile and thiocyanate products may also be formed from glucosinolate precursors. The former requires the action of an epithiospecifier protein and Fe^{++} on the unstable 2-hydroxy-3-butenyl intermediate (*33,34,36–42*). The mechanism of thiocyanate formation is presently unknown (*15*), but it has been suggested that a labile thiocyanate-forming factor similar to epithiospecifier protein may be present in the limited number of genera from which the thiocyanates have been isolated (*21*).

Biological Activity

Glucosinolates and their hydrolysis products have been shown to exhibit both acute and chronic toxic effects in mammalian systems. Early studies were focused on goitrogenic effects of the Cruciferae (*43,44*). Thiocyanates (*45*) and L-5-vinyl-2-thiooxazolidone (goitrin) (*46*) produced by glucosinolate hydrolysis (*47*) were identified as the goitrogenic factors. Goitrin was determined to be the more important compound since its activity could not be reversed by added dietary iodine (*45,48*). *In vitro* experiments demonstrated that progoitrin and *epi*-progoitrin were hydrolyzed to *R*- and *S*-goitrin by native gut flora and were thus toxic even in the absence of thioglucosidase activity (*49–52*). Embryotoxicity has been reported for a number of hydrolysis products (*53*). Acute rodent LD_{50} values for glucosinolate hydrolysis products have been determined (*53,54*), and hepatic, renal, and pancreatic lesions have been reported following ingestion of OZTs (*50*) and nitriles (*53,55–58*).

   Longer-term adverse effects of glucosinolates and their hydrolysis products are more difficult to identify. Syrian hamsters fed a dried cabbage diet had increased pancreatic carcinoma formation (*59*), and cabbage fed to 1,2-dimethylhydrazine-treated mice produced a higher incidence of adenocarcinomas (*60*). Indole-3-carbinol fed to 1,2-dimethylhydrazine-treated rats during initiation caused an increase in carcinogenesis (*61*). Diets containing 10 and 20% cabbage enhanced tumorigenicity of 1,2-dimethylhydrazine, whereas diets of 40% cabbage provided a protective effect in mice (*62*). Sauerkraut incubated with simulated gastric fluid displayed alkylating activity which was enhanced when nitrite was added to the test solution (*63*). Lüthy *et al.* (*64,65*) reported that goitrin is nitrosated by treatment with nitrite under stomach conditions to form N-nitroso-5-vinyl-2-oxazolidone, a mutagen in the Ames *Salmonella* assay.

Albert-Puleo (*65*) provides a review of some of the medicinal uses of Cruciferae, with special emphasis on the anticancer activity of the family. Several epidemiological studies on diet and cancer have been performed which suggest a decrease in colon cancer risk associated with frequent ingestion of vegetables, especially cabbage and other cruciferous vegetables (*67–71*).

Hepatic tumorigenicity of aflatoxin B_1 in rats can be reduced by dietary cabbage and cauliflower (*72,73*). Likewise, the incidence of dimethylbenz[*a*]anthracene-induced mammary tumors in rats can be reduced by oral administration of cabbage and cauliflower (*74*), and I3C and 3,3'-diindolylmethane (*75*). Both of these compounds and I3A also inhibit benzo[*a*]pyrene-induced neoplasia of the forestomach in mice (*76*). Benzo[*a*]pyrene-treated mice gavaged with benzyl isothiocyanate developed fewer forestomach and lung tumors, and diethylnitrosamine-treated mice had fewer stomach tumors (*77*). Benzyl isothiocyanate (BI) (*78*) and benzyl glucosinolate (BG) (*21,79*) inhibited formation of mammary tumors in 7,12-dimethylbenz[*a*]anthracene-treated rats.

On the biochemical level, compounds that stimulate phase 1 or phase 2 mammalian xenobiotic-metabolizing enzymes may possess cancer chemoprotective activity since these systems are important in the metabolism and detoxification of carcinogens (*80*). Welch (*81*), however, has pointed out that although the mixed-function oxidase (MFO) system (phase 1) may detoxify foreign compounds, it is also capable of activating biologically inert substances and may explain some of the conflicting results of studies cited above (*60–62*). Phase 2 system induction may provide a greater protective effect against dietary carcinogenesis (*82*). For further details of these mechanisms, readers are referred to recent reviews (*83–86*).

A number of these xenobiotic-metabolizing enzyme systems have been found to be stimulated by glucosinolates and their hydrolysis products. Rats fed cabbage, I3C, I3A, and diindolylmethane had increased hepatic ethoxyresorufin deethylation and cytochrome P-450 activities (*75*). *Brassica* indoles (*87–91*), brussels sprouts, and cabbage (*92*) have been shown to increase aryl hydrocarbon hydroxylase levels in rodents. Brussels sprouts (but not cabbage) fed to benzo[*a*]-pyrene-treated mice caused significant increases in hepatic epoxide hydrolase activity, and ethanol extracts of brussels sprouts administered to benzo[*a*]pyrene-and aflatoxin B_1-treated mice produced significant increases in cytochrome P-450 levels in the females (*92*). Binding of aflatoxin B_1 to rat hepatic DNA was decreased, whereas hepatic and intestinal glutathione-S-transferase, microsomal epoxide hydroxylase, intestinal aryl hydrocarbon hydroxylase, and ethoxycoumarin O-deethylase activities were increased with a cabbage diet (*91*). Bradfield and Bjeldanes (*93*) reported that 1-methoxyindole-3-carbaldehyde was a potent inducer of monooxygenase. Induction of MFO activity in the small intestine was increased by cabbage, whereas brussels sprouts increased this activity in intestines and liver, and glucobrassicin (but not sinigrin, progoitrin, or glucotropaeolin) induced increased MFO activity (*94*).

Several investigators have recently reported the effects of glucosinolates and their breakdown products on the phase 2 enzyme glutathione-S-transferase. This system was stimulated in liver by BI (*95*), R-goitrin (*96,97*), allyl isothiocyanate, and brussels sprouts (*96*). Wallig and Jeffery (*98*) reported increased pancreatic glutathione-S-transferase following a single dose of 1-cyano-2-hydroxy-3-butene (CHB).

Prochaska *et al.* (*82*) screened vegetable extracts for the induction of quinone reductase (QR) *in vitro*, and found that extracts of Cruciferae were consistent and potent inducers of this phase 2 enzyme. Zhang *et al.* (*99*) isolated and identified sulforaphane [(-)-1-isothiocyanato-(4*R*)-(methyl-sulfinyl)butane] from broccoli on the basis of its QR-inducing effect.

Neither I3C nor its acid reaction products inhibited covalent binding of aflatoxin B_1-8,9-Cl_2 to calf thymus DNA *in vitro*, although the reaction products did inhibit the microsome-mediated binding of the aflatoxin to DNA *in vitro* (*100*). Some of the I3C-mediated protection against hepatocarcinogenisis may, therefore, be due to direct inhibition (by reaction products of I3C and stomach acid) of microsome-activated DNA alkylation rather than to the induction of xenobiotic-metabolizing enzymes.

Analytical Methods

Reviews of the methodology of glucosinolate analysis have been provided by McGregor *et al.* (*101*) and Dietz (*102*). The two most commonly used techniques are the determination of total glucosinolate content (used primarily for rapeseed oil and rapeseed meal) and determination of individual glucosinolate levels.

Thioglucosidase hydrolysis yields glucose in a 1:1 ratio of glucosinolate:sugar and glucosinolate:sulfate. Total glucosinolate content can thus be determined by measuring glucose (*103–111*), sulfate (*112,113*), H^+ (*114*), OZT (*46,115*), or isothiocyanate (*116*) release following the controlled hydrolysis by exogenous myrosinase. When converted to thiourea derivatives, the isothiocyanates may be determined by measurement of their specific UV absorbance (*105*). The glucose techniques are useful for the analysis of plant material or extracts which are glucose free (seeds), but substantial error may result from attempts to apply them to glucose-containing materials such as leaves and flowering heads (*117*).

Various methods have been used for the determination of total nonhydrolyzed glucosinolates (*118–125*). Unfortunately, these general methods all suffer from drawbacks that render them less than universally applicable; although capable of providing data on total content, they are unable to distinguish between individual glucosinolates and are also subject to interferences from other plant constituents.

Paper and thin-layer chromatography have been used for the separation of oxazolidine-2-thiones, isothiocyanates, thiourea derivatives of isothiocyanates, and intact glucosinolates (*22,126–132*). These methods provide qualitative information.

The development of gas chromatographic (GC) instrumentation permitted the separation and quantitation of the volatile mustard oils produced by myrosinase hydrolysis with greater accuracy and precision (*36,103,133–139*). Mass spectrometry (MS) has been used to characterize glucosinolates (*140–143*) and their myrosinase hydrolysis products (*8,36,135,144–147*).

Lack of volatility and the heat lability of the parent glucosides renders them unsuitable for GC analysis without derivatization. Trimethylsilyl (TMS) derivatization of glucosinolates affords the volatility required for GC analysis (*148*). Currently, the most widely used GC methods are those developed by Thies (*149–151*) and Heaney and Fenwick (*152*). These techniques avoid the potential sources of error inherent in the thioglucosidase hydrolysis assays by inactivating myrosinase and analyzing enzymatically desulfated, TMS-derivatized glucosinolates. The method is highly sensitive and the derivatized compounds may be determined in the usual manner. When structural data are required, GC separation may be followed by online matrix isolation-Fourier transform infrared (GC/MI/FT-IR) spectroscopy (*153*) or MS (*140,154–156*). Disadvantages are that the separated compounds are chemically altered (rendering this method unsuitable for isolating compounds for physiological studies) and that the ion-exchange, enzymatic desulfation, and derivatization steps are more time-consuming than other methods (*157*). Extensive surveys of the occurrence of glucosinolates in various cruciferous vegetables have been performed by GC determination of myrosinase hydrolysis products (*103,137,139*). Similar surveys using enzymatic desulfation and GC determination of TMS derivatives have also been performed (*158–161*).

Desulfoglucosinolate standards have been separated by supercritical fluid chromatography (*162*), but the method has not yet been applied to plant material.

High performance liquid chromatography (HPLC) is generally the method of choice for the separation of water-soluble, nonvolatile analytes. A number of investigators have applied HPLC methods to nonderivatized glucosinolates which had been desulfated by techniques previously developed for gas chromatography (*163–169*). This method eliminates the TMS derivatization step required for GC analysis, but the time-consuming ion-exchange and desulfation steps remain and the separated compounds are not suitable for physiological studies (*169*). Glucosinolates separated by HPLC are monitored by detection of UV absorbance rather than by the destructive method of flame-ionization detection (FID) used in GC analysis, thereby enabling it to be used as a preparative technique for obtaining pure compounds. Several investigators have used HPLC for separation of intact, nonderivatized glucosinolates (*118,119,170,171*). Compounds separated by HPLC have been positively identified by comparison of their retention times and UV spectra (obtained by photodiode array detection (PDAD)) with those of authentic compounds (*172*), by collection of the separated compounds followed by MS, and by HPLC/MS (*173– 175*).

Experimental

Our own HPLC method combines gradient ion-pair reverse phase chromatography with a rapid cleanup and glucosinolate concentration step. This procedure allows the rapid determination of intact glucosinolates in vegetative material such as leaves and floral parts, and can be scaled up for production of analytical standards and pure compounds for biological evaluation.

All solvents and reagents were HPLC grade. Progoitrin and *epi*-progoitrin were the generous gifts of G. Spencer of the U.S. Department of Agriculture Northern Regional Research Laboratory, Peoria, IL. Glucotropaeolin, gluconapin, glucosinalbin, gluconasturtiin, and glucobrassicin were generously provided by J. Lewis of the Agricultural and Food Research Council Institute of Food Research, Norwich, NR47UA, UK. Sinigrin was purchased from Aldrich Chemical Co., Inc., Milwaukee, WI. Glucocheirolin and glucoiberin were purchased from Atomergic Chemetals Corp., Farmingdale, NY. *epi*-Progoitrin was purchased from Biocatalysts Ltd., Mid Glamorgan, CF37 5BR, UK. All standards were purified, as necessary, by HPLC.

Frozen and fresh broccoli (*Brassica oleracea* var. *botrytis* L. subvar. *cymosa* Lam.) were purchased at a local supermarket. Extraction of glucosinolates from plant material was similar to that described by Heaney and Fenwick (*160*). Thirty grams of plant material (10 g for freeze-dried material) was added to 200 ml boiling 70% methanol. This material was cooled and then homogenized in an explosion-proof blender. The resulting slurry was heated for 2.5 min, cooled, and filtered. The residue was washed twice with 50 ml 70% methanol, and the filtrate was concentrated to approximately 5 ml on a rotary evaporator. The concentrated extract was diluted to 25 ml with water (broccoli test solution, BTS) and stored at -40°C.

Glucosinolates were concentrated and interfering substances were removed from this solution by use of 6 ml C_{18} disposable solid-phase extraction (SPE) columns (J.T. Baker, Phillipsburg, NJ) (use of brand names does not constitute endorsement by the Food and Drug Administration, the Department of Health and Human Services, or the U.S. government). Columns were washed with 5 ml methanol and then conditioned with 5 ml 0.01 M aqueous tetrabutylammonium sulfate (IPC-A) (Alltech Associates, Deerfield, IL). A 5 ml aliquot of broccoli test solution was added to the column, and the column was washed with 1 ml water. Glucosinolates were eluted with 1 ml methanol/water (55:45) (LC test solution, LCTS).

Chromatographic analyses were performed using a Waters Model 600E pump (Milford, MA) equipped with a column oven (45°C) and a Rheodyne 7010 injector (Cotati, CA) (20 µl loop). Separation was performed on a Beckman Ultrasphere C_{18} ion-pair analytical HPLC column (Fullerton, CA), 5 µm packing (ϕ4.6 x 250 mm) equipped with a ϕ4.6 x 50 mm guard column. Detection was accomplished with a Hewlett-Packard Model 1040A photodiode array detector (Palo Alto, CA). Quantitation was performed by recording area-under-the-curve (AUC) values at 237 nm for each glucosinolate standard. Ultraviolet spectra of separated compounds were recorded by using the photodiode array detector. A binary gradient was used for the separation, in which mobile phase A consisted of 12% methanol/88% aqueous 0.005 M IPC-A and mobile phase B was composed of 34% methanol/66% aqueous 0.005 M IPC-A. The gradient ran from 100% A at 0 min to 100% B at 20 min (gradient curve shape 4) and was held at 100% B for an additional 15 min. Flow rate was 1.3 ml/min.

The effects of different types of food processing on the glucosinolate content of cruciferous vegetables were evaluated. Because it has been suggested that glucosinolates in steamed or boiled vegetables are lost, in part by leaching into the cooking water (*159,176*), our first experiment examined the effects of boiling water on glucosinolates. A mixture of aqueous standard solutions (each 0.02 mg/ml) was analyzed by HPLC after no treatment, boiling for 15 min, boiling for 30 min, and treating with 0.01 N HCl for 1 h at 37°C (to approximate conditions in the stomach). Glucosinolates in plant material were determined by HPLC of LCTS obtained from broccoli after no treatment (raw), stir-frying (5 min), adding to boiling water and boiling (5 or 30 min), and freeze-drying (after freezing in liquid nitrogen). Frozen broccoli purchased from a local supermarket was examined by preparing the BTS from still-frozen florets after removal from a freshly opened package.

Results

The recoveries of individual glucosinolates from the SPE columns varied considerably because of the elution conditions required by the diverse side chains. Elution with less than 55% methanol resulted in incomplete elution of glucosinolates. Higher recoveries were achieved with methanol concentrations greater than 55%, but elution of interfering plant pigments with the analyte offset any gains in recovery. Glucosinolate recoveries with the SPE method described above averaged 85%. Glucosinolates isolated from broccoli were identified by comparison of UV spectra obtained by using the photodiode array detector with those of standard compounds and with those obtained for broccoli extracts spiked with individual standards.

Examination of Figure 2a reveals that the major glucosinolates found in the broccoli (fresh, raw) used in this study were glucoiberin and glucobrassicin. Progoitrin, *epi*-progoitrin, and gluconasturtiin, which were present in lesser quantities, were also identified in the fresh florets. These results agree with those in previous reports (*18,158*). Some peaks appearing in the chromatogram could not be identified because reference standards for several glucosinolates were not available. Thus, minor glucosinolates apparently absent in this study despite having been reported by other investigators as constituents of broccoli (Table I), may actually have been present.

The glucosinolates (with the exception of progoitrin) were found to be relatively heat labile when processed within the food matrix (Figure 2b, 2c, Table II) or in aqueous solution (Table III). Stir-frying for 5 min (Figure 2c) was the least destructive of the cooking methods examined. As expected, raw, freeze-dried, and frozen broccoli contained the highest glucosinolate levels. In addition, the longer the broccoli was cooked, the greater the glucosinolate loss.

Table II. Effect of Processing on Glucosinolates[a] in Broccoli

	GI	Pro	*epi*-Pro	Sin	GB	GN
Raw fresh						
	15.4 [100]	2.5 [100]	3.2 [100]	1.6 [100]	42.6 [100]	8.1 [100]
Stir-fried fresh						
	10.3 [67]	2.2 [88]	—	0.4 [25]	32.3 [76]	3.3 [41]
Boiled (5 min) fresh						
	7.8 [51]	2.5 [100]	—	—	13.4 [32]	2.0 [25]
Boiled (30 min) fresh						
	4.7 [31]	2.5 [100]	—	—	5.5 [13]	—
Freeze-dried fresh (calculated on wet-weight basis)						
	21.0 [136]	1.7 [68]	—	2.3 [143]	73.9 [174]	—
Frozen (uncooked)						
	14.3 [93]	1.1 [44]	—	0.8 [50]	65.6 [154]	—

[a] μg/g broccoli [% of raw, fresh], GI = glucoiberin; Pro = progoitrin; *epi*-Pro = *epi*-progoitrin; Sin = sinigrin; GB = glucobrassicin; GN = gluconasturtiin.

Hydrolysis products of glucobrassicin (I3C and I3A) and progoitrin (goitrin) were also identified in the broccoli. Quantitation of these compounds was not attempted because it was beyond the scope of the original research plan. These products may have resulted from hydrolysis catalyzed by the release of endogenous myrosinase during chopping of the broccoli into pieces of manageable size for cooking or freezing. The data in Table III indicate that the aliphatic glucosinolates are labile in boiling water and in 0.1 N HCl. No attempts were made to determine the fate of the processed standards. In addition, we did not possess sufficient quantities of progoitrin, *epi*-progoitrin, and glucobrassicin to include these compounds in this experiment.

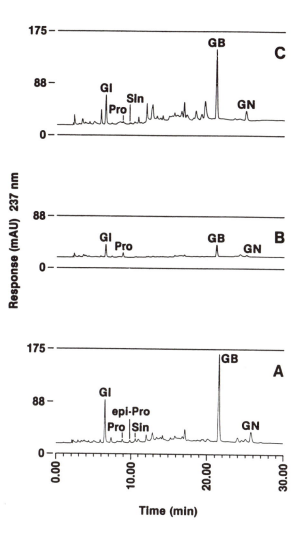

Figure 2. Representative chromatograms of A) fresh raw, B) boiled 5 min, and C) stir fried broccoli. GI = glucoiberin; Pro = progoitrin; *epi*-Pro = *epi*-pro-goitrin; Sin = sinigrin; GB = glucobrassicin; GN = gluconasturtiin.

**Table III. Effect of Processing on
the Concentration of Glucosinolate Standards (0.02 mg/ml)**

Glucosinolate	% Change		
	Boil 15 min	Boil 30 min	0.1 N HCl (1 h)
Glucoiberin	-5	-24	-33
Glucocherolein	-7	-21	-28
Sinigrin	-9	-37	-30
Indole-3-carbinol	-13	-82	-100

Conclusions

Broccoli was chosen for this investigation because it is widely consumed both raw and cooked and because of its indole glucosinolate content. Our original purpose in examining the glucosinolate content of crucifers was related to potential public health aspects. Human toxicological issues related to the consumption of cruciferous vegetables (except in quantities which greatly exceed those found in the "normal" diet) have been largely discounted. The potential chemoprotective effects attributed to a number of vegetables, however, are attracting considerable attention from the public health community. For this reason, a rapid and relatively simple method for separating and quantitating these compounds was needed in order to monitor their decomposition during various types of food processing. The biological activity of the plant depends to a great extent on which of the hydrolysis products reach the target organs. Because the final products of hydrolysis are dependent upon many complex variables (and controlling these variables is extremely difficult when performing *in vivo* experiments), and because is unlikely that all hydrolysis products in the complex milieu have been identified, a simpler approach to the problem is to monitor the intake of intact glucosinolates. In addition, critical examination of the literature along with a chromatographic survey of Cruciferae for indole compounds has led Wall *et al.* (*177*) to conclude that content and occurrence of indole-3-carbinol in plants has been heavily overestimated and that its role as a dietary factor in chemoprevention may be suspect. Results of studies on non-indole glucosinolates may indicate a broader role for these compounds than has been previously suspected (*98,99*).

Complicating the possible link between the glucosinolate content of cruciferous vegetables and cancer chemoprevention are reports on the isolation of antimutagenic nonglucosinolate compounds from these plants. A heat and pronase labile protein (MW 48,000) that inhibited the *Salmonella typhimurium* mutagenic activity of amino acid pyrolysates was isolated from cabbage (*178–181*). β-Sitosterol and 15-nonacosanone isolated from dried Savoy Chieftan cabbage inhibited 2-amino-anthracene- or N-methyl-N-nitrosourea-induced mutagenesis in the Ames assay. These compounds, nonacosane, and pheophytin *a* (also from cabbage) inhibited 2-amino-anthracene-induced mutagenicity in V79 mammalian cells, whereas all but pheophytin were active against the mutagenicity of N-methyl-N-nitrosourea (*182*).

In conclusion, if carcinogenic processes are inhibited by cruciferous vegetables, the phenomenon would appear to be extremely complex and to depend on a number of interconnecting mechanisms. Although glucosinolates and their hydrolysis products may be important components of the postulated chemo-protective effect of this plant family, synergistic effects have yet to be addressed. The total milieu of phytochemicals is probably necessary for full dietary protection. Finally, the glucosinolates and their products possess demonstrated toxicity, and overuse syndromes are a potential consumer hazard.

Literature Cited

1. Edlin, H.L. *Plants and Man. The Story of Our Basic Food*; Natural History Press: Garden City, NY, 1969; p 110.
2. Nieuwhof, M. *Cole Crops, Botany, Cultivation and Utilization*; Leonard Hill Books: London, UK, 1969; pp 1–11.
3. Toxopeus, H. *Proceedings Eucarpia 'Cruciferae 1974' Conference*; Proceedings Congress Eucarpia 5th; Scottish Horticultural Research Institute: Edinburgh, 1974; pp 1–7.
4. Van Marrewijk, N.P.A.; Toxopeus, H. *Proceedings Eucarpia 'Cruciferae 1979' Conference*; Proceedings Congress Eucarpia 10th; European Association for Research on Plant Breeding: Wageningen, 1979; pp 47–56.
5. Zeven, A.C.; Brandenburg, W.A. *Econ. Bot.* **1986**, *40*, 397–408.
6. Ettlinger, M.G.; Lundeen, A.J., *J. Am. Chem. Soc.* **1956**, *78*, 4172–4173.
7. Clapp, R.C.; Long, L., Jr.; Dateo, G.P.; Bissett, F.H.; Hasselstrom, T. *J. Am. Chem. Soc.* **1959**, *81*, 6278–6281.
8. Bailey, S.D.; Bazinet, M.L.; Driscoll, J.L.; McCarthy, A.I. *J. Food Sci.* **1961**, *26*, 163–170.
9. Friis, P.; Kjaer, A. *Acta Chem. Scand.* **1966**, *20*, 698–705.
10. MacLeod, A.J.; MacLeod, G. *J. Food Sci.* **1970**, *35*, 734–738.
11. Wallbank, B.E.; Wheatley, G.A. *Phytochemistry* **1976**, *15*, 763–766.
12. Buttery, R.G.; Guadagni, D.G.; Ling, L.C.; Seifert, R.M.; Lipton, W. *J. Agric. Food Chem.* **1976**, *24*, 829–832.
13. Olsen, O.; Rasmussen, K.W.; Sørensen, H. *Phytochemistry* **1981**, *20*, 1857–1861.
14. Linscheid, M.; Wendisch, D.; Strack, D. *Z. Naturforscher* **1980**, *35 c*, 907–914.
15. Heaney, R.K.; Fenwick, G.R. In *Natural Toxicants in Foods: Progress and Prospects;* Watson, D.H., Ed.; Ellis Horwood Series in Food Science and Technology; Ellis Horwood, Ltd.: Chichester, UK, 1987; pp 76–109.
16. Kjaer, A. In *Chemistry in Botanical Classification*; Bendy, G.; Santesson, J., Eds.; Academic Press: London, UK, 1974; pp 229–234.
17. Ettlinger, M.G.; Dateo, G.P., Jr.; Harrison, B.W.; Mabry, T.J.; Thompson, C.P. *Proc. Natl. Acad. Sci. USA* 1961, *47*, 1875–1880.
18. VanEtten, C.H.; Tookey, H.L. In *Handbook of Naturally Occurring Food Toxicants*; Rechcigl, M., Jr., Ed.; CRC Series in Nutrition and Food; CRC Press, Inc.: Boca Raton, FL, 1983; pp 15–30.
19. Challenger, F. *Aspects of the Organic Chemistry of Sulfur*; Butterworths: London, UK, 1959; pp 115–161.
20. Kjaer, A. *Fortschr. Chem. Org. Naturst.* **1960**, *18*, 122–176.

21. Fenwick, G.R.; Heaney, R.K.; Mullin, W.J. *CRC Crit. Rev. Food Sci. Nutr.* **1983**, *18*, 123–201.
22. Josefsson, E. *Phytochemistry* **1967**, *6*, 1617–1627.
23. Josefsson, E. *J. Sci. Food Agric.* **1970**, *21*, 98–103.
24. MacLeod, A.J.; Nussbaum, M.L. *Phytochemistry* **1977**, *16*, 861–865.
25. MacLeod, A.J.; Pikk, H.E. *Phytochemistry* **1978**, *17*, 1029–1032.
26. Miller, K.W.; Boyd, J.N.; Babish, J.G.; Lisk, D.J.; Stoewsand, G.S. *J. Food Safety* **1983**, *5*, 131–143.
27. Goodrich, R.M.; Parker, R.S.; Lisk, D.J.; Stoewsand, G.S. *Food Chem.* **1988**, *27*, 141–150.
28. Gmelin, R.; Virtanen, A.I. *Acta Chem. Scand.* **1960**, *14*, 507–510.
29. Phelan, J.R.; Vaughan, J.G. *J. Exp. Bot.* **1980**, *31*, 1425–1433.
30. Phelan, J.R.; Allen, A.; Vaughan, J.G. *J. Exp. Bot.* **1984**, *35*, 1558–1564.
31. Pocock, K.; Heaney, R.K.; Wilkinson, A.P.; Beaumont, J.E.; Vaughan, J.G.; Fenwick, G.R. *J. Sci. Food Agric.* **1987**, *41*, 245–257.
32. Björkman, R. In *The Biology and Chemistry of the Cruciferae*; Vaughan, J.G.; MacLeod, A.J.; Jones, B.M.G., Eds.; Academic Press: New York, NY, 1976; pp 191–205.
33. Cole, R.A. *Phytochemistry* **1976**, *15*, 759–762.
34. Tookey, H.L.; VanEtten, C.H.; Daxenbichler, M.E. In *Toxic Constituents of Plant Foods*; Liener, I.E., Ed.; Academic Press: New York, NY, 1980, pp 103–142.
35. Ettlinger, M.G.; Lundeen, A.J. *J. Am. Chem. Soc.* **1957**, *79*, 1764–1765.
36. Daxenbichler, M.E.; VanEtten, C.H.; Spencer, G.F. *J. Agric. Food Chem.* **1977**, *25*, 121–124.
37. VanEtten, C.H.; Daxenbichler, M.E.; Peters, J.E.; Wolff, I.A.; Booth, A.N. *J. Agric. Food Chem.* **1966**, *14*, 426–430.
38. Pihakaski, K.; Pihakaski, S. *J. Exp. Bot.* **1978**, *29*, 335–346.
39. Gmelin, R.; Virtanen, A.I. *Ann. Acad. Sci. Fenn. Ser. AII* **1961**, *107*, 1–25.
40. MacLeod, A.J.; Islam, R. *J. Sci. Food Agric.* **1976**, *27*, 909–912.
41. Lüthy, J.; Benn, M.H. *Phytochemistry* **1979**, *18*, 2028–2029.
42. Tookey, H.L. *Can. J. Biochem.* **1973**, *51*, 1654–1660.
43. Chesney, A.M.; Clawson, T.A.; Webster, B. *Johns Hopkins Hosp. Bull.* **1928**, *43*, 261–277.
44. Webster, B.; Chesney, A.M. *Am. J. Pathol.* **1930**, *6*, 275–284.
45. Greer, M.A. *Physiol. Rev.* **1950**, *30*, 513–548.
46. Astwood, E.B.; Greer, M.A.; Ettlinger, M.G. *J. Biol. Chem.* **1949**, *181*, 121–130.
47. Greer, M.A. *J. Am. Chem. Soc.* **1956**, *78*, 1260–1261.
48. Greer, M.A.; Kendall, J.W.; Smith, M. In *The Thyroid Gland*; Pitt-Rivers, R.; Trotter, W.R., Eds.; Butterworths: Washington, DC, 1964; pp 357–389.
49. Greer, M.A. In *Recent Progress in Hormone Research*; Pincus, G., Ed.; Academic Press: New York, NY, 1962; pp 187–219.
50. VanEtten, C.H.; Gagne, W.E.; Robbins, D.J.; Booth, A.N.; Daxenbichler, M.E.; Wolff, I.A. *Cereal Chem.* **1969**, *46*, 145–155.
51. Oginsky, E.L.; Stein, A.E.; Greer, M.A. *Proc. Soc. Exp. Biol. Med.* **1965**, *119*, 360–364.
52. Tani, N.; Ohtsuru, M.; Hata, T. *Agric. Biol. Chem.* **1974**, *38*, 1623–1630.
53. Nishie, K.; Daxenbichler, M.E. *Food Cosmet. Toxicol.* **1980**, *18*, 159–172.

54. VanEtten, C.H.; Daxenbichler, M.E.; Wolff, I.A. *J. Agric. Food Chem.* **1969**, *17*, 483–491.
55. Gould, D.H.; Fettman, M.J.; Daxenbichler, M.E.; Bartuska, B.M. *Toxicol. Appl. Pharmacol.* **1985**, *78*, 190–201.
56. Wallig, M.A.; Gould, D.H.; Fettman, M.J.; Willhite, C.C. *Food Chem. Toxicol.* **1988**, *26*, 149–157.
57. Wallig, M.A.; Gould, D.H.; Fettman, M.J. *Food Chem. Toxicol.* **1988**, *26*, 137–147.
58. Dietz, H.M.; Panigrahi, S.; Harris, R.V. *J. Agric. Food Chem.* **1991**, *39*, 311–315.
59. Birt, D.F.; Tibbels, M.G.; Pour, P.M. *Fed. Proc.* **1986**, *45*, 1076.
60. Temple, N.J.; Basu, T.K. *J. Natl. Cancer Inst.* **1987**, *79*, 1131–1134.
61. Pence, B.; Buddingh, F.; Yang, S. *J. Natl. Cancer Inst.* **1986**, *77*, 269–276.
62. Srisangham, C.; Hendricks, D.G.; Sharma, R.P.; Salunkhe, D.K.; Mahoney, A.W. *J. Food Safety* **1980**, *4*, 235–245.
63. Groenen, P.J.; Busink, E. *Food Chem. Toxicol.* **1988**, *26*, 215–225.
64. Lüthy, J.; Carden, B.; Bachmann, M.; Friederich, U.; Schlatter, C. *Mitt. Geb. Lebensmittelunters. Hyg.* **1984**, *75*, 101–109.
65. Lüthy, J.; Carden, B.; Friederich, U.; Bachmann, M. *Experientia* **1984**, *40*, 452–453.
66. Albert-Puleo, M. *J. Ethnopharmacol.* **1983**, *9*, 261–272.
67. Graham, S.; Dayal, H.; Swanson, M.; Mittelman, A.; Wilkinson, G. *J. Natl. Cancer Inst.* **1978**, *61*, 709–714.
68. *Diet, Nutrition and Cancer*; National Academy of Sciences: Washington, DC, 1982; p 11.
69. Graham, S.; Mettlin, C. *Am. J. Epidemiol.* **1979**, 1–20.
70. Haenszel, W.; Locke, F.B.; Segi, M. *J. Natl. Cancer Inst.* **1980**, *64*, 17–22.
71. Manousos, O.; Day, N.E.; Trichopoulos, D.; Gerovassilis, F.; Tzonou, A.; Polychronopoulou, A. *Int. J. Cancer* **1983**, *32*, 1–5.
72. Boyd, J.N.; Babish, J.G.; Stoewsand, G.S. *Food Chem. Toxicol.* **1982**, *20*, 47–52.
73. Stoewsand, G.S.; Babish, J.G.; Wimberly, H.C. *J. Environ. Pathol. Toxicol.* **1978**, *2*, 399–406.
74. Wattenberg, L.W. *Cancer Res.* **1983**, *43*, 2448s–2453s.
75. McDanell, R.; McLean, A.E.M.; Hanley, A.B.; Heaney, R.K.; Fenwick, G.R. *Food Chem. Toxicol.* **1987**, *25*, 363–368.
76. Wattenberg, L.W.; Loub, W.D. *Cancer Res.* **1978**, *38*, 1410–1413.
77. Wattenberg, L.W. *Carcinogenesis* **1987**, *8*, 1971–1973.
78. Wattenberg, L.W. *J. Natl. Cancer Inst.* **1977**, *58*, 395–398.
79. Wattenberg, L.W.; Hanley, A.B.; Barany, G.; Sparnins, V.L.; Lam, L.K.T.; Fenwick, G.R. In *Diet, Nutrition and Cancer*; Hayashi, Y.; Nagao, M.; Sugimura, T.; Takayama, S.; Tomatis, L.; Wattenberg, L.W.; Wogan, G.N., Eds.; Japan Sci. Soc. Press: London, 1986; pp 193–203.
80. McDanell, R.; McLean, A.E.M.; Hanley, A.B.; Heaney, R.K.; Fenwick, G.R. *Food Chem. Toxicol.* **1988**, *26*, 59–70.
81. Welch, R.M. *Pharmac. Rev.* **1979**, *30*, 457–467.
82. Prochaska, H.J.; Santamaria, A.B.; Talalay, P. *Proc. Natl. Acad. Sci. USA* **1992**, *89*, 2394–2398.

83. Kada, T.; Inoue, T.; Namiki, M. In *Environmental Mutagenesis, Carcinogenesis, and Plant Biology*; Klekowski, Jr., E.J., Ed.; Praeger Publishers: New York, NY, 1982; Vol. 1, pp 133–151.
84. Hartman, P.E.; Shankel, D.M. *Environ. Mol. Mutagen.* **1990**, *15*, 145–182.
85. Davis, D.L. *Environ. Res.* **1989**, *50*, 322–340.
86. Wattenberg, L.W. *Cancer Res.* **1983**, *43*, 2448s–2453s.
87. Wattenberg, L.W. *Cancer* **1971**, *28*, 99–102.
88. Loub, W.D.; Wattenberg, L.W.; Davis, D.W. *J. Natl. Cancer Inst.* **1975**, *54*, 985–988.
89. Babish, J.G.; Stoewsand, G.S. *Food Cosmet. Toxicol.* **1977**, *16*, 151–155.
90. Schertzer, H.G. *Food Chem. Toxicol.* **1983**, *21*, 31–35.
91. Whitty, J.P.; Bjeldanes, L.F. *Food Chem. Toxicol.* **1987**, *25*, 581–587.
92. Hendrich, S.; Bjeldanes, L.F. *Food Chem. Toxicol.* **1983**, *21*, 479–486.
93. Bradfield, C.A.; Bjeldanes, L.F. *J. Agric. Food Chem.* **1987**, *35*, 896–900.
94. McDanell, R.; McLean, A.E.M.; Hanley, A.B.; Heaney, R.K.; Fenwick, G.R. *Food Chem. Toxicol.* **1989**, *27*, 289–293.
95. Vos, R.M.E.; Snoek, M.C.; van Berkel, W.J.H.; Muller, F.; van Bladeren, P.J. *Biochem. Pharmacol.* **1988**, *37*, 1077–1082.
96. Chang, Y.; Bjeldanes, L.F. *Food Chem. Toxicol.* **1985**, *23*, 905–909.
97. Bogaards, J.J.P.; van Ommen, B.; Falke, H.E.; Willems, M.I.; van Bladeren, P.J. *Food Chem. Toxicol.* **1990**, *28*, 81–88.
98. Wallig, M.A.; Jeffery, E.H. *Fundam. Appl. Toxicol.* **1990**, *14*, 144–159.
99. Zhang, Y.; Talalay, P.; Cho, C.-G.; Posner, G.H. *Proc. Natl. Acad. Sci. USA* **1992**, *89*, 2394–2398.
100. Fong, A.T.; Swanson, H.I.; Dashwood, R.H.; Williams, D.E.; Hendricks, J.D.; Bailey, G.S. *Biochem. Pharmacol.* **1990**, *39*, 19–26.
101. McGregor, D.I.; Mullin, W.J.; Fenwick, G.R. *J. Assoc. Off. Anal. Chem.* **1983**, *66*, 825–849.
102. Dietz, H.M. *Food Lab. News* **1989**, *16*, 30–36.
103. VanEtten, C.H.; Daxenbichler, M.E.; Williams, P.H.; Kwolek, W.F. *J. Agric. Food Chem.* **1976**, *24*, 452–455.
104. VanEtten, C.H.; Daxenbichler, M.E.; Kwolek, W.F.; Williams, P.H. *J. Agric. Food Chem.* **1979**, *27*, 648–650.
105. Wetter, L.R.; Youngs, C.G. *J. Am. Oil Chem. Soc.* **1976**, *53*, 162–164.
106. Bjørkman, R. *Acta Chem. Scand.* **1972**, *26*, 1111–1116.
107. Lein, K.-A.; Schön, W.J. *Angew. Bot.* **1969**, *43*, 87–92.
108. Thies, W. *Fette, Seifen, Anstrichm.* **1985**, *87*, 347–350.
109. Gardrat, C.; Mesnard, S. *Rev. Fr. Corps Gras* **1990**, *37*, 91–96.
110. Heaney, R.K.; Fenwick, G.R. *Z. Pflanzenzücht.* **1981**, *87*, 89–95.
111. Kershaw, S.J.; Johnstone, R.A.W. *J. Am. Oil Chem. Soc.* **1990**, *67*, 821–826.
112. Mustakas, G.C.; Kirk, L.D.; Griffin Jr., E.L. *J. Am. Oil Chem. Soc.* **1976**, *53*, 12–16.
113. Fiebig, H.-J.; Jörden, M.; Aitzetmüller, K. *Fett Wiss. Technol.* **1990**, *92*, 173–178.
114. Leoni, O.; Iori, R.; Palmieri, S. *J. Agric. Food Chem.* **1991**, *39*, 2322–2326.
115. Kreula, M.; Kesvaara, M. *Acta Chem. Scand.* **1959**, *13*, 1375–1382.
116. Appelqvist, L.-A.; Josefsson, E. *J. Sci. Food Agric.* **1967**, *18*, 510–519.
117. VanEtten, C.H.; Daxenbichler, M.E. *J. Assoc. Off. Anal. Chem.* **1977**, *60*, 946–949.

118. Møller, P.; Plöger, A.; Sørensen, H. In *Advances in the Production and Utilization of Cruciferous Crops*; Sørensen, H., Ed.; Martinus Nijhoff: Dortrecht, 1985; pp 97–110.
119. Schung, V.E.; Haneklaus, S. *Fett Wiss. Technol.* **1987**, *89*, 32–36.
120. Schung, V.E.; Haneklaus, S. *Fett Wiss. Technol.* **1990**, *92*, 57–61.
121. Schung, V.E.; Haneklaus, S. *Fett Wiss. Technol.* **1990**, *92*, 101–106.
122. Biston, R.; Dardenne, P.; Cwikowski, M.; Marlier, M.; Severin, M.; Wathelet, J.-P. *J. Am. Oil Chem. Soc.* **1988**, *65*, 1599–1600.
123. Tkachuk, R. *J. Am. Oil Chem. Soc.* **1981**, *58*, 819–822.
124. Starr, C.; Suttle, J.; Morgan, A.G.; Smith, D.B. *Agric. Sci. Camb.* **1985**, *104*, 317–323.
125. Wong, R.S.-C. In *Biotechnology of Plant Fats and Oils*; Rattray, J.B., Ed.; American Oil Chemist's Society: Champaign, IL, 1991; pp 52–57.
126. Kjaer, A.; Rubinstein, K. *Acta Chem. Scand.* **1953**, *7*, 528–536.
127. Rodman, J.E. *Phytochem. Bull.* **1978**, *11*, 6–31.
128. Olsen, O.; Sørensen, H. *Phytochemistry* **1979**, *18*, 1547–1552.
129. Olsen, O.; Sørensen, H. *J. Am. Oil Chem. Soc.* **1981**, *58*, 857–865.
130. Wagner, H.; Horhammer, L.; Nufer, H. *Arzneim. Forsch.* **1965**, *15*, 453– 457.
131. Rakow, D.; Gmelin, R.; Thies, W. *Z. Naturforscher* **1981**, *36c*, 16–22.
132. Woggon, H.; Macholz, R.; Jehle, D. *Nahrung* **1990**, *6*, 515–525.
133. Kjaer, A.; Jart, A. *Acta Chem. Scand.* **1957**, *11*, 1423.
134. Cole, R.A. *Phytochemistry* **1975**, *14*, 2293–2294.
135. Cole, R.A. *J. Sci. Food Agric.* **1980**, *31*, 549–557.
136. Olsen, O.; Sørensen, H. *Phytochemistry* **1980**, *19*, 1783–1787.
137. Daxenbichler, M.E.; VanEtten, C.H.; Williams, P.H. *J. Agric. Food Chem.* **1979**, *27*, 34–37.
138. Hasapis, X.; MacLeod, A.J.; Moreau, M. *Phytochemistry* **1981**, *20*, 2355–2358.
139. Carlson, D.G.; Daxenbichler, M.E.; VanEtten, C.H.; Tookey, H.L.; Williams, P.H. *J. Agric. Food Chem.* **1981**, *29*, 1235–1239.
140. Eagles, J.; Fenwick, G.R.; Gmelin, R.; Rakow, D. *Biomed. Mass Spectrom.* **1981**, *8*, 265–269.
141. Bojesen, G.; Larsen, E. *Biol. Mass Spectrom.* **1991**, *20*, 286–288.
142. Fenwick, G.R.; Eagles, J.; Gmelin, R.; Rakow, D. *Biomed. Mass Spectrom.* **1980**, *7*, 410–412.
143. Fenwick, G.R.; Eagles, J.; Self, R. *Org. Mass Spectrom.* **1982**, *17*, 544-546.
144. Kjaer, A.; Ohashi, M.; Wilson, J.M.; Djerassi, C. *Acta Chem. Scand.* **1963**, *17*, 2143–2154.
145. Bach, E.; Kjaer, A.; Shapiro, R.H.; Djerassi, C. *Acta Chem. Scand.* **1965**, *19*, 2438–2440.
146. Spencer, G.F.; Daxenbichler, M.E. *J. Sci. Food Agric.* **1980**, *31*, 359–367.
147. Gil, V.; MacLeod, A.J. *Phytochemistry* **1980**, *19*, 227–231.
148. Underhill, E.W.; Kirkland, D.F. *J. Chromatogr.* **1971**, *57*, 47–54.
149. Thies, W. *Fette, Seifen, Anstrichm.* **1976**, *78*, 231–234.
150. Thies, W. *Z. Pflanzenzücht.* **1977**, *79*, 331–335.
151. Thies, W. *Naturwissenschaften* **1979**, *66*, 364–365.
152. Heaney, R.K.; Fenwick, G.R. *J. Sci. Food Agric.* **1980**, *31*, 593–599.
153. Mossoba, M.M.; Shaw, G.J.; Andrzejewski, D.; Sphon, J.A.; Page, S.W. *J. Agric. Food Chem.* **1989**, *37*, 367–372.

154. Christensen, B.W.; Kjaer, A.; Ogaard Madsen, J.; Olsen, C.E.; Olsen, O.; Sørensen, H. *Tetrahedron* **1982**, *38*, 353–359.

155. Shaw, G.J.; Andrzejewski, D.; Roach, J.A.G.; Sphon, J.A. *J. Agric. Food Chem.* **1989**, *37*, 372–378.

156. Shaw, G.J.; Andrzejewski, D.; Roach, J.A.G.; Sphon, J.A. *J. Agric. Food Chem.* **1990**, *38*, 616–619.

157. Quinsac, A.; Ribaillier, D. In *Advances in the Production and Utilization of Cruciferous Crops*; Sørensen, H., Ed.; Martinus Nijhoff/Junk: Dortrecht, 1985; pp 11–13.

158. Lewis, J.; Fenwick, G.R. *Food Chem.* **1987**, *25*, 250–268.

159. Sones, K.; Heaney, R.K.; Fenwick, G.R. *J. Sci. Food Agric.* **1984**, *35*, 712–720.

160. Heaney, R.K.; Fenwick, G.R. *J. Sci. Food Agric.* **1980**, *31*, 785–793.

161. Heaney, R.K.; Fenwick, G.R. *J. Sci. Food Agric.* **1980**, *31*, 794–801.

162. Lafosse, M.; Rollin, P.; Elfakir, C.; Morin–Allory, L.; Martens, M.; Dreux, M. *J. Chromatogr.* **1990**, *505*, 191–197.

163. Spinks, E.A.; Sones, K.; Fenwick, G.R. *Fette, Seifen, Anstrichm.* **1984**, *86*, 228–231.

164. Minchinton, I.R.; Sang, J.P.; Burke, D.G.; Truscott, R.J.W. *J. Chromatogr.* **1982**, *247*, 141–148.

165. Truscott, R.J.W.; Burke, D.G.; Minchinton, I.R. *Biochem. Biophys. Res. Commun.* **1982**, *107*, 1258–1264.

166. Truscott, R.J.W.; Minchinton, I.R.; Burke, D.G.; Sang, J.P. *Biochem. Biophys. Res. Commun.* **1982**, *107*, 1368–1375.

167. Truscott, R.J.W.; Minchinton, I.; Sang, J. *J. Sci. Food Agric.* **1983**, *34*, 247–254.

168. Quinsac, A.; Ribaillier, D.; Elfakir, C.; Lafosse, M.; Dreux, M. *J. Assoc. Off. Anal. Chem.* **1991**, *74*, 932–939.

169. Fiebig, H.G. *Fett Wiss. Technol.* **1991**, *93*, 264–267.

170. Helboe, P.; Olsen, O.; Sørensen, H. *J. Chromatogr.* **1980**, *197*, 199–205.

171. Castor-Normandin, F.; Gauchet, C.; Prevot, A.; Sørensen, H. *Rev. Fr. Corps Gras* **1986**, *33*, 119–126.

172. Björkqvist, B.; Hase, A. *J. Chromatogr.* **1988**, *435*, 501–507.

173. Mellon, F.A.; Chapman, J.R.; Pratt, J.A.E. *J. Chromatogr.* **1987**, *394*, 209–222.

174. Kokkonen, P.S.; van der Greef, J.; Niessen, W.M.A.; Tjaden, U.R.; ten Hove, G.J.; van der Werken, G. *Biol. Mass Spectrom.* **1991**, *20*, 259–263.

175. Lange, R.; Petrzika, M. *Fett Wiss. Technol.* **1991**, *93*, 284–288.

176. Mullin, W.J.; Sahasrabudhe, M.R. *J. Inst. Can. Sci. Technol. Aliment.* **1978**, *11*, 50–52.

177. Wall, M.E.; Taylor, H.; Perera, P.; Mansukh, C.W. *J. Nat. Prod.* **1988**, *51*, 129–135.

178. Morita, K.; Hara, M.; Kada, T. *Agric. Biol. Chem.* **1978**, *42*, 1235–1238.

179. Morita, K.; Kada, T.; Inoue, T. *Jpn. Kokai Tokkyo Koho* **1979**, *79,122,715*

180. Morita, K.; Kada, T.; Inoue, T. *Ger. Offen.* **1979**, *2,810,293*.

181. Kada, T.; Morita, K.; Inoue, T. *Mutat. Res.* **1978**, *53*, 351–353.

182. Lawson, T.; Nunnally, J.; Walker, B.; Bresnick, E.; Wheeler, D.; Wheeler, M. *J. Agric. Food Chem.* **1989**, *37*, 1363–1367.

RECEIVED October 4, 1993

Limonoids and Phthalides

Chapter 15

Biochemistry of Citrus Limonoids and Their Anticarcinogenic Activity

Shin Hasegawa[1], Masaki Miyake[2], and Yoshihiko Ozaki[2]

[1]Fruit and Vegetable Chemistry Laboratory, U.S. Department of Agriculture, Agricultural Research Service, 263 South Chester Avenue, Pasadena, CA 91106
[2]Wakayama Agricultural Biological Research Institute, Tsukatsuki, Momoyama, Wakayamaken 649–61, Japan

Citrus limonoids, one of the bitter principles in *Citrus*, have been shown to possess anticarcinogenic activity in laboratory animals and antifeedant activity against insects and termites. The demand for limonoids has increased significantly in recent years. Citrus seeds and byproducts of juice processing are major sources of these limonoids. The biochemistry of limonoids in *Citrus* is reviewed here, especially detailing how, where and when these limonoids are biosynthesized and accumulate in various tissues.

Limonoids are a group of chemically related triterpene derivatives found in the Rutaceae and Meliaceae families. Citrus limonoids are one of the bitter principles in citrus juices. They are also present as glucose derivatives in mature fruit tissues and seeds, and are one of the major secondary metabolites present in *Citrus*.

Recently, limonoids have been found to possess anticarcinogenic activity in laboratory animals (*1–4*). The furan moiety attached to the D-ring is most likely responsible for the induction of the detoxifying enzyme system glutathione *S*- transferase (GST). Among 38 limonoids isolated from *Citrus* and its hybrids, eight of them have been tested and all have been found to stimulate GST activity in mice (*2*). Limonin (**1**, Figure 1) and nomilin (**4**), two major limonoids in *Citrus*, inhibited the formation of tumors induced by benzo[*a*]pyrene in mice (*1*). Compound **1** and limonin 17-β-D-glucopyranoside (limonin glucoside, **3**) also inhibited the formation of oral tumors induced by 7,12-dimethylbenz[*a*]anthracene in hamsters (*3,4*).

Limonoids also act as antifeedants against insects and termites (*5–10*). This subject has been widely and intensely studied. One of the biological functions of limonoids *in vivo* could be their antifeedant activity. In fact, citrus seedlings, young leaves and immature fruit tissues, which are vulnerable to insect attack, contain relatively high concentrations of limonoid aglycones. Mature leaves, on the other hand, contain low concentrations of aglycones. Moreover, limonoid aglycones in the fruit tissues gradually disappear during the late stages of fruit growth and maturation (*11,12*). They are converted to their corresponding glucosides, which are considered to be biologically inactive.

0097–6156/94/0546–0198$06.00/0

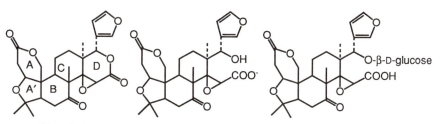

Limonin (1)   Limonoate A-ring lactone (2)   Limonin 17-β-D-glucopyranoside (3)

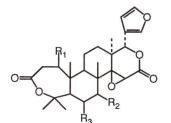

Nomilin (4) R₁=OAc, R₂=O, R₃=H
Deacetylnomilin (5) R₁=OH, R₂=O, R₃=H
6-keto-7β-Deacetylnomilol (6) R₁=OH, R₂=OH, R₃=O

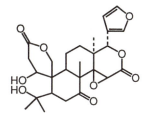

Ichangin (7)

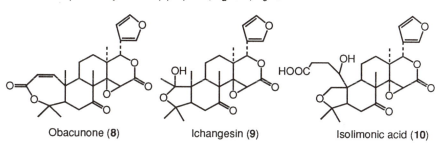

Obacunone (8)   Ichangesin (9)   Isolimonic acid (10)

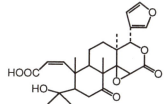

Obacunoic acid (11)

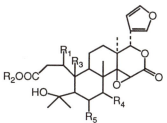

Nomilinic acid (12) R₁=OAc, R₂=H, R₃=CH₃, R₄=O, R₅=H
Deacetylnomilinic acid (13) R₁=OH, R₂=H, R₃=CH₃, R₄=O, R₅=H
Calamin (14) R₁=OH, R₂=CH₃, R₃=CH₃, R₄=OH, R₅=O
19-Hydroxydeacetylnomilinic acid (15)
R₁=OH, R₂=H, R₃=CH₂OH, R₄=O, R₅=H

Figure 1. Structures of major limonoids in *Citrus*.

Because of these biological activities, the demand for limonoids has increased significantly in recent years. Citrus seeds contain high concentrations of both limonoid aglycones and glucosides (*13,14*), while fruit tissues and by-products of juice processing such as peels and molasses are excellent sources of limonoid glucosides (*11,12*). We have made a substantial progress in the field of limonoid biochemistry in *Citrus*. In this chapter, we will discuss how, where and when limonoids are metabolized, catabolized and stored in *Citrus*.

Limonoid Aglycones and Glucosides

Thirty-eight limonoid aglycones, 23 neutral and 15 acidic, have been isolated from *Citrus* and its hybrids (Table I). Eight of them have been tested for anticarcinogenic activity and they were all found capable of stimulating the GST activity in mice (*2*). They include: 1,4-deacetylnomilin (**5**), obacunone (**8**), isoobacunoic acid, limonol, deoxylimonin and ichangin (**7**). Compounds **1** and **4** are two of the major limonoids in *Citrus* and both contribute to limonoid bitterness in citrus juices (*13,14*). Compound **4** has been shown to be more effective in reducing tumor formation in mice (*1*), but **1** is more effective preventing oral cancer in hamsters (*3*).

Table I. Limonoids in *Citrus* and its Hybrids

Neutral limonoids

Limonin	Nomilin
Obacunone	Deacetylnomilin
Ichangin	Deoxylimonin
Deoxylimonol	7α-Limonol
Limonyl acetate	7α-Obacunol
7α-Obacunyl acetate	Ichangensin
Citrusin	Calamin
Retrocalamin	Cyclocalamin
Methyl deacetylnomilinate	Isocyclocalamin
6-keto-7β-Deacetylnomilol	6-keto-7β-Nomilol
Methyl isoobacunoate diosphenol	
1-(10-19)Abeo-obacun-9(11)-en-7α-yl acetate	
1-(10-19)Abeo-7α-acetoxy-10β-hyddroxyisoobacunoic acid 3,10-lactone	

Acidic limonoids

Deacetylnomilinate	Nomilinate
Isoobacunoate	Epiisoobacunoate
trans-19-Hydroxyobacunoate	Isolimonate
Limonoate A-ring lactone	Deoxylimonate
17-Dehydrolimonoate A-ring lactone	Calaminate
Retrocalaminate	Isoobacunoate diosphenol
Obacunoate	19-Hydroxydeacetylnomilinate
Cyclocalaminate	

The neutral limonoids are not ideal forms to be used as additives for functional foods because they are insoluble in H_2O and some limonoids such as **1**,

4 and **7**, are bitter in taste (*15,16*). The acidic limonoids are, on the other hand, soluble in H_2O and they are, except for nomilinic acid (**12**), tasteless. The neutral limonoids, however, are predominant in *Citrus* and its hybrids. Acidic limonoids comprise approximately 20% of the total aglycones in citrus seeds (*13*).

Seventeen limonoid glucosides have been isolated from *Citrus* and its hybrids (Table II). Each of these compounds contains one D-glucose molecule attached via a β-glucosidic linkage to the C-17 position of the corresponding aglycone (*17,18*). The structure of limonin 17-β-D-glucopyranoside (**3**), known as limonin glucoside, is shown in Figure 1. Nine limonoid glucosides contain one carboxlic acid group and eight contain two carboxylic acid groups. They are all soluble in H_2O and MeOH, but not very soluble in other organic solvents. They appear to be stable between pH 2 and 8. One known exception is nomilin glucoside, which is converted to obacunone glucoside above pH 7.5 and to nomilinic acid glucoside below pH 3.5. Since they are all tasteless, limonoid glucosides would be ideal forms to be used as food additives. Only two of them, **3** and nomilin glucoside, have been tested for anticancer activity (*4*). Compound **3** was found to inhibit the oral cancer formation induced by 7,12-dimethylbenz[*a*]anthracene in hamster buccal pouch.

Table II. Limonoid Glucosides in *Citrus*

Monocarboxylic acids
   17-β-D-Glucopyranosides of

Limonin	Nomilin
Obacunone	Deacetylnomilin
Ichangin	Ichangensin
Calamin	6-keto-7β-Deacetylnomilol
Methyl deacetylnomilinate	

Dicarboxylic acids
   17-β-D-Glucopyranosides of

Nomilinic acid	Deacetylnomilinic acid
Obacunoic acid	*trans*-Obacunoic acid
Isoobacunoic acid	Epiisoobacunoic acid
Isolimonic acid	19-Hydroxydeacetylnomilinic acid

Occurrence

Limonoids are present in three different forms in *Citrus*. For example, the dilactone **1** is also present as the open D-ring form (monolactone), the limonoate A-ring lactone (**2**) and also as the glucoside form, **3**, in citrus seeds (Figure 1). Since 38 limonoid aglycones have been isolated from *Citrus*, it is very possible that each aglycone has two naturally occurring forms, monolactone and dilactone, and also a corresponding glucoside derivative form. Therefore a total of at least 114 limonoids (3 x 38) may be present in *Citrus*. Only two forms, the monolactones and glucosides, are present in fruit tissues.

Citrus seeds are a major source of both limonoid aglycones and glucosides (*14*). As shown in Table III, the concentration of total aglycones in the seeds of eight species is generally higher than that of the glucosides, on average; the ratio of total aglycones to total glucosides being approximately 2.2. Grapefruit seeds are the richest sources of limonoids. Compounds **1** and **4** can be easily isolated by crystallization from an extract of grapefruit seeds.

Table III. Limonoid Aglycones and Glucosides in Citrus Seeds

Seeds	Glucosides (%)	Aglycones (%)	Total (%)	Aglycones/ Glucosides
Grapefruit	0.698	2.39	3.09	3.4
Lemon	0.637	1.26	1.90	2.0
Tangerine	0.526	1.23	1.76	2.3
Valencia	0.871	1.48	2.35	1.7
Fukuhara	0.766	1.59	2.36	2.1
Hyuganatsu	0.312	0.904	1.22	2.9
Sanbokan	0.416	0.679	1.10	1.6
Shimamikan	0.677	1.23	1.91	1.8
Average	0.613	1.35	1.96	2.2

For most species, **1** is the predominant limonoid aglycone, followed by **4, 5** and **8** (*13*). The seeds of *C. ichangensis*, however, contain an unique limonoid, ichangensin (**9**), which is the predominant aglycone (*19,20*), and *C. aurantium* (sour orange or bitter orange) seeds contain relatively high concentrations of **7** and isolimonic acid (**10**) (*21*). The seeds of calamondin (*C. reticulata var. austera* x *Fortunella* sp.) contain a group of limonoids inherited from *Fortunella* sp. This group consists of calamin (**14**), cyclocalamin, retrocalamin and methyl iso-obacunoate diosphenol (*22*).

Unlike the fruit tissues and juices where **3** is predominant (*23,24*), the major limonoid glucoside in most citrus seeds is nomilin glucoside, followed by the glucosides of **8, 12** and **5** in order of decreasing concentration (*14*).

Commercial citrus juices in the United States contain high concentrations of limonoid glucosides (*23*) (Table IV). Orange juice has the highest level, averaging 320 ppm, followed by grapefruit juice averaging 190 ppm and lemon juice averaging 83 ppm. Mandarin juices in Japan also contain high concentrations of limonoid glucosides, averaging 225 ppm at 9.0° Brix (*25*).

Compound **3** is the predominant limonoid glucoside found in all juice samples. In orange juice it comprises 56% of the total limonoid glucosides present, while in grapefruit and lemon juices, it comprises an

Table IV. Limonoid Glucosides in Commercial Citrus Juices

Juice	Number of samples	Total (ppm)
Orange	15	320 ± 48
Grapefruit	8	190 ± 36
Lemon	4	83 ± 10

average of 63% and 66%, respectively. Orange juices also have the glucosides of **12, 4**, deacetylnomilinic acid (**13**) and **8** in order of decreasing concentration (*24*). Citrus peels and molasses also contain high concentrations of limonoid glucosides (*11,12*). In contrast, citrus juices contain very low concentrations of limonoid aglycones. Orange juice, for example, generally contains about 2 ppm (*15*).

Limonoid glucosides are one of the major secondary metabolites in citrus fruit tissues. The presence of such high concentrations of limonoid glucosides in citrus juices and fruit tissues may have important health significance. The aglycone moiety of limonoid glucosides is considered to be the biologically active group. Commercial β-glucosidases, including naringinases, do not hydrolyze the glucosides. When consumed, however, these glucosides are most likely hydrolyzed by intestinal floral organisms and subsequently, the aglycones will be liberated. A species of bacterium that is capable of hydrolyzing the glucosides has been isolated from soil (*17*).

Biosynthesis of Limonoids in Citrus and Its Hybrids

This subject has been recently reviewed in detail (*26*). The biosynthetic pathways of limonoids in *Citrus* and its hybrids are shown in Figure 2. These pathways have been proposed on the basis of data obtained from radioactive tracer work.

Most citrus species and its hybrids accumulate **1** as the major limonoid. The pathway from **4** to **1** via **8**, obacunoate (**11**) and **7** is most likely present in all citrus species and hybrids (*27,28*). In addition, limonoids involved in this pathway such as **4, 8** and **1** are the major limonoids present in most citrus species. Therefore, this pathway is considered to be the major biosynthetic pathway of limonoids in *Citrus* and its hybrids.

C. aurantium (sour orange or bitter orange) contains relatively high concentrations of **10** and **7** (*21,29*). Recently, we found that this species possesses an alternative pathway to synthesize these limonoids via 19-hydroxydeacetyl-nomilinic acid (**15**) (*30*). Compound **7** has been shown to be an effective GST activity inducer (*2*). It is of interest that the A′-ring of **7** is open, which may play a key part in the role of the compound as an effective enzyme inducer.

C. ichangensis (Ichang Chie) is grown in China and is used as herbal medicine. This species accumulates a nonbitter limonoid, ichangensin (**9**), as the major limonoid (*19*). As shown in Figure 2, **9** is biosynthesized from **4** via **5** and **13** (*20*). Compound **9** is unique to this species and its hybrids such as *C. sudachi, C. junos* and *C. sphaerocarpa* (*31*). There is no A-ring lactone present in the **9** molecule. Since the A-ring and A′-ring of limonoids have significant influence on the anticancer activity (*2*), it would be worthwhile to test this unique limonoid for its activity.

Calamondin (*C. reticulata var. austera* x *Fortunella* sp.) is a very popular citrus fruit in the Philippines. In addition to the common limonoids, this species possesses the calamin group of limonoids: calamin (**14**), cyclocalamin, retrocalamin and methyl isoobacunoate diosphenol (*22*). All of these limonoids are methyl esters and all are oxygenated at C-6. The calamin group also contains 6-keto-7β-nomilol and 6-keto-7β-deacetylnomilol (**6**). A biosynthetic pathway of these compounds starting from **13** has been proposed (*32,33*) (Figure 2). This group of limonoids has not been tested for anticancer activity. There are significant differences in the molecular structures around the A- and B-rings compared to that of other limonoids.

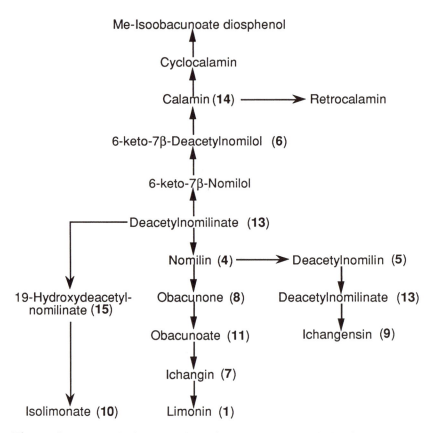

Figure 2. Proposed biosynthetic pathways of limonoids in *Citrus* and its hybrids.

Studies on the biosynthesis of minor limonoids in *Citrus* are very limited. Deoxylimonin, deoxylimonate and 17-dehydrolimonoate A-ring lactone have been shown to be biosynthesized from limonin (*34,35*). Based on their structures, limonol and limonyl acetate are also most likely synthesized from **1**. 7α-Obacunol and 7α-obacunyl acetate appear to be synthesized from **8**.

Biosynthesis of Limonoid Glucosides

This subject has also been reviewed recently (*36*). Limonoid glucosides are present only in mature fruit tissues and seeds (*17,18,31,37,38*). They are biosynthesized by the glucosidation of corresponding aglycones. The glucosidation of limonoid aglycones (monolactones) begins during the late stages of fruit growth and maturation (*11,12*). This conversion begins at the same time in both seeds and fruit tissues, but is delayed in peels by about one month. The limonoid glucoside content increases steadily thereafter.

This conversion [between limonoate A-ring lactone (2) and limonin 17-β-D-glucopyranoside (3)] is catalyzed by UDP-D-glucose transferase (*39*). The enzyme activity has been directly detected in mature fruit tissues and seeds by radioactive tracer work. The activity was also confirmed in cell free extracts of navel orange albedo tissues. Other tissues, including stems and leaves, and immature fruit tissues and seeds do not possess the glucosides or the transferase activity.

Sites of Limonoid Biosynthesis

Citrus seedlings are excellent tools for the preparation of ^{14}C-labelled **4** substrates (*40*). Up to 5% of ^{14}C labelled acetate fed to lemon seedling stems is converted to [14]C-**4**. The phloem region of the stems is the site of **4** biosynthesis, starting from acetate, mevalonate or farnesyl pyrophosphate (*41*). Other tissues including leaves, seeds and fruit tissues are incapable of biosynthesizing limonoids from these substrates. Compound **4** is biosynthesized in stems, which are directly attached to fruit, but not in stems attached to mature leaves located far away from fruit.

Compound **4** is then translocated from the stem to other locations such as leaves, fruit tissues and seeds where it is further converted to other limonoids such as **8, 11** and **1** (*42*). Radioactive tracer work has confirmed that all tissues are capable of biosynthesizing other limonoids from **4** (*43*). In the fruit tissue, aglycones remain as biologically active monolactones. In the seed, some monolactones are converted to biologically inactive dilactones.

During the late stages of fruit growth and maturation, limonoid monolactones are further metabolized to their corresponding glucosides. The glucosidation occurs only in fruit tissues and seeds.

Biosynthesis of Limonoids in Citrus Seeds

It was generally believed that the limonoids in the seed are translocated from the fruit tissues to be stored (Figure 3). Our data, however, have shown that the limonoids are biosynthesized in the seeds independently and are not translocated from the fruit tissues. There are several pieces of evidence to support this. Limonoids are present in the three forms: monolactones, dilactones and glucosides in the seeds, while in the fruit tissues there are only two forms: monolactones and glucosides. Seeds accumulate both aglycones and glucosides, while fruit tissues accumulate mainly glucosides. Compound **3** is the predominant limonoid glucoside in the fruit tissue while in the seed, nomilin glucoside is the major glucoside with the concentration of **3** being very low. Moreover, seeds possess their own enzyme systems for the metabolism and catabolism of limonoids.

Conclusion

Substantial progress has been made in the field of limonoid biochemistry in *Citrus*. We now know how, when and where limonoids and their glucoside derivatives are biosynthesized and stored. Because of their biological importance in roles such as anticancer activity in animals and antifeedant activity against insects and termites, the demand for limonoids has increased significantly in recent years. Citrus seeds, peels and molasses, which are byproducts of juicing, are excellent sources of limonoids. Routine procedures for the extraction and isolation of both aglycones and glucosides have been established on a pilot scale.

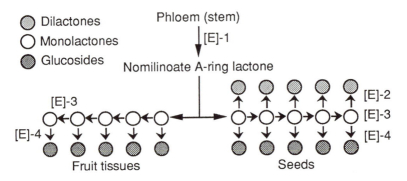

[E]-1   The group of enzymes involved in the biosynthesis of nomilin from acetate via mevalonate and farnesyl pyrophosphate. They are present only in the phloem region of stems, and not in the fruit tissues, seeds or leaves.

[E]-2   Enzymes involved in the biosynthesis of other limonoids from nomilin. The activities of this group occurs in leaves, fruit tissues and seeds regardless of maturity.

[E]-3   Limonoid D-ring lactone hydrolase, which catalyzes the lactonization of the open D-ring of monolactones to form dilactones. This activity occurs only in the seeds.

[E]-4   UDP-D-glucose transferase, which catalyzes the glucosidation of limonoid aglycones. The activity occurs only in fruit tissues and seeds during late stages of fruit growth and maturation.

Figure 3. Schematic diagram of limonoid biosynthesis in *Citrus*.

Further investigation of the anticarcinogenic activity is in progress. We are particularly interested in finding the correlation between activity and structural configuration. Also, we need more information on the anticarcinogenic activity of limonoid glucosides, such as how the glucosides are absorbed through intestines and if the aglycones are liberated from the glucosides by intestinal flora before absorption.

One of the disadvantages of using limonoids as a food additive is that neutral limonoid aglycones are insoluble in H_2O. Studies on the solubilization of neutral limonoids by chemical and/or biological means, such as glucosidation or conversion to acidic forms like 17-dehydrolimonoate A-ring lactone and deoxy-limonic acid, are in progress.

Literature Cited

1. Lam, L. K. T.; Hasegawa, S. *Nutrition and Cancer* **1989**, *12*, 43.
2. Lam, L. K. T.; Li, Y.; Hasegawa, S. *J. Agric. Food Chem.* **1989**, *37*, 878 .
3. Miller, E. G.; Fanous, R.; Rivera-Hidalgo, F.; Binnie, W. H.; Hasegawa, S.; Lam, L. K. T. *Carcinogenesis* **1989**, *10*, 1535.
4. Miller, E. G.; Gonzales-Sanders, A. P.; Couvillon, A.M.; Wright, J. M.; Hasegawa, S.; Lam, L. K. T. *Nutrition and Cancer* **1992**, *17*, 1.
5. Klocke, J. A.; Kubo, I. *Ent. Exp. & Appl.* **1982**, *32*, 299.
6. Alford, A. R.; Bentley, M. D. *J. Econ. Entomol.* **1986**, *79*, 35.

7. Alford, A. R.; Cullen, J. A.; Storch, R.H.; Bentley, M.D. *J. Econ. Entomol.*, **1987**, *80*, 575.
8. Bentley, M. D.; Rajab, M. S.; Alford, A. R.; Mendel, M. J.; Hassanali, A. *Entomol. Exp. Appl.* **1989**, *49*, 189.
9. Bentley, M. D.; Rajab, M. S.; Mendel, M. J.; Alford, A. R. *J. Agric. Food Chem.* **1990**, *38*, 1400.
10. Liu, Y. B.; Alford, A. R.; Rajab, M. S.; Bentley, M. D. *Physiol. Entomol.* **1990**, *15*, 37.
11. Hasegawa, S.; Ou, P.; Fong, C. H.; Herman, Z.; Coggins, C. W. Jr.; Atkin, D. A. *J. Agric. Food Chem.* **1991**, *39*, 262.
12. Fong, C. H.; Hasegawa, S.; Coggins, C. W., Jr.; Atkin, D. W.; Miyake, M. *J. Agric. Food Chem.* **1992**, *40*, 1178.
13. Hasegawa, S.; Bennett, R. D.; Verdon, C. P. *J. Agric. Food Chem.* **1980**, *28*, 922.
14. Ozaki, Y.; Fong, C. H.; Herman, Z.; Maeda, H.; Miyake, M.; Ifuku, Y.; Hasegawa, S. *Agric. Biol. Chem.* **1991**, *55*, 137.
15. Maier, V. P.; Bennett, R. D.; Hasegawa, S. In *Citrus Science and Technology* Nagy, S.; Shaw, P. E.; Veldhuis, M. K., Eds.; Avi Publishing: Westport, CT, 1977; Vol. 1, p 355.
16. Rouseff, R. L.; Matthews, R. F. *J. Food Sci.* **1984**, *49*,777.
17. Hasegawa, S.; Bennett, R. D.; Herman, Z.; Fong, C. H.; Ou, P. *Phytochemistry* **1989**, *28*, 1717.
18. Bennett, R. D.; Hasegawa, S.; Herman, Z. *Phytochemistry* **1989**, *28*, 2777.
19. Bennett, R. D.; Herman, Z.; Hasegawa, S. *Phytochemistry* **1988**, *27*, 1543.
20. Herman, Z.; Hasegawa, S.; Fong, C. H.; Ou, P. *J. Agric. Food Chem.* **1989**, *37*, 850.
21. Bennett, R. D.; Hasegawa, S. *Phytochemistry* **1980**, *19*, 2417.
22. Bennett R. D.; Hasegawa, S. *Tetrahedron* **1981**, *37*, 17.
23. Fong, C. H.; Hasegawa, S.; Herman, Z.; Ou, P. *J. Food Sci.* **1989**, *54*, 1505.
24. Herman, Z.; Fong, C. H.; Ou, P.; Hasegawa, S. *J. Agric. Food Chem.* **1990**, *38*, 1860.
25. Ozaki, Y.; Ayano, S.; Miyake, M.; Maeda, H.; Ifuku, Y.; Hasegawa, S. *Biosci. Biotech. Biochem.* **1992**, *56*, 836.
26. Hasegawa, S.; Herman, Z. In *Biosynthesis and Metabolism of Secondary-Metabolites Natural Products* Petroski, R. J.; McCormick, S. P., Eds.; ACS Series, in press.
27. Hasegawa, S.; Herman, Z. *Phytochemistry* **1985**, *24*, 1973.
28. Herman, Z.; Hasegawa, S. *Phytochemistry* **1985**, *24*, 2911.
29. Miyake, M.; Ayano, S.; Ozaki, Y.; Herman, Z.; Hasegawa, S. *Nippon Nogeikagaku Kaishi* **1992**, *66*, 31.
30. Bennett, R. D.; Miyake, M.; Ozaki, Y.; Hasegawa, S. *Phytochemistry* **1991**, *30*, 3803.
31. Ozaki,Y.; Miyake, M.; Maeda, H.; Ifuku, Y.; Bennett, R. D.; Herman, Z.; Fong, C. H.: Hasegawa, S. *Phytochemistry* **1991**, *30*, 2659.
32. Herman, Z.; Bennett, R. D.; Ou, P.; Fong, C. H.; Hasegawa, S. *Phytochemistry* **1987**, *26*, 2247.
33. Hasegawa, S.; Herman, Z.; Ou, P.; Fong, C. H. *Phytochemistry* **1988**, *27*, 1349.
34. Hasegawa, S.; Bennett, R. D.; Verdon, C. P. *Phytochemistry* **1980**, *19*, 1445.
35. Hasegawa, S.; Maier, V. P.; Bennett, R. D. *Phytochemistry* **1974**, *13*, 103.

36. Hasegawa, S.; Fong, C. H.; Herman, Z.; Miyake, M. In *Flavor Precursors* Teranishi, R.; Takeoka, G. R.; Guntert, M., Eds.; ACS Symposium Series 490; American Chemical Society: Washington, DC, 1992; p 87.

37. Maeda, H.; Ozaki, Y.; Miyake, M.; Ifuku, Y.; Hasegawa, S. *Nippon Nogeikagaku Kaishi* **1990**, *64*, 1231.

38. Bennett, R. D.; Miyake, M.; Ozaki, Y.; Hasegawa, S. *Phytochemistry* **1991**, *30*, 3803.

39. Herman, Z.; Fong, C. H.; Hasegawa, S. *Proc. Citrus Research Conference, Pasadena, CA* **1990**.

40. Hasegawa, S.; Bennett , R. D.; Maier, V. P. *Phytochemistry* **1984**, *23*, 1601.

41. Ou, P.; Hasegawa, S.; Herman, Z.; Fong, C. H. *Phytochemistry* **1988**, *27*, 115.

42. Hasegawa, S.; Herman, Z.; Orme, E. D.; Ou, P. *Phytochemistry* **1986**, *25*, 2783.

43. Fong, C. H.; Hasegawa, S.; Herman, Z.; Ou, P. *J. Sci. Food Agric.* **1991**, *54*, 393.

RECEIVED October 8, 1993

Chapter 16

Inhibition of Chemically Induced Carcinogenesis by Citrus Limonoids

Luke K. T. Lam[1,2], Jilun Zhang[2], Shin Hasegawa[3], and Herman A. J. Schut[4]

[1]Department of Medicinal Chemistry, University of Minnesota, Minneapolis, MN 55455
[2]LKT Laboratories Inc., Minneapolis, MN 55413
[3]Fruit and Vegetable Chemistry Laboratory, U.S. Department of Agriculture, Agricultural Research Service, 263 South Chester Avenue, Pasadena, CA 91106
[4]Department of Pathology, Medical College of Ohio, Toledo, OH 43614

Limonin and nomilin are two of the bitter principles found in citrus fruits such as lemon, lime, orange and grapefruit. Both citrus limonoids have been found to induce increased activity of the detoxifying enzyme glutathione *S*-transferase. The increased enzyme activity was correlated with the ability of these compounds to inhibit chemically induced carcinogenesis in laboratory animals. Administration of nomilin by gavage to ICR/Ha mice reduced the incidence and number of forestomach tumors per mouse induced by benzo[*a*]-pyrene (BP). Addition of nomilin and limonin to the diet at various concentrations inhibited BP-induced lung tumor formation in A/J mice. The inhibition of lung tumors was correlated with inhibition of the formation of BP-DNA adducts in the lung. Topical application of the limonoids was found to inhibit both the initiation and the promotion phases of carcinogenesis in the skin of SENCAR mice. Nomilin appeared to be more effective during the initiation stage while limonin was more potent as an inhibitor during the promotion phase of carcinogenesis. These findings suggest citrus limonoids may be useful as cancer chemopreventive agents.

Citrus limonoids are one of the two main classes of compounds responsible for the bitter taste in citrus fruits. The most prevalent limonoids are limonin and nomilin. They are present in Rutaceous plants, which include the commonly eaten fruits lemon, lime, orange, and grapefruit (*1*). The bitterness due to limonin is of major concern and economic impact (*2*). The characteristic of this class of compounds is a furan moiety substituted at the 3 position (*3*). The point of substitution is the D-ring lactone of a highly oxygenated triterpene. Recently, it has been determined that citrus limonoids have certain biological activities that may allow their use as chemopreventive agents.

0097–6156/94/0546–0209$06.00/0
© 1994 American Chemical Society

Glutathione *S*-transferase Induction

Many furan-containing natural products have been shown to induce the detoxifying enzyme system, glutathione *S*-transferase (GST) (*4–7*). The GST system is a major detoxifying enzyme system that catalyzes the conjugation of glutathione with electrophiles, including activated carcinogens (*8*). The conjugates are generally less reactive and more water soluble, facilitating their excretion. An increase of GST activity by a substance, therefore, usually enhances the protection mechanism against the noxious effects of xenobiotics, including carcinogens. Many chemicals that are GST inducers have been found to inhibit chemically induced carcinogenesis.

The structures of the naturally occurring furanoids that have been found to induce GST activity range from the simple 2-alkyl substituted compound, 2-*n*-heptyl furan, and the sulfur analog, 2-*n*-butyl thiophene, formed during the roasting of meat (*7*), to the more complex molecules such as kahweol and cafestol, isolated from green coffee beans (*4*), and salannin, identified in the seed of the neem tree of India (*9*). The presence of the furan moiety appears to be essential for enzyme induction. When the furan ring in cafestol is saturated by hydrogenation, its activity as GST enzyme inducer is lost (*5*).

In the case of limonin and nomilin, the triterpene structure also plays a part in the relative GST inducing activity of these compounds. The activity of limonin was found to be much less than that of nomilin. The A ring in limonin is orthogonal to the plane of the molecule. On the other hand, if the A or A' ring is open such that the A ring is no longer above the plane of the molecule, as in the case of ichangin and isoobacunoic acid, the inducing activity is higher than that of limonin (*6*).

Table I shows the effects of limonin and nomilin on GST activity in the liver and small intestinal mucosa of mice. Butylated hydroxyanisole, a known inducer of GST (*10*), was used as a positive marker in this experiment. The test compounds were either dissolved or suspended in cottonseed oil and were given by gavage at the indicated doses once every other day over a period of seven days. On the eighth day the animals were sacrificed and the tissues removed. GST assay was performed using the universal substrate chlorodinitrobenzene (CDNB) (*11*).

BHA in this experiment induced GST activity to three times the control level, which is comparable to previous observations at this dose level (*10*). Limonin was found to be inactive in the liver. Nomilin, on the other hand, was quite active even at the low dose of 5 mg. The relative inducing activity in the small bowel mucosa was higher than that in the liver for both limonin and nomilin. No appreciable elevation of GST activity was detected in the forestomach.

Inhibition of Benzo[a]pyrene-induced Forestomach Tumors

The high GST inducing activity of the limonoids suggested that they might be chemopreventive agents. The first experiment to test their effectiveness was carried out with female ICR mice. The limonoids were given by gavage three times a week at two dose levels for 4 weeks. Two additional administrations before and one after the carcinogen were given. The carcinogen, benzo[*a*]pyrene (BP), was given at 1 mg/animal twice a week on the days when the inhibitors were not given. Eighteen weeks after the first dose of BP, the animals were sacrificed. The forestomach was excised and the tumors were counted. The tumors were confirmed histopathologically.

Table II shows a significant decrease in tumor incidence as a result of BHA and nomilin treatment. BHA was used as an inhibitor control because it has been found to reduce forestomach tumor formation in similar experiments. The level of inhibition by BHA in this experiment was similar to that observed previously (*12*). Nomilin given at 10 mg/dose was found to reduce the number of tumors/animal by greater than 50%. Limonin under these conditions was somewhat less effective (*13*).

These results indicated that the inhibitory effects of limonin and nomilin follow the same trend as their ability to induce GST activity. Nomilin, a much better inducer of GST than limonin, was more active as an inhibitor of carcinogenesis than limonin. The inability of limonoids to elevate GST activity in the forestomach did not appear to be an important factor in the inhibition of carcinogenesis in this target tissue. A significant enhancement of GST activity in the major detoxifying center, namely the liver, appeared to be positively correlated with the inhibitory potential of limonoids not only in the portal of entry but also at a remote site such as the lung.

Inhibition of Benzo[*a*]pyrene-induced Lung Tumors

Besides forestomach tumors, BP also induces lung tumors in laboratory animals (*14*). In general, A strain mice are used to test inhibitors that may inhibit BP-induced lung tumor formation. In this experiment, the inhibition of BP-induced lung tumorigenesis by dietary limonoids in A/J mice was determined (*15*).

Limonin and nomilin at various concentrations were added to the diet of A/J mice before and during BP administration. The carcinogen was given by gavage at 1 mg/animal twice a week for 4 weeks. Three days after the last dose of BP, the animals were returned to normal lab chow diet. The experiment was terminated 18 weeks after the first dose of BP.

The results are shown in Table III. The weight gain of the animals at the end of the experiment was similar for all experimental groups. Under these experimental conditions, 100% of the animals developed pulmonary adenoma with the exception of the 0.27% nomilin group, which had 83% incidence, which was not statistically different from the control group. The number of tumors per animal, however, was significantly reduced for both the limonin and nomilin groups. The reduction was dose dependent. At 0.5%, the limonin group had 6.1 tumors /animal compared to 11.9 in the control group, a 50% reduction. At 0.25%, there were 8.1 tumors/animal, a 32% reduction, and at 0.125% no reduction was obtained.

The dosages used for nomilin were half those of limonin because we anticipated it would be more active judging from the previous forestomach tumor experiment (*13*). At 0.27%, 2.9 tumors/animals were found in this group, compared to 11.9 in the control group, which is an inhibition of 75%. In the 0.135% group, 5.7 tumors/animal were found (a 52% reduction) and no protection was obtained at 0.068%, the lowest dose. At 0.27%, it is estimated that the animals were consuming approximately 10 mg nomilin/day. At this level a small reduction of forestomach tumor was also found which was similar to that found in the previous experiment using ICR mice.

Like the inhibition of forestomach tumors, the protection against lung carcinogenesis by limonoids was positively correlated with the induction of GST activity.

Table I. Effects of Limonin, Nomilin and BHA on the Activity of GlutathioneS-transferase in the Tissues of Female ICR/Ha Mice

Compound[a] (dose in mg)	Liver		Small Intestine Mucosa		Forestomach	
	GST activity[b]	Test/ control	GST activity[b]	Test/ control	GST activity[b]	Test/ control
Control (—)	1.15 ± 0.19	—	0.33 ± 0.04	—	0.66 ± 0.08	—
BHA[c] (7.5)	$5.24 \pm 0.60***$	4.55	$1.08 \pm 0.03***$	3.25	0.84 ± 0.18	1.29
Limonin (5)	1.29 ± 0.35	1.12	$0.45 \pm 0.08*$	1.33	0.69 ± 0.13	1.05
Limonin (10)	1.24 ± 0.23	1.08	$0.45 \pm 0.09*$	1.36	0.60 ± 0.12	0.92
Nomilin (5)	$2.86 \pm 0.60**$	2.48	$1.00 \pm 0.03***$	3.00	0.70 ± 0.16	1.07
Nomilin (10)	$3.96 \pm 0.81**$	3.44	$1.39 \pm 0.15***$	4.17	0.76 ± 0.15	1.16

SOURCE: Reprinted with permission from ref. 13. Copyright 1989.
Significantly different from control, $*p<0.05$, $**p<0.005$, $***p<0.001$.
[a]Limonin and nomilin were given as fine suspensions and BHA in solution in 0.3 ml cottonseed oil. Control was given cottonseed oil only.
[b]The GST activity was determined according to the method of Havig et al. using CNDB as the substrate (11). Mean $\pm$ SD (n = 3).
[c]3-tert-Butyl-4-hydroxyanisole (Sigma Chemical Co., recrystallized from hexanes-acetone).

Table II. Effects of Limonin and Nomilin on BP-induced Neoplasia of the Forestomach of ICR/HA Mice

Compound[a] (dose in mg)	Number of mice	Weight gain (g)	Number (percent) of mice with tumors	Tumors/mouse[b] (Tumors per tumor-bearing mouse)	
Control (—)	18	9.0	18 (100)	3.6 ± 0.8	(3.6 ± 0.8)
BHA (7.5)	16	8.7	11 (69**)	$1.7 \pm 0.4*$	(2.5 ± 0.4)
Limonin (5)	16	10.2	14 (88)	3.1 ± 0.8	(3.5 ± 0.8)
Limonin (10)	19	9.0	16 (84)	2.5 ± 0.6	(3.0 ± 0.7)
Nomilin (5)	18	9.4	15 (83)	2.3 ± 0.4	(2.7 ± 0.4)
Nomilin (10)	19	9.4	13 (72**)	$1.7 \pm 0.4*$	(2.3 ± 0.4)

SOURCE: Reprinted with permission from ref. 13. Copyright 1989.
One mg BP in 0.2 ml corn oil was administered by p.o. intubation twice a week for 4 weeks. Tumors greater than 0.5 mm in diameter were counted.
[a] Limonin and nomilin were given as fine suspension and BHA in solution in 0.3 ml cottonseed oil. Control was given cottonseed oil only. All were administered by p.o intubation 3 times a week for 4 weeks.
[b] Mean $\pm$ SE. Significantly different from control, $*p<0.05$, $**p<0.025$.

Inhibition of the Formation of BP-DNA Adducts in the Lung

If increased activity of the detoxifying enzyme GST is responsible for the inhibition of BP-induced lung tumor formation, then one of the possible mechanisms of action is the rapid detoxification of BP-diol epoxide that is the ultimate carcinogenic form of BP (16,17). Since the carcinogenic action of BP-diol epoxide is the binding to DNA (18–20), limonoid treatment is expected to reduce BP-DNA binding in the target tissue by enhancing the rapid conjugation of BP-diol epoxide with GSH that is catalyzed by GST (21,22).

Randerath (23) developed a very sensitive method to determine carcinogen-DNA binding, which uses ^{32}P to label the carcinogen bound mono-nucleotide. The carcinogen-nucleotide adducts are then separated by two dimensional TLC and the radioactivity determined.

Limonin at either 10 or 20 mg/dose by gavage did not produce noticeable reduction of adduct formation as shown in Table IV (15). On the other hand, nomilin, the better inhibitor of lung carcinogenesis was found to reduce adduct formation in the liver and lung. No reduction of adduct formation was found in the forestomach of A/J mice. These results correlate well with the relative effectiveness of the two limonoids, which suggested that inhibition of BP-DNA adduct formation in the liver and in the target tissue may be one of the critical mechanisms of action of the limonoids.

Inhibition of Two-stage Mouse Skin Carcinogenesis

In addition to the inhibition of BP-induced tumorigenesis in mice, it has been determined that the two limonoids are also capable of inhibiting DMBA-induced carcinogenesis in hamster cheek pouches (24). The experimental conditions used in the hamster study suggested that limonoids may be inhibiting at the promotion stage of carcinogenesis as well. To test this hypothesis, the two-stage skin tumor model — DMBA initiation and TPA promotion (25) — was used. Both limonin and nomilin were tested in this experiment (26).

In this experiment a total of twelve groups were used. Group 1 to 3 were the control groups. Group 1 received solvent only throughout the experiment. In groups 2 and 3, 1 mg limonin or nomilin was applied to the back of the animals throughout the experimental period to check whether these compounds alone cause any unusual responses from the animal. Group 4 was the positive control; 100 mmol DMBA was applied and followed after ten days by 5 mg TPA twice a week for 12 weeks.

Groups 5 to 8 were designed to test the inhibition of initiation. In these groups the limonoids were applied one day before and on the same day of DMBA treatment, then the animals were promoted with TPA alone. Group 5 received 1 mg limonin per treatment; group 6, 0.25 mg limonin; group 7, 1 mg nomilin; group 8, 0.25 mg nomilin.

Groups 9 to 12 were designed to test the inhibition of promotion. Initiation was DMBA alone, then the limonoids were applied one hour before each TPA treatment for the entire promotion period. Group 9 received 1 mg limonin per treatment; group 10, 0.25 mg limonin; group 1, 1 mg nomilin; group 12, 0.25 mg nomilin.

Tumors on the dorsal skin of the animals were recorded each week. The animals in the three solvent and inhibitor control groups (groups 1–3) did not

Table III. Effects of Limonin and Nomilin on BP-induced Neoplasia of the Lung and Forestomach in A/J Mice

Compound[a]	BP[b]	Number of mice	Weight gain (g)	Lung Tumor incidence	Lung Tumors/ mouse	Forestomach Tumor incidence	Forestomach Tumors/ mouse
None	–	14	4.4	21	0.5	0	0.0
Limonin 0.5%	–	14	3.6	21	0.2	0	0.0
Nomilin 0.27%	–	15	4.1	13	0.1	0	0.0
None	+	17	4.0	100	11.9	82	2.7
Limonin 0.125%	+	19	4.9	100	10.8	95	3.2
Limonin 0.25%	+	16	3.8	100	8.1	100	2.6
Limonin 0.5%	+	18	4.3	100	6.1	100	2.7
Nomilin 0.068%	+	18	4.4	100	14.9	100	4.2
Nomilin 0.135%	+	16	4.4	100	5.7	100	3.2
Nomilin 0.27%	+	12	4.1	83	2.9	92	1.8

[a] Citrus limonoids were formulated into pelleted semipurified diet at the percentages indicated. Experimental diets were given one week before, during, and 3 days after BP administration.

[b] BP (1 mg in 0.2 ml corn oil) was given by gavage twice a week for 4 weeks.

Table IV. Effects of Limonin and Nomilin on BP-DNA Adduct Formation in a ^{32}P Postlabelling Assay

Compound[b]	Relative adduct labelling[a] x 10 Liver	Lung	Forestomach
None	N.D.	N.D.	N.D.
Limonin 20 mg	N.D.	N.D.	N.D.
Nomilin 20 mg	N.D.	N.D.	N.D.
BP	0.33	0.50	0.28
BP-Limonin 10 mg	0.39	0.50	0.28
BP-Limonin 20 mg	0.39	0.67	0.33
BP-Nomilin 10 mg	0.25	0.27	0.34
BP-Nomilin 20 mg	0.21	0.19	0.32

[a] Not corrected for intensification factor. N.D. = not detected.

[b] Limonin and nomilin were given by gavage in 0.3 ml cottonseed oil every other day for a total of 3 administrations. BP (1 mg in 0.2 ml corn oil) was given by gavage once, 24 h after the last dose of limonoid. The animals were killed 24 h after BP administration.

develop any tumors. The positive control group (group 4) was found to have the highest number of tumors/animal from the beginning to the end of the experiment.

The results from the anti-initiation groups are shown in Figure 1. Limonin given topically at 0.25 or 1 mg per application showed equal, slight inhibition of the number of tumors per mouse, but the difference from control was not significant. Nomilin at 1 mg/dose, showed more inhibition of skin tumor formation than limonin, while 0.25 mg/dose did not not, but again, neither was statistically significant.

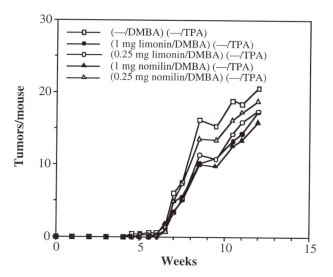

Figure 1. Effect of topical limonin and nomilin on tumor initiation. Limonoids were applied topically before and during initiation.

Figure 2 shows the inhibition by limonoids at the promotion phase of carcinogenesis. Limonin at 1 mg/dose applied one hour before TPA inhibited the number of tumors/mouse throughout the course of the experiment. At 0.25 mg/dose, inhibition was also found to be significant. Nomilin, which showed higher inhibitory activity at the initiation stage and in the other tumor model experiments, was found to be less effective than limonin in reducing tumors at the promotional phase. No significant difference was found at the low or high dose.

Table V shows the number of tumors/animal at sacrifice. The number of tumor/animal was 20.6 for the DMBA alone/TPA alone group. The data show that nomilin was more effective as an inhibitor at the initiation stage of carcinogenesis and limonin was more effective at the promotion stage.

Summary

The two most prevalent limonoids from citrus have been determined to induce GST enzyme activity. The inducing property was correlated with the inhibition of tumor formation when the compounds were administered to the animals by gavage or by addition to the diet. The concentration of limonin and nomilin in citrus juices is fairly low (27,28). The overall consumption of these two specific limonoids,

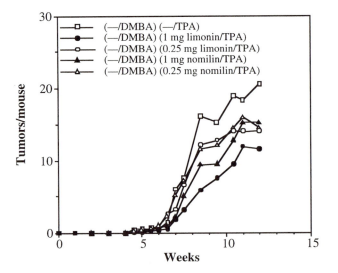

Figure 2. Effect of topical limonin and nomilin on tumor promotion. Limonoids were applied topically during promotion.

Table V. Reduction of Skin Tumors by Citrus Limonoids Applied at the Initiation and Promotion Phases of Carcinogenesis

Group[a] (Initiation) (Promotion)	Number of mice	Number (percent) of mice with tumors	Tumors per mouse	Percent inhibition
1 (—/—) (—/—)	15	0 (0)	0	—
2 (1 mg limonin/—) (1 mg limonin/—)	15	0 (0)	0	—
3 (1 mg nomilin/—) (1 mg nomilin/—)	15	0 (0)	0	—
4 (—/DMBA) (—/TPA)	20	19 (95)	20.6 ± 2.4	—
5 (1 mg limonin/DMBA) (—/TPA)	20	20 (100)	17.4 ± 2.3	15.5
6 (0.25 mg limonin/DMBA) (—/TPA)	19	19 (100)	17.4 ± 1.7	15.5
7 (1 mg nomilin/DMBA) (—/TPA)	19	19 (100)	15.9 ± 2.2	22.8
8 (0.25 mg nomilin/DMBA) (—/TPA)	20	20 (100)	18.8 ± 2.0	8.7
9 (—/DMBA) (1 mg limonin/TPA)	20	20 (100)	11.6 ± 2.0**	43.7
10 (—/DMBA) (0.25 mg limonin/TPA)	20	19 (95)	14.0 ± 2.2*	32.0
11 (—/DMBA) (1 mg nomilin/TPA)	20	19 (95)	15.2 ± 2.1	26.2
12 (—/DMBA) (0.25 mg nomilin/TPA)	20	18 (90)	14.4 ± 2.0	30.1

[a] Limonin, nomilin, DMBA (100 nmol) and TPA (5 µg), dissolved in 200 µl acetone, were applied topically.
[b] Values are the mean ± SE. Statistically different from control, *$p<0.10$, **$p<0.01$.

consequently, is not high. Recent findings suggest, however, that many other structurally related limonoids such as ichangin, isoobacunoic and obacunone are also good GST inducers (*6*). In addition, it was discovered that limonoids exist as their glycosidic form at very high concentration in juices (*29,30*). Thus the total exposure to these compounds is expected to be substantial. The cumulative effects of citrus limonoids on cancer chemoprevention remains to be determined.

Acknowledgment

This work was supported by a grant from the American Institute for Cancer Research.

Literature Cited

1. Hasegawa, S.; Bennett, R. D.; Verdon, C. P. *J. Agric. Food Chem.* **1980**, *28*, 922–925.
2. Maier, V. P.; Bennett, R. D.; Hasegawa, S. In *Citrus Science and Technology;* Nagy, S.; Shaw, P.; Velduis, M. K., Eds.; Avi Publishing: Westport, CT, 1977; Vol. I, 335–396.
3. Dreyer, D. L. *Fortsch. Chem. Org. Naturst.* **1968**, *26*, 190–244.
4. Lam, L. K. T.; Sparnins, V. L.; Wattenberg, L. W. *Cancer Res.* **1982**, *42*, 1193–1198.
5. Lam, L. K. T., Sparnins, V. L.; Wattenberg, L. W. *J. Med. Chem.* **1987**, *30*, 1399–1403.
6. Lam, L. K. T.; Li, Y.; Hasegawa, S. *J. Agric. Food Chem.* **1989**, *37*, 878–880.
7. Lam, L. K. T.; Zheng, B. L. *Nutr. Cancer* **1992**, *17*, 19–26.
8. Chasseaud, L. F. *Adv. Cancer Res.* **1979**, *29*, 175–274.
9. Lam, L. K. T., unpublished results.
10. Lam, L. K. T.; Sparnins, V. L.; Hochalter, J. B.; Wattenberg, L. W. *Cancer Res.* **1981**, *41*, 3940–3943.
11. Habig, W. H.; Pabst, M. J.; Jakoby, W. B. *J. Biol. Chem.* **1974**, *249*, 7130–7139.
12. Wattenberg, L. W.; Coccina, J. B.; Lam, L. K. T. *Cancer Res.* **1980**, *40*, 2820–2823.
13. Lam, L. K. T.; Hasegawa, S. *Nutr. Cancer* **1989**, *12*, 43–47.
14. Speier, J. L.; Lam, L. K. T.; Wattenberg, L. W. *J. Nat. Cancer Inst.* **1978**, *60*, 605–609.
15. Lam, L. K. T.; Schut, H. A. J., manuscript in preparation.
16. Buening, M. K.; Wislocki, P. G.; Levin, W.; Thakker, D. R.; Akagi, H.; Koreeda, M.; Jerina, D. M.; Conney, A. H. *Proc. Natl. Acad. Sci. USA* **1978**, *75*, 5358–5361.
17. Slaga, T. J.; Bracken, W. J.; Gleason, G.; Levin, W.; Yagi, H.; Jerina, D. M.; Conney, A. H. *Cancer Res.* **1979**, *39*, 67–71.
18. Cooper, C. S.; Grover, L. L.; Sims, P. *Prog. Drug Metab.* 1983, *7*, 295–396.
19. Gräslund, A.; Jernström, B. Q. *Rev. Biophys.* **1989**, *22*, 1–37.
20. Jeffrey, A. M.; Weinstein, I. B.; Jeanette, K. W. Grzeskowiak, K. Nakanishi, K.; Harvey, R. G.; Autrup, H.; Harris, C. C. *Nature* **1977**, *269*, 348–350.
21. Copper, C. S. Hewer, A.; Ribeiro, O.; Grover, P. L.; Sims, P. *Carcinogenesis* **1980**, *1*, 1075–1080.
22. Hesse, S.; Jernström, B.; Martinez, M.; Moldeus, P.; Christodoulides, L.; Ketterer, B. *Carcinogenesis* **1982**, *3*, 757–760.

23. Gupta, R. C.; Reddy, M. V.; Randerath, K. *Carcinogenesis* **1982**, *3*, 1081–1092.
24. Miller, G. G.; Fanous, R.; Rivera-Hidalgo, F.; Binnie, W. H.; Hasegawa, S.; Lam, L. K. T. *Carcinogenesis* **1989**, *10*, 1535–1537.
25. Slaga, J. J.; Fischer, S. M.; Weeks, C. E.; *et al. Compr. Survey* **1982**, *7*, 71–75.
26. Lam, L. K. T.; Zhang, J.; Hasegawa, S., manuscript in preparation.
27. Guadagui, D. G.; Maier, V. P.; Turnbaugh J. *Sci. Food Agric.* **1973**, *24*, 1277–1288.
28. Rouseff, R. L.; Matthews, R. T. *J. Food Sci.* **1984**, *49*, 777–779.
29. Fong, C. H.; Hasegawa, S.; Herman, Z. *Phytochem.* **1989**, *28*, 2777–2781.
30. Herman, Z.; Fong, C. H.; Ou, P.; Hasegawa, S. *J. Agric. Food Chem.* **1990**, *38*, 1860–1861.

RECEIVED October 8, 1993

Chapter 17

Inhibition of Oral Carcinogenesis by Green Coffee Beans and Limonoid Glucosides

E. G. Miller[1], A. P. Gonzales-Sanders[1], A. M. Couvillon[1], J. M. Wright[1], Shin Hasegawa[2], Luke K. T. Lam[3], and G. I. Sunahara[4]

[1]Department of Biomedical Sciences, Baylor College of Dentistry, Dallas, TX 75246
[2]Fruit and Vegetable Chemistry Laboratory, U.S. Department of Agriculture, Agricultural Research Service, 263 South Chester Avenue, Pasadena, CA 91106
[3]Department of Medicinal Chemistry, University of Minnesota, Minneapolis, MN 55455
[4]Nestle Research Centre, CH−1800, Vevey, Switzerland BP 353

The hamster cheek pouch model was used to test green coffee beans, defatted green coffee beans, green coffee bean oil, and three limonoid glucosides for antineoplastic activity. The cancer chemopreventive activity of green coffee beans was found to be due to chemicals in the oil, probably kahweol and cafestol, and to one or more chemicals in the defatted portion of the bean. Using a technique developed in this laboratory, it was found that the oil induced increased glutathione S-transferase activity; the defatted beans did not. In a separate experiment, one of the limonoid glucosides, limonin 17-β-D-glucopyranoside, was found to inhibit, by approximately 55%, the development of oral tumors. This chemical had no affect on the glutathione S-transferase activity of oral epithelial cells.

This research on the cancer chemopreventive activity of green coffee beans is an outgrowth of previous work in Dr. Lee Wattenberg's laboratory at the University of Minnesota. Their results showed that the addition of green coffee beans (20%) to the diet of experimental animals inhibited by approximately 60% the development of 7,12-dimethylbenz[a]anthracene (DMBA)-induced mammary tumors (*1*). Further studies (*2*) led to the isolation of two cancer chemopreventive agents from green coffee beans. The compounds, kahweol and cafestol, are diterpenoids that are usually found as esters of fatty acids. The free alcohols are chemically similar, differing from each other by one extra double bond that is found in one of the six-sided rings in kahweol (*3*).

The antineoplastic activity of these two coffee bean constituents was demonstrated in a rat model for mammary carcinogenesis. It was found that oral intubation of the two diterpenes once a day for 3 days prior to a single dose of the carcinogen resulted in a diminished (40%) neoplastic response (*4*). Each dose

0097−6156/94/0546−0220$06.00/0

contained 60 mg of the diterpene. The kahweol and cafestol were dissolved in cottonseed oil.

Additional studies with kahweol and cafestol suggest that both compounds are type A blocking agents that induce increases in glutathione *S*-transferase (GST) activity (5). Since kahweol is the more potent inducer of GST activity, it is assumed that kahweol is the more potent inhibitor of carcinogenesis. The experiments with kahweol (5) also showed that certain features in its chemical structure are critical to its ability to stimulate GST activity. One of these structural features is a furan ring, the other is the extra double bond.

This laboratory has continued the research on the cancer chemopreventive activity of green coffee beans (6,7). Using the hamster cheek pouch model for oral carcinogenesis, it was found that a diet containing 20% green coffee beans (Colombian) inhibited the formation of DMBA-induced epidermoid carcinomas by more than 95% (6). With the help of Dr. Würzner, Dr. Liardon, and Mr. Bertholet at the Nestlé Research Centre, we were able to obtain large quantities of a 50:50 mixture of kahweol and cafestol. Using this mixture we prepared a diet that approximated the kahweol and cafestol content of a diet containing 20% green coffee beans (Colombian). This experiment showed that when the modified diet was fed to hamsters, it inhibited the development of DMBA-induced oral tumors by approximately 35–40% (7).

Green Coffee Beans and Green Coffee Bean Fractions

In summary, the data from these experiments (6,7) indicate that green coffee beans might contain additional cancer chemopreventive agents. To test this possibility, we went back to the green coffee bean and to two fractions of the green coffee bean, green coffee bean oil and the defatted green coffee bean. In the Colombian green coffee bean, the green coffee bean oil accounts for approximately 15% (w/w) of the whole bean. The oil contains kahweol and cafestol as well as other plant oils and lipids. After the oil is removed, the remainder of the green coffee bean (85%) is the defatted green coffee bean. This fraction is essentially devoid of kahweol and cafestol.

For the experiment, 64 female Syrian golden hamsters (Lak:LVG strain) weighing 80–90 g were purchased from the Charles River Breeding Laboratories (Wilmington, MA). The animals were housed in stainless steel cages in a temperature-controlled room (22°C) with a 12 hour light-dark cycle. Water and food were furnished *ad libitum.*

The hamsters were given 10 days to adjust to their new surroundings. During this time they were fed a Purina Lab Chow specifically formulated for small rodents. The animals were then weighed and separated into four equal groups. The hamsters in group 1 were fed the Purina Lab Chow. The hamsters in groups 2, 3, and 4 were fed the same diet supplemented with 15% green coffee beans (group 2), 12.75% defatted green coffee beans (group 3), or 2.25% green coffee bean oil (group 4).

After the animals adjusted to their respective diets, the treatments with the carcinogen were initiated. The left buccal pouches of all of the animals in each group were inverted and painted 3 x weekly with a 0.5% solution of DMBA dissolved in heavy mineral oil. Each application with a camel hair brush places approximately 0.05 ml of the DMBA solution on the surface of the pouch. After a total of 36 applications the treatments were discontinued. One week later all of the animals were sacrifice by inhalation of an ethyl ether overdose. The pouches were excised. Tumors, when present grossly, were counted and measured (length, width,

and height). The sum of these three measurements divided by six was used to calculate an average radius for each tumor. By using the formula for the volume of a sphere, $4/3\pi r^3$, an approximation of the volume of each tumor was determined. A simple sum of the volume of each tumor in one pouch was defined to be the animal's tumor burden (7–9). Once the measurements had been taken the excised pouches were fixed in 10% formalin, embedded in paraffin, processed by routine histological techniques, and stained with hematoxylin and eosin. The microscopic and macroscopic observations were used to determine tumor incidence, number, mass, and type. Student's *t*-test was used to assess the significance of the data.

Tumor Data. Two animals died before the end of the experiment. Both of these animals were excluded from the study. At the end of the experiment, the number of hamsters in the four groups was 16 (group 1), 15 (group 2), 15 (group 3), and 16 (group 4). All of the hamsters in groups 1 and 4, 12 of the 15 hamsters in group 2, and 13 of the 15 hamsters in group 3 had visible tumors. Most of the animals, 15/16 in group 1, 10/15 in group 2, 13/15 in group 3, and 14/16 in group 4 had multiple tumors. The total number of tumors ranged from a low of 39 for groups 2 and 3 to a high of 75 for group 1. The values for tumor radii, which were calculated from the three measurements, ranged from 0.5 to 5.0 mm. The values for tumor volume ranged from 0.5 to 524 mm^3.

The data for average tumor number, average tumor burden, and average tumor mass are given in Table I. As illustrated, the treatment with green coffee beans (group 2) led to a 45% reduction in average tumor number and a 55% reduction in average tumor mass. Tumor burden was reduced by 70%. A similar comparison between groups 1 and 3 showed that the diet containing defatted green coffee beans reduced average tumor number by 45% and the average tumor mass by 25%. The decrease in average tumor burden was 55%. The same comparison for groups 1 and 4 (green coffee bean oil diet) yielded a 25% reduction in average tumor number, a 55% reduction in average tumor mass, and a 60% reduction in average tumor burden.

Table I. Effect of Dietary Green Coffee Beans and Green Coffee Bean Fractions on Oral Carcinogenesis

Group	Number of Animals	Number of Tumors[a]	Tumor Burden[a] (mm^3)	Tumor Mass[b] (mm^3)
1	16	4.7 ± 0.5	145 ± 41	30.9
2	15	2.6 ± 0.6****	43 ± 15***	16.5
3	15	2.6 ± 0.4****	62 ± 34*	23.8
4	16	3.5 ± 0.6*	57 ± 15**	16.3

SOURCE: Adapted from ref. 10.

[a] Values are means ± S.E.

[b] Values for average tumor mass were calculated by dividing the values for average tumor burden by corresponding values for average tumor number.

Statistically different from Group 1: *, $p < 0.10$; **, $p < 0.05$; ***, $p < 0.025$; ****, $p < 0.005$.

Histologically all tumors were epidermoid carcinomas. At the end of the experiment, the average weight of the animals in the four groups was 187 ± 5 g (group 1), 200 ± 6 g (group 2), 191 ± 5 g (group 3), and 193 ± 4 g (group 4). The differences in weight were not significant; the hamsters receiving the diets containing green coffee beans or the green coffee bean fractions actually gained more weight during the experiment than the animals on the regular diet.

The data on the extraction of the oil showed that 98% of the oil was removed. The defatted green coffee beans contained the remaining 2%. These data, when coupled with our earlier data with kahweol and cafestol (7), support the possibility that green coffee beans might contain additional bioactive agents. One or more of these unidentified agents appears to be present in the defatted portion of the bean.

Effects on GST Activity. Additional support for these conclusions can be found in an experiment recently completed in our laboratory. For this study 32 female Syrian golden hamsters (Lak:LVG) were weighed and separated into four equal groups, and placed on one of 4 diets. The animals in group 1 remained on a diet of Purina Lab Chow. The hamsters in groups 2, 3, and 4 received the same diet supplemented with either 15% green coffee beans (group 2), 12.75% defatted green coffee beans (group 3), or 2.25% green coffee bean oil (group 4).

The animals remained on their respective diets for 25 days. The hamsters were then sacrificed, and the right buccal pouches were excised, frozen, and stored at -80°C. Using two known procedures, one for the separation of intact oral epithelial sheets (11) and the second for the isolation of the enzyme fraction (12), we have developed a technique to measure the GST activity of oral epithelial cells. Using this technique 8–10 frozen pouches per day can be processed. The GST activity of the cytosolic fractions was determined spectrophotometrically at 30°C with 1-chloro-2,4-dinitrobenzene (CDNB) as the substrate (13). The complete assay mixture without the cytosolic fraction was used as a control. The assays were run in duplicate. The activity of the enzyme was expressed as µmol of CDNB metabolized per minute per mg of protein. The bicinchoninic acid protein assay was used to determine the protein concentration of the cytosolic fractions (14).

The effect of green coffee beans and the two green coffee bean fractions on the GST activity of buccal pouch epithelial cells is given in Table II. As can be seen, the addition of green coffee beans and green coffee bean oil to the diets of the experimental animals (groups 2 and 4) led to an increase in GST activity. Even though this increase was relatively small (approximately 15%), a statistical comparison (Student's t-test) between groups 1 and 4 yielded a probability value less than 0.025. A similar comparison, group 1 vs. group 2, yielded a probability value less than 0.100. From the Table it can also be seen that the addition of defatted green coffee beans to the diets of the experimental animals (group 3) increased GST activity by only 2%. This increase was not significant.

These data also support the conclusion that green coffee beans contain additional cancer chemopreventive agents. The two diets containing kahweol and cafestol (groups 2 and 4) produced an increase in GST activity. This supports earlier data (5) indicating that these chemicals are type A blocking agents. With defatted green coffee beans very little, if any, effect on GST activity was seen. This suggests that the cancer chemopreventive activity associated with the defatted portion of the bean is not due to the small amounts of kahweol and cafestol remaining in the defatted green coffee beans.

Table II. Effect of Green Coffee Beans and Green Coffee Bean Fractions on the GST Activity of Oral Epithelial Cells

Group	Treatment (diet)	Enzyme activity[a] (μmol/min/mg of protein)
1	Control diet	0.295 ± 0.045
2	15% Green coffee beans	0.334 ± 0.057*
3	12.75% Defatted green coffee beans	0.302 ± 0.023
4	2.25% Green coffee bean oil	0.342 ± 0.042**

[a] Values are means $\pm$ S.D.
Statistically different from Group 1: *, $p<0.10$; **, $p<0.025$.

Further research will be needed to identify the bioactive agent or agents in the defatted green coffee beans. Additional experiments will also be required to determine how these chemical(s) might affect the action of DMBA. It is interesting to note that the data in Table I indicate that the treatment with defatted green coffee beans affected tumor number more than tumor mass. This suggests that this treatment mainly affected the initiation phase of carcinogenesis.

These results could have significance for humans. It is known that very little kahweol and cafestol actually reach the coffee cup; during brewing these chemicals are trapped by the filter. Most of the chemicals that enter the coffee cup come from the defatted portion of the bean. If these as yet unidentified cancer chemopreventive agents can withstand the roasting process, then it is possible that they may be of benefit to humans.

Limonoid Glucosides

The structural features in kahweol which appear to be associated with cancer chemopreventive activity are also found in other plant oils, such as the citrus limonoids. Two of these compounds, limonin and nomilin, are highly oxidized triterpenes containing a furan ring and additional double bonds throughout the rest of the multi-ring structure.

Recent studies with limonin and nomilin (9,15) now indicate that one or both of these citrus chemicals might be cancer chemopreventive agents. In the first experiment (15), nomilin was found to inhibit benzo[a]pyrene-induced neoplasia in the forestomach of mice. Limonin was considerably less potent as an inhibitor of tumorigenesis in this animal model. In a similar study (Lam, L. K. T., LKT Laboratories, Inc., unpublished data), limonin and nomilin were both effective in inhibiting the development of cancer in a two-stage model for skin carcinogenesis in mice. In the first experiment, the limonoids were dissolved in cottonseed oil and given by intubation. For the second study, the limonoids were dissolved in acetone and applied topically to the skin. In our laboratory, topical applications of limonin and to a lesser extent nomilin were found to inhibit the formation of DMBA-induced carcinomas in the buccal pouch of hamsters (9).

The results with limonin and nomilin suggest that these limonoids, which are by-products of the citrus industry, may eventually prove useful as cancer chemopreventive agents. Several factors, however, could limit how these chemicals might be utilized as inhibitors of carcinogenesis in humans. The primary problem is that these chemicals are intensely bitter. This severely restricts the possibility of using limonin or nomilin as a food additive or as a food supplement. A second factor is that limonin and nomilin are only soluble in organic solvents. This limits the types of products to which these chemicals might be added. The third factor is human consumption. Every effort has been made to lower the concentration of limonin and nomilin in citrus products. The success of the citrus industry, however, leaves us with some basic questions. How much limonin and nomilin can be given to humans? Are elevated concentrations of limonin and nomilin toxic?

The negative features associated with limonin and nomilin are not found in a series of compounds that have recently been isolated and identified (*16,17*). In each case the compound is a modified citrus limonoid that is formed by the addition of one molecule of glucose to the limonoid. The glucose is attached at C-17 by a β-glucosidic linkage. The addition of glucose opens the D ring which is part of the limonoid nucleus. The furan ring is not affected by this modification. As compared to limonin and nomilin the concentration of these limonoid glucosides in commercial citrus juices is extremely high (*18,19*). The average concentration in orange, grapefruit, and lemon juices was 320, 190, and 82 ppm. The predominant limonoid glucosides in commercial orange juice are limonin 17-β-D-glucopyranoside (LG), nomilinic acid 17-β-D-glucopyranoside (NAG), and nomilin 17-β-D-glucopyranoside (NG). In all orange juice samples LG was the predominant limonoid glucoside with concentrations ranging from 130–220 ppm. Taste tests indicate that the limonoid glucosides have a slightly astringent taste as compared to the intense bitterness of the corresponding aglycone (*18*). The limonoid glucosides are water soluble.

Because the limonoid glucosides are natural products that are consumed in relatively high concentrations by humans, we tested LG, NG, and NAG for cancer chemopreventive activity. Like the earlier study with limonin and nomilin (*9*), solutions of the limonoid glucosides were applied topically to the hamster cheek pouches. In the earlier experiment (*9*), we used 2.5% solutions of limonin and nomilin. To evaluate stoichiometric changes, 3.5% solutions of the three limonoid glucosides were used for this experiment.

Sixty-four female Syrian golden hamsters (Lak:LVG) were obtained from the Charles River Breeding Laboratories (Wilmington, MA). The animals were housed as described earlier. Water and food (Purina Lab Chow) were given *ad libitum*. After arriving, the hamsters were given 10 days to adjust to the new surroundings. Following this period of time, the animals were divided into 4 equal groups. The average weight of the hamsters in each group was 104 g.

The left buccal pouches of the hamsters in each group were pretreated with two separate daily applications of water (group 1), or a 3.5% solution of LG (group 2), NG (group 3), or NAG (group 4) dissolved in water. Following pretreatment, the left buccal pouches of the animals were painted 5 x per week. On an alternating week-to-week basis, the pouches were treated with two or three applications of a 0.5% solution of the carcinogen, DMBA, dissolved in heavy mineral oil. On alternate days, the pouches were treated with water (group 1) or the 3.5% solution of LG (group 2), NG (group 3) or NAG (group 4).

The solutions of the limonoid glucosides were prepared on the day of use. All of the solutions were painted on the pouches with a camel hair brush. The 3 limonoid glucosides were prepared by Dr. Hasegawa (*16*) and their structures confirmed by nuclear magnetic resonance spectral analysis.

After a total of 71 treatments, 35 with DMBA and 36 with water or the limonoid glucosides, the hamsters were sacrificed by inhalation of an ethyl ether overdose. The left buccal pouches were excised and the tumors were counted and measured (length, width, and height). These measurements were used to calculate the tumor data. The excised pouches were then fixed in 10% formalin, embedded in paraffin, and processed by routine histological techniques. Student's t-test was used to analyze the significance of the tumor data. For each analysis, the data from one of the groups treated with a limonoid glucoside (groups 2–4) were compared with the data for group 1, the group treated with water.

Tumor Data. Three animals died before the end of the experiment. All of the hamsters that died prematurely were excluded from the study. At the end of the experiment, there were 16 hamsters in group 1 and 15 hamsters in groups 2, 3, and 4. All of the animals in groups 1 and 4, 12 of the 15 animals in group 2 and 14 of the 15 hamsters in group 3 had visible tumors. Multiple tumors were common (14/16 in group 1, 10/15 in group 2, 11/15 in group 3, and 13/15 in group 4). The high and low values for the total number of tumors were 73 (group 4) and 53 (group 3). The values for tumor radii ranged from a low of 0.6 to a high of 4.9 mm.

As illustrated in Table III, there was very little difference in the average number of tumors for the animals in all four groups. The figures ranged from a low of 3.5 (group 3) to a high of 4.9 (group 4). These differences were not significant. The high and low values for average tumor burden were 147 mm^3 (group 1) and 65 mm^3 (group 2). This difference between groups 1 and 2 was significant, $p<0.05$. Since tumor burden takes into account tumor number and size, it would appear that the treatment with LG mainly affected the size of the tumors not the number. The data for average tumor mass support this possibility. The data for average tumor burden for groups 3 and 4 indicate that NG and NAG had very little affect on the development of the DMBA-induced tumors.

Table III. Effect of LG, NG, and NAG on Oral Carcinogenesis

Group	Number of Animals	Number of Tumors[a]	Tumor Burden[a] (mm^3)	Tumor Mass[b] (mm^3)
1	16	4.4 ± 0.5	147 ± 41	33.4
2	15	3.7 ± 0.7	65 ± 15*	17.6
3	15	3.5 ± 0.6	121 ± 36	34.6
4	15	4.9 ± 0.7	139 ± 40	28.4

SOURCE: Adapted from ref. 20.
[a] Values are means ± S.E.
[b] Values for average tumor mass were calculated by dividing the values for average tumor burden by corresponding values for average tumor number.
Statistically different from Group 1: *, $p<0.05$.

Histologically, all tumors were epidermoid carcinomas. At the end of the experiment, the average weight of the animals in the four groups was 200–203 g. Throughout the course of the experiment, the animals appeared normal.

The data for tumor number, burden and mass for LG are very similar to the data from our earlier studies with limonin (9). In that experiment, the topical treatments with limonin produced a 20% reduction in tumor number, a 50% reduction in tumor mass, and a 60% reduction in tumor burden. Topical application of LG reduced the tumor number, mass and burden by 15%, 45% and 55% respectively. The strong similarity in these two sets of tumor data suggest that the addition of glucose and the opening of the D ring does not alter the cancer chemopreventive activity of limonin. Another possibility would be the removal of the glucose and the reformation of the D ring in the tissues. These reactions would allow for LG to be converted to limonin.

Effects on GST Activity. Studies with limonin (15,21) suggest that this chemical is at best a very poor inducer of GST activity. Even though LG is structurally similar to limonin, we wanted to see what effect, if any, this citrus chemical might have on the GST activity of oral epithelial cells. For this experiment, 16 female Syrian golden hamsters (Lak:LVG strain) were weighed and separated into 2 equal groups. The hamsters in group 1 served as controls. On alternating weekdays (Monday, Wednesday and Friday, or Tuesday and Thursday) the left buccal pouches of these animals were painted with water. On the same days, the left buccal pouches of the hamsters in group 2 were painted with a 3.5% (w/w) solution of LG. The limonoid glucoside was dissolved in water. The pattern that was used for the paintings is similar to the pattern that was used in our earlier experiment on tumorigenesis. The only difference was that the pouches were not treated with the carcinogen, DMBA.

After a total of 10 treatments, all of the hamsters were sacrificed. The left buccal pouches were excised, frozen, and stored at -80°C. Using the technique developed in our laboratory, the epithelial sheets were isolated and the cytosolic fractions were prepared (11,12) and assayed for GST activity (13,14).

The effect of topical treatments of LG on the GST activity of buccal pouch epithelial cells is given in Table IV. From the table, it is apparent that LG did not produce an increase in GST activity. Using Student's *t*-test it was found that the differences in these two sets of values were not significant.

Table IV. Effect of LG on the GST Activity of Oral Epithelial Cells

Group	Treatment (topical)	Enzyme activity[a] (μmol/min/mg of protein)
1	Water	0.284 ± 0.012
2	3.5% LG solution	0.263 ± 0.044

[a] Values are means $\pm$ S.D.

These results were not surprising considering the earlier data with limonin (15,21), but these two sets of results with limonin and LG leave us with a very basic question. What is the mechanism of action? One possible clue comes from the data with limonin and LG. In the two studies, the primary effect was on tumor mass not tumor number. One reason for this effect was readily apparent during the experiments. In both experiments, there was a lag of approximately one to two weeks in the appearance of the tumors. This suggests that these two citrus limonoids might be affecting the promotion phase of carcinogenesis.

Another aspect of this research has to do with the connection between green coffee beans and citrus limonoids. Experiments with kahweol (5) showed that certain features in its chemical structure were critical to its ability to act as an inducer of GST activity. The most important structural feature was the furan ring. The data from these studies suggested that there might be a family of furan-containing natural products with antineoplastic activity. It was also assumed that these bioactive agents would be type A blocking agents, inducers of increased GST activity. The data now indicate that six furan-containing chemicals can inhibit tumorigenesis in laboratory animals (4,7,9,10,15,20,22). The chemicals are cafestol and kahweol from coffee beans; limonin, nomilin and LG from citrus; and 2-*n*-heptylfuran, a byproduct formed during the roasting of meat. These results certainly strengthen the idea that there may be a family of furan-containing natural products with antineoplastic activity; support for the second assumption that these bioactive agents would all be type A blocking agents, however, is lessening. The evidence indicates that cafestol, kahweol, nomilin, and 2-*n*-heptylfuran can induce increases in GST activity (2,5,15,22). The data with limonin and now LG indicate that these chemicals have very little, if any, effect on the activity of this enzyme. Thus the distinct possibility of not one, but two families of furan-containing natural products with cancer chemopreventive activity exists. One family would be a group of type A blocking agents, the other would be a group of antineoplastic agents whose mechanism of action has yet to be determined.

Acknowledgments

The authors would like to thank Mary Freeman Johnson for her help in the preparation of the manuscript, and Josephine Taylor and Henry Burns, Jr. for technical assistance. This research was funded by a grant from the Hillcrest Foundation in Dallas, TX. A. M. Couvillon and A. P. Gonzales-Sanders were supported by the National Institutes of Health (Bethesda, MD) Traineeship DE07188-02. A. P. Gonzales-Sanders was also the recipient of an American Association of Dental Student Research Fellowship from Johnson & Johnson.

Literature Cited

1. Wattenberg, L. W. *Cancer Res.* **1983**, *43*, 2448s–2453s.
2. Lam, L. K. T.; Sparnins, V. L.; Wattenberg, L. W. *Cancer Res.* **1982**, *42*, 1193–1198.
3. Lam, L. K. T.; Yee, C.; Chung, A.; Wattenberg, L. W. *J. Chromatog.* **1985**, *328*, 422–424.
4. Wattenberg, L. W.; Lam, L. K. T. *Banbury Rep.* **1984**, *17*, 137–145.
5. Lam, L. K. T.; Sparnins, V. L.; Wattenberg, L. W. *J. Med. Chem.* **1987**, *30*, 1399–1403.
6. Miller, E. G.; Formby, W. A.; Rivera-Hidalgo, F.; Wright, J. M. *Oral Surg.* **1988**, *65*, 745–749.
7. Miller, E. G.; McWhorter, K.; Rivera-Hidalgo, F.; Wright, J. M.; Hirsbrunner, P.; Sunahara, G. I. *Nutr. Cancer*, **1991**, *15*, 41–46.
8. Niukian, K.; Schwartz, J.; Shklar, G. *Nutr. Cancer* **1987**, *9*, 171–176.

9. Miller, E. G.; Fanous, R.; Rivera-Hidalgo, F.; Binnie, W. H.; Hasegawa, S.; Lam, L. K. T. *Carcinogenesis* **1989**, *10*, 1535–1537.
10. Miller, E. G.; Couvillon, A. M.; Gonzales-Sanders, A. P.; Binnie, W. H.; Würzner, H. P.; Sunahara, G. I. In *14th International Scientific Colloquium on Coffee*; Association Scientifique Internationale du Café: Paris, 1992; pp 46–51.
11. Harris, R. R.; MacKenzie, I. C.; William, R. A. D. *J. Invest. Dermatol.* **1980**, *74*, 402–406.
12. Lam, L. K. T.; Sparnins, V. L.; Hochalter, J. D.; Wattenberg, L. W. *Cancer Res.* **1981**, *41*, 3940–3943.
13. Habig, W. H.; Pabst, M. J.; Jakoby, W. B. *J. Biol. Chem.* **1974**, *249*, 7130–7139.
14. Redinbaugh, M. G.; Turley, R. D. *Anal. Biochem.* **1986**, *153*, 267–271.
15. Lam, L. K. T.; Hasegawa, S. *Nutr. Cancer* **1989**, *12*, 43–47.
16. Hasegawa, S.; Bennett, R. D.; Herman, Z.; Fong, C.H.; Ou, P. *Phytochemistry* **1989**, *28*, 1717–1720.
17. Bennett, R. D.; Hasegawa, S.; Herman, Z. *Phytochemistry* **1989**, *28*, 2777–2781.
18. Fong, C. H.; Hasegawa, S.; Herman, Z.; Ou, P. *J. Food Sci.* **1990**, *54*, 1505–1506.
19. Herman, Z.; Fong, C. H.; Ou, P.; Hasegawa, S. *J. Agric. Food Chem.* **1990**, *38*, 1860–1861.
20. Miller, E. G.; Gonzales-Sanders, A. P.; Couvillon, A. M.; Wright, J. M.; Hasegawa, S.; Lam, L. K. T. *Nutr. Cancer*, **1992**, *17*, 1–7.
21. Lam, L. K. T.; Li, Y. Hasegawa, S. *J. Agric. Food Chem.* **1989**, *37*, 878–880.
22. Lam, L. K. T.; Zheng, B. L. *Nutr. Cancer* **1992**, *17*, 19–26.

RECEIVED May 28, 1993

Chapter 18

Stimulation of Glutathione *S*-Transferase and Inhibition of Carcinogenesis in Mice by Celery Seed Oil Constituents

G.-Q. Zheng, Jilun Zhang, P. M. Kenney, and Luke K. T. Lam

LKT Laboratories Inc., 2233 University Avenue West, St. Paul, MN 55114

Bioassay-directed fractionation of celery seed oil from the plant *Apium graveolens* L. (Umbelliferae) led to the isolation of five natural products, including limonene (**1**), *p*-mentha-2,8-dien-1-ol (**2**), *p*-mentha-8(9)-en-1,2-diol (**3**), 3-*n*-butyl phthalide (**4**), and sedanolide (**5**). Their structures were determined by spectral analysis. Compounds **2**, **4** and **5** exhibited high activity (up to 5-fold over the control) to induce the detoxifying enzyme glutathione *S*-transferase in the liver, forestomach and small intestinal mucosa of female A/J mice. They were then tested for their ability to inhibit benzo[*a*]pyrene-induced carcinogenesis in mouse forestomach. After pretreatment of mice with phthalides **4** and **5**, the number of mice with tumor was reduced from 68% to 30% and 11%, respectively. The number of tumors per mouse was also significantly decreased. No tumor inhibition was observed with compound **2**. The phthalides **4** and **5**, which determine the characteristic odor of celery, are present in relatively high amount in celery seed oil that is used for seasoning and flavoring purposes.

A short-term enzyme assay based on the induction of glutathione *S*-transferase (GST) activity has been successfully used for the isolation of anticarcinogenic natural products from edible plants. GST is a detoxifying enzyme that catalyzes the reaction of glutathione with electrophiles to form less toxic and more water-soluble conjugates for elimination by excretion (*1–3*). Because most reactive ultimate carcinogenic forms of chemical carcinogens are electrophiles, induction of increased GST activity is believed to be a major mechanism for carcinogen detoxification (*3–7*).

Compounds that stimulate GST may be considered as potential inhibitors of carcinogenesis. Positive correlation between the inhibitory activity of anticarcinogens and their ability to increase GST activity has been observed (*8–10*). Thus, examining the capacity to increase GST activity can be used as a method for preliminary screening of natural anticarcinogens. The active compounds thus isolated may be further tested for their inhibitory activity in the animal tumorigenesis assays.

0097–6156/94/0546–0230$06.00/0

Recently, using the GST assay as a guide, we have isolated five compounds from celery seed oil (*Apium graveolens* L., Umbelliferae). Three of them showed high GST-inducing activity. In this study, these compounds were tested for their capacity to inhibit benzo[*a*]pyrene (BP)-induced forestomach tumor formation in mice.

Isolation and Identification of Celery Seed Oil Constituents

GST-directed fractionation led to the isolation of five known compounds previously found in celery seed oil (*11–13*)(Table I). Fractionation of the oil was carried out by using preparative liquid chromatography with the solvent system of hexane-ethyl acetate. A 10 g portion of the oil was subjected to separation on two connected silica gel cartridges eluted with hexane, hexane-EtOAc (99:1), hexane-EtOAc (95:5), and EtOAc. Fractions were collected and analyzed by silica gel TLC.

Fractions with the same R_f values were combined to yield 9 major fractions (A-I), which were tested with the GST assay. Fraction B (eluted with hexane) showed GST-inducing activity and was evaporated *in vacuo* to afford *d*-limonene (**1**, 6.87 g, 68.7%). Fractions F, G and H (all eluted with hexane:EtOAc, 95:5) exhibited higher GST-inducing activity and were further purified by silica gel column chromatography (hexane-EtOAc as the solvent) to afford 3-*n*-butyl phthalide (**4**, from fraction F, 0.30 g, 3.0%), sedanolide (**5**, from fraction G, 0.39 g, 3.9%), *p*-mentha-2,8-dien-1-ol (**2**, from fraction H, 0.19 g, 1.9%), and *p*-mentha-8(9)-en-1,2-diol (**3**, from fraction H, 0.07 g, 0.7%). The rest of the fractions were inactive, and thus not subjected to further investigation.

Table I. Properties of the Isolated Celery Seed Oil Constituents

d-Limonene or *p*-mentha-1,8-diene (**1**): Colorless oil
    High resolution EI-MS: *m/z* 136.1250 (calcd for $C_{10}H_{16}$, 136.1252, M$^+$)
p-Mentha-2,8-dien-1-ol (**2**): Colorless oil
    High resolution EI-MS: *m/z* 152.1192 (calcd for $C_{10}H_{16}O$, 152.1201, M$^+$)
p-Mentha-8(9)-en-1,2-diol (**3**): White solid crystallized from MeOH
    High resolution EI-MS: *m/z* 170.1289 (calcd for $C_{10}H_{18}O_2$, 170.1307, M$^+$)
3-*n*-Butyl phthalide (**4**): Colorless oil
    High resolution EI-MS: *m/z* 190.0997 (calcd for $C_{12}H_{14}O_2$, 190.0994, M$^+$)
Sedanolide (**5**): Colorless oil
    High resolution EI-MS: *m/z* 194.1313 (calcd for $C_{12}H_{18}O_2$, 194.1307, M$^+$)

The structures (Figure 1) of these compounds were determined by spectral analysis (NMR, MS, IR) and further by comparison of their spectral data with those published in the literature (*14-18*). They can be divided into two groups, limonene-type monoterpenes and butyl phthalides.

Induction of GST Activity

Female A/J mice, six weeks of age, were obtained from Harlan Sprague Dawley Co. (Indianapolis, IN). Animals were housed in temperature controlled animal quarters with a 12 h light/dark cycle. They were acclimated for one week after arrival, then fed semipurified diet (ICN Nutritional Biochemicals) until the end of the experiment. The mice were divided into experimental and control groups with 4 mice per group. The experimental groups were given the test compounds by gavage once every two days for a total of 3 doses. The control group was given cottonseed oil alone.

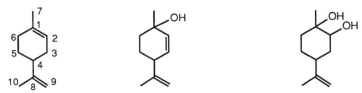

d-Limonene (**1**) p-Mentha-2,8-dien-1-ol (**2**) p-Mentha-8(9)-en-1,2-diol (**3**)

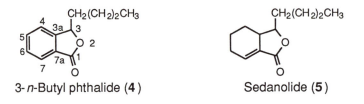

3-n-Butyl phthalide (**4**)   Sedanolide (**5**)

Figure 1. Structures of compounds isolated from celery seed oil.

Twenty-four hours after the last administration of test compound, the mice were killed by cervical dislocation. The liver, forestomach, and the mucosa layer of the proximal 1/3 of the small intestine were removed and homogenized in cold 1.15% KCl solution (pH 7.4). The homogenates were centrifuged at 9,000 x g for 20 min and the supernatants were centrifuged at 100,000 x g for 1 h. The cytosolic fractions were kept frozen at -80°C until use.

The activity of cytosolic GST was determined according to the method of Habig and co-workers using CDNB as the substrate (*1*). Assays were performed at 30°C in 0.1 M phosphate buffer (pH 6.5), in the presence of 5 mM GSH, 1 mM CDNB, and 20 l of the cytosol. The reaction was monitored at 340 nm on a Beckman DU-65 spectrophotometer equipped with a temperature-controlled cell compartment. Complete assay mixture without the cytosolic enzyme was used as the control. The protein concentration of these samples was determined by using the method of Lowry and co-workers (*19*).

The effects of compounds **1–5** on the GST activity in the mouse tissues are shown in Table II. These compounds appear to be responsible for the high GST-inducing activity of the oils observed in the preliminary screening. The phthalides **4** and **5**, at 20 mg per dose, increased GST activity 4.5–6.0 times the controls in liver. At the same dose, compounds **1** and **2** stimulated GST activity to 2.8 and 3.7 times

Table II. Effects of Compounds from Celery Seed Oil on GST Activity in Target Tissues of Female A/J Mice

Compound[a]	Liver		Small bowel mucosa		Forestomach	
	GST Activity[b]	Ratio[c]	GST Activity[b]	Ratio[c]	GST Activity[b]	Ratio[c]
Control	0.92 ± 0.17		0.50 ± 0.15		0.50 ± 0.07	
d-Limonene (1)	2.56 ± 0.53**	2.78	0.82 ± 0.54	1.64	0.57 ± 0.11	1.14
3-n-Butyl phthalide (4)	5.43 ± 0.59**	5.90	2.14 ± 0.37**	4.28	1.00 ± 0.11**	2.00
Sedanolide (5)	4.20 ± 0.67**	4.57	1.63 ± 0.29**	3.26	0.73 ± 0.13*	1.46
Control	1.39 ± 0.11		0.34 ± 0.01		0.56 ± 0.05	
p-Mentha-2,8-dien-1-ol (2)	5.17 ± 1.04**	3.72	1.27 ± 0.55*	3.74	0.90 ± 0.15**	1.61
3-n-Butyl phthalide (4)	7.36 ± 1.05**	5.29	1.76 ± 0.17**	5.18	0.76 ± 0.15*	1.36
Sedanolide (5)	6.56 ± 0.99**	4.72	1.53 ± 0.19**	4.50	0.74 ± 0.15	1.32
Control	0.79 ± 0.06		0.33 ± 0.03		1.37 ± 0.12	
p-Mentha-8(9)-en-1,2-ol (3)	1.37 ± 0.26**	1.73	0.90 ± 0.14**	2.73	1.22 ± 0.07	0.89

[a] Mice were administered 20 mg of test compound in 0.3 ml cottonseed oil per dose by p.o. intubation every two days for a total of three doses. The control mice were given only 0.3 ml cottonseed oil.
[b] Values are the mean μmol/min/mg of protein $\pm$ S.E. from 4 samples. Statistically different from control using the Student's t-test, *p < 0.05; **p < 0.005.
[c] Test/control.

the controls, respectively, while **3** enhanced GST activity only 1.70 times the control. In the small intestinal mucosa, the GST activity induced by **4** and **5** was 3.2–5.2 times the controls. In this tissue, compounds **1–3** also showed significant GST-inducing activity. In the forestomach, compounds **2**, **4** and **5** still exhibited significant activity as enzyme inducers, but no activity was observed with **1** and **3**.

These data indicate that the phthalides are more active than limonene-type monoterpenes. The five-membered lactone ring of phthalides appeared to be important for the high enzyme-inducing activity, while the aromatic ring may not be required for the maintenance of such high activity. Partial saturation of the phenyl ring with an intact lactone group as in **5** still retained high enzyme-inducing activity.

Inhibition of BP-Induced Forestomach Tumorigenesis

The tumor inhibition experiment was carried out according to previously published procedures (*20, 21*). Female A/J mice were housed as above and fed semipurified diet. The 128 mice were divided into 8 groups. The carcinogen BP control and the BP plus inhibitor groups each contained 20 mice. The vehicle control and the inhibitor-only controls each had 12 mice per group. BP was given by oral intubation twice a week for 4 weeks at 1 mg in 0.2 ml cottonseed oil per dose. The inhibitors (10 mg suspended in 0.3 ml of cottonseed oil per dose) were given by oral intubation three times a week on days other than the carcinogen treatment. The last dose of inhibitors was given on the day after the last dose of carcinogen. Two additional doses of inhibitors were administered before the first dose of carcinogen. Three days after the last dose of BP, the animals were returned to normal laboratory chow until the termination of the experiment. Eighteen weeks after the first dose of BP, the mice were killed and the forestomachs were removed and fixed in formalin. Tumors of the forestomach were counted under a dissecting microscope with a millimeter scale. Tumors that were 0.5 mm or larger were recorded and checked histopathologically.

The tumor inhibition data indicate that **4** and **5** significantly inhibit carcinogenesis in the mouse forestomach when given 1 week prior to and during carcinogen administration (Table III). The inhibition was manifested by a reduction in both tumor incidence (percentage of mice with tumors) and number of tumors per mouse. In the BP-treated group, 13 of 19 animals (68%) developed forestomach tumors. After treatment with **4** and **5** at 10 mg per dose, the tumor incidence was greatly reduced to 30% and 11%, respectively. This indicated more than 55% of inhibition of tumorigenesis after treatment of mice with the phthalides. No significant statistical difference was observed between the control mice and the mice treated with **2**. The number of tumors per mouse was also significantly reduced as the results of treatment with the phthalides. The greatest inhibitory effect was obtained with **5**. It resulted in more than 83% reduction of the mean number of tumors per mouse. About 67% reduction of the occurrence of forestomach tumor in mice by **4** was observed. The monoterpene **2** was almost inactive, which produced only small or no significant reduction of forestomach tumor formation.

The animal growth curve indicated that at the doses employed in this experiment all the inhibitor and carcinogen-treated groups experienced weight losses during the four-week treatment period (Figure 2). After treatment, however, most of the animals regained enough weight so that the final average body weight approached the control level.

Comparison of the data from the vehicle control with those of the inhibitor-only controls indicated that neither **4** nor **5** was an inducer of forestomach tumorigenesis in mice. In addition, the average animal body weight gains (Table III) of the BP plus inhibitors suggest that both compounds have little toxic effects in mice when given at the 20 mg dose level.

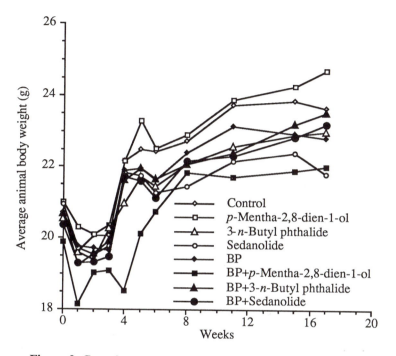

Figure 2. Growth curves of animals for the experimental period.

Possible Mechanism for the Inhibition by Phthalides

Correlation of induction of GST activity with the inhibition of carcinogenesis by anticarcinogens has been well documented. The results of this study further confirm the correlation although the relative inhibitory potency may be slightly different from the magnitude of enzyme-inducing activity. Phthalides **4** and **5**, which are very active enzyme inducers in the liver and other target tissues, were effective inhibitors against BP-induced tumorigenesis. Stimulation of GST activity by the phthalides appears to be a major mechanism for the inhibition of BP or other carcinogens that may be detoxified in the same manner. The induced enzyme GST could catalyze the reaction of glutathione with the activated carcinogenic forms of BP such as BP diol-epoxide to form less toxic conjugates which are readily eliminated by excretion. The present results suggest that with the advantages of simplicity and short term the GST assay may be used as a guide for the rapid screening of a large number of samples and isolation of potential anticarcinogenic

Table III. Effects of Compounds from Celery Seed Oil on BP-induced Neoplasia in the Forestomach of Female A/J Mice

Compound[a]	Number of mice[b]	Weight gain	Number of mice with tumors[d]	Percent of mice with tumors	Tumors per mouse[e]
Control	12	2.92	0	0	0
p-Mentha-2,8-dien-1-ol (2)	11	3.69	0	0	0
3-n-Butyl phthalide (4)	11	2.47	0	0	0
Sedanolide (5)	11	0.84[c]	0	0	0
BP	19	2.07	13	68	1.16 ± 0.23
BP + p-Mentha-2,8-dien-1-ol (2)	14	2.13	7	50	0.64 ± 0.19
BP + 3-n-Butyl phthalide (4)	20	2.83	6*	30	0.40 ± 0.16*
BP + Sedanolide (5)	18	2.85	2**	11	0.17 ± 0.12**

[a] Mice were administered 1 mg BP in 0.2 ml cottonseed oil per dose by p.o. intubation, two doses/wk for 4 wks, and/or 10 mg of each test compound in 0.2 ml cottonseed oil per dose, 3 doses/wk for 4 wks. Two additional doses of test compounds were administered before the first dose of BP. The control mice were only given 0.2 ml cottonseed oil.

[b] Indicates effective number of mice at the end of the experiment.

[c] One animal in this group had an extremely low weight compared to the remainder of the group; if the weight of that animal is excluded, the average weight gain for the group is 1.56 g.

[d] All tumors $\geq$ 0.5 mm were included. Statistically different from BP control using χ^2 analysis, *p < 0.025; **p < 0.005.

[e] Values are mean ± SE. Statistically different from BP control using Student's t-test, *p < 0.025; **p < 0.005.

natural products from plants. The active leads thus selected can be further tested for their inhibitory activity in the long-term animal tumorigenesis assays.

Phthalides as Potential Chemopreventive Agents

As a class of bioactive natural products, phthalides occur widely in umbelliferous plants (*14–16,22–25*). They are present in celery seed oil in relatively high amounts (~7%) (*14*) as well as in fresh celery in ppm quantities (*26,27*). These two phthalides are known to determine the characteristic odor and flavor of celery (*14,28*). Limonene was previously reported to inhibit chemical carcinogenesis in mice (*31*).

Celery is regularly consumed as a vegetable by humans. Celery seed oil is also used frequently in food flavorings and seasonings, as well as in perfumery and pharmaceutical preparation (*29*). Since edible plants are already consumed by humans on a daily basis, anticarcinogenic natural products isolated from the same sources should be readily acceptable as diet supplements if the toxicity and side effects are minimal. Although individual phthalide is present in small quantities in celery, the cumulative consumption of this class of compound from other edible umbelliferous plants such as parsley (*30*) may be substantial. The results of the present study indicate that phthalides naturally occurring in edible umbelliferous plants may be useful chemopreventive agents against BP-induced tumorigenesis, which is of significance for the environmental exposure of humans to this type of polycyclic aromatic hydrocarbon.

Acknowledgments

This study was supported by a grant from the National Cancer Institute (USPHS CA 47720).

Literature Cited

1. Habig, W. H.; Pabst, M. J.; Jakoby, W. B. G. *J. Biol. Chem.* **1974**, *249*, 7130.
2. Jakoby, W. B.; Habig, W. H. In *Enzymatic Basis of Detoxification*; Jakoby, W. B., Ed.; Academic Press: New York, 1980; pp 63–94.
3. Chasseaud, L. F. *Adv. Cancer Res.* **1979**, *29*, 175.
4. Talalay, P.; DeLong, M. J.; Prochaska, H. J. In *Cancer Biology and Therapeutics*; Cory, J. G.; Szentivanyi, A., Eds.; Plenum Press: New York, 1987; pp 197–219.
5. Prochaska, H. J.; DeLong, M. J.; Talalay, P. *Proc. Natl. Acad. Sci. USA* **1985**, *82*, 8232.
6. Benson, A. M.; Batzinger, R. P.; Ou, S. L.; Bueding, E.; Cha, Y. N. *Cancer Res.* **1978**, *38*, 4486.
7. Sparnins, V. L.; Barany, G.; Wattenberg, L. W. *Carcinogenesis* **1988**, *9*, 131.
8. Lam, L. K. T.; Zheng, B. L. *Nutr. Cancer* **1992**, *17*, 19.
9. Sparnins, V. L.; Wattenberg, L. W. *J. Natl. Cancer Inst.* **1981**, *66*, 769.
10. Sparnins, V. L.; Venegas, P. L.; Wattenberg, L. W. *J. Natl. Cancer Inst.* **1982**, *68*, 493.
11. Van Wassenhove, P.; Dirinck, P.; Schamp, N. *Med. Fac. Landbouww. Rijksuniv. Gent.* **1988**, *53*, 85.
12. Herrmann, H. Z. *Lebensm. Unters. Forsch.* **1978**, *167*, 262.
13. Salzer, U. J. *CRC Crit. Rev. Food Sci. Nutr.* **1977**, *9*, 345.

14 Bjeldanes, L. F.; Kim, I. S. *J. Org. Chem.* **1977**, *42*, 2333.
15. Chulia, A. J.; Garcia, J.; Mariotte, A. M. *J. Nat. Prod.* **1986**, *49*, 514.
16. Kaouadji, M.; Pouget, C. *J. Nat. Prod.* **1986**, *49*, 184.
17. Breitmaier, E.; Voelter, W. *Carbon-13 NMR Spectroscopy*; Verlagsgesellschaft VCH: Weinheim, FRG, 1987; pp 231–232 (for phthalides) and 328 (for monoterpenes).
18. Sadtler Research Laboratories, Inc. *Standard Spectra Collection, 1980 Cumulative Alphabetical Index*; Sadtler Research Laboratories, Inc.: Philadelphia, 1980; *d*-Limonene: #2852 (^{1}H NMR) and #1266 (^{13}C NMR).
19. Lowry, O. H.; Rosebrough, N. J.; Farr, A. L.; Randall, R. J. *J. Biol. Chem.* **1951**, *193*, 265.
20. Lam, L. K. T.; Hasegawa, S. *Nutr. Cancer* **1989**, *12*, 43.
21. Speier, J. L.; Lam, L. K. T.; Wattenberg, L. W. *J. Natl. Cancer Inst.* **1978**, *60*, 605.
22. Kaouadji, M.; Mariotte, A. M.; Reutenauer, H. *Z. Naturforsch.* **1984**, *39c*, 872.
23. Kaouadji, M.; DePachtere, F.; Pouget, C.; Chulia, A. J. *J. Nat. Prod.* **1986**, *49*, 872.
24. Cichy, M.; Wray, V.; Hofle, G. *Liebigs. Ann. Chem.* **1984**, 394.
25. Banerjee, S. K.; Gupta, B. D.; Sheldrick, W. S.; Hofle, G. *Liebigs.Ann. Chem.* **1982**, 699.
26. MacLeod, A. J.; MacLeod, G.; Subramanian, G. *Phytochemistry* **1988**, *27*, 373.
27. Tang, J.; Zhang, Y.; Hartman, T. G.; Rosen, R. T.; Ho, C. T. *J. Agric. Food Chem.* **1990**, *38*, 1937.
28. Uhlig, J. W.; Chang, A.; Jen, J. J. *J. Food Sci.* **1987**, *52*, 658.
29. Formacek, V.; Kubeczka, K. H. *Essential Oils Analysis by Capillary Gas Chromatography and Carbon-13 NMR Spectroscopy*; John Wiley & Sons: New York, 1982; pp 37–39.
30. Shaath, N. A.; Griffin, P.; Dedeian, S.; Paloympis, L. In *Flavors and Fragrances: A World Perspective. Proceedings of the 10th International Congress of Essential Oils, Fragrances and Flavors*; Lawrence, B. M.; Mookherjee, B. D.; Willis, B. J., Eds.; Washington, DC, 1986; pp 715–729.
31. Wattenberg, L. W.; Sparnins, V. L.; Barany, G. *Cancer Res.* **1989**, *49*, 2689.

RECEIVED October 8, 1993

PHYTOCHEMICALS FROM FRUITS AND VEGETABLES

Chapter 19

Citrus Juice Flavonoids with Anticarcinogenic and Antitumor Properties

John A. Attaway

Florida Department of Citrus, Scientific Research Department, Citrus Research and Education Center, 700 Experiment Station Road, Lake Alfred, FL 33850

Citrus flavonoids have a broad spectrum of biological activity including anticarcinogenic and antitumor activities. Quercetin inhibits carcinogenesis in a number of models and is able to selectively inhibit a variety of tumor cell growth. Polymethoxylated flavonoids, such as tangeretin and nobiletin, are more potent inhibitors of tumor cell growth than hydroxylated flavonoids. They also possess potent anti-invasive and anti-metastatic activities. In this chapter the biological activity of citrus flavonoids is briefly overviewed. Emphasis is on their anticarcinogenic and antitumor activities.

It is well-known that citrus fruits and juices are healthful foods. The Committee on Diet and Health, Food and Nutrition Board, of the National Research Council (*1*) recommended that one should "[e]very day eat five or more servings of a combination of vegetables and fruits, especially green and yellow vegetables and citrus fruits." With the exception of vitamin C, however, the source of these healthful benefits is not well understood

For over 30 years, the Florida Department of Citrus has sponsored research directed to documenting new health benefits of citrus fruit and juices. The Department's slogan describing this program is "Citrus Beyond Vitamin C." It has three major thrusts: (a) The study of water-soluble B-complex vitamins, specifically folate and thiamin (vitamin B_1), (b) the evaluation of citrus pectins as a dietary fiber, and (c) the investigation of the therapeutic value of the citrus flavonoids. The potential value of flavonoids as anticarcinogenic and antitumor agents are described in this chapter. Emphasis is placed on the findings of the investigators funded by the Department of Citrus, but contributions of other investigators in the field will also be mentioned.

Flavonoids

Flavonoids are a broad class of compounds, which occur widely in many fruits and vegetables. Since studies by Szent-Gyorgyi (*2,3*) indicated that citrus bioflavonoids have vitamin-like activity, there have been literally hundreds of individual plant flavonoids isolated and characterized, and subdivided into some 15 classes including flavones, isoflavones, flavonones, flavanones, flavanols, chalcones, dihydrochalcones and catechins. Their properties have been found to include antitumor activity, antiviral activity, antiinflammatory activity, activity against allergic type reactions, effects on capillary fragility and the ability to inhibit human platelet aggregation.

This interest in the health-related properties of flavonoids has resulted in an explosion in the literature resulting in a series of symposia designed to bring together the world's leading authorities in this complex field. For those interested in digging deeply into the mass of published material, it is recommended that you include The Flavonoids, Parts I and II, edited by J. B. Harborne, T. J. Mabry and Helga Mabry and published by Academic Press, Inc. in 1975; The Flavonoids — Advances in Research Since 1980, edited by J. B. Harborne and published by Chapman and Hall in 1988; Plant Flavonoids in Biology and Medicine: Biochemical, Pharmacological, and Structure-Activity Relationships, edited by V. Cody, E. Middleton, Jr., and J. B. Harborne and published by Alan R. Liss, Inc. (the proceedings of a symposium held in Buffalo, New York, July 22–26, 1985); Plant Flavonoids in Biology and Medicine II: Biochemical, Cellular and Medicinal Properties, edited by V. Cody, E. Middleton, Jr., J. B. Harborne and A. Beretz and published by Alan R. Liss, Inc (the proceedings of a symposium held in Strasbourg, France, August 31–September 3, 1987); and Plant Flavonoids in Biology and Medicine III: Current Issues in Flavonoid Research, edited by N. P. Das and published by the National University of Singapore Press (the proceedings of a symposium held in Singapore, November 13–17, 1989).

Three types of flavonoids, over 61 of which have been identified and characterized, occur in citrus (4)(See Figure 1 for several structures). These are flavanones, flavones and anthocyanins (the latter are important only in blood oranges). Flavanones are the most abundant, as the flavanone glycoside hesperidin is the principal flavonoid of orange juice, and the flavanone glycoside naringin is the principal flavonoid of grapefruit juice. On the other hand, the flavone aglycones such as nobiletin, tangeretin, and sinensetin are much more interesting for their biological activity, even though they occur in much lower concentrations. The highly methoxylated flavones found in food plants are almost unique to citrus (5), and exhibit much higher levels of biological activity than their hydroxylated counterparts.

Department of Citrus Supported Research

Since 1960, the Florida Department of Citrus has supported research by different investigators on the health benefits of flavonoids which occur in citrus juices. These scientists include Elliott Middleton, Jr., MD, State University of New York at Buffalo; Marc Bracke, MD, PhD, Laboratory of Experimental Cancerology, Ghent University, Ghent, Belgium and Dr. Ralph Robbins, Department of Food Science and Human Nutrition, University of Florida, Gainesville, Florida.

Since the late 1970s, Middleton has made major contributions to the study of antiinflammatory, anti-allergy properties of flavonoids, and in the mid-1980s, he enlarged his program to include the study of anticancer and antiviral properties.

Bracke has recently discovered that the citrus flavonoid tangeretin has the ability to prevent invasion of normal tissue by cancer cells, thus possibly blocking the onset of the "metastatic cascade effect" through which a primary cancerous tumor spreads to secondary and tertiary sites. This activity will be discussed in detail.

During the 1960s and 1970s, Robbins (5) found that citrus flavonoids affect capillary fragility, can function as antiplatelet agents and may be important in preventing blood cell clumping which can lead to coronary thrombosis.

Studies specifically on the anticancer activity of flavonoids found in citrus juices have been conducted by Middleton and coworkers and Bracke, Mareel and coworkers, both groups with partial support from the Florida Department of Citrus.

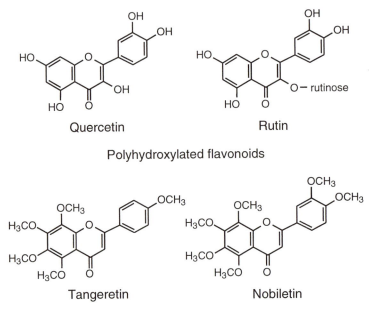

Figure 1. Structures of some citrus flavonoids.

Anticarcinogenic Activity of Citrus Flavonoids

One of the earliest reports of anticarcinogenic activity by citrus flavonoids was by Wattenberg and coworkers (6–8) who found that the polymethoxyflavones, tangeretin and nobiletin, were active inducers of benzo[a]pyrene (BP) hydroxylase activity in the liver and lung of the rat. In the Wattenberg studies, these were the only two naturally-occurring constituents of edible foods that had this effect, leading to the landmark observation that dietary constituents might conceivably alter the response of man or animals to exposure to BP and other polycyclic aromatic hydrocarbons known to be carcinogens. The hydroxylated flavonoids (e.g., quercetin and myricetin) usually inhibit cytochrome P-450 monooxygenase activity, whereas polymethoxylated flavonoids (e.g., tangeretin and nobiletin) are inducers and activators of phase I enzyme monooxygenase activity (9). The effect of polymethoxylated flavonoids on carcinogenesis, however, has not been studied yet.

Some studies on the inhibitory effects of citrus flavonoids on carcinogenesis, tumor cell growth and the invasion of tumor cell into normal cell are summarized in Table I. Research relating to flavonoids as possible chemopreventive agents in chemical carcinogenesis was also reviewed by Fujiki *et al.* (24), who postulated that in the 2-step, initiator and promoter sequence, flavonoids act as inhibitors of the tumor promoter step, as shown by Glusker and Russi (25) and Bracke *et al.* (26,27).

While the bioflavonoid quercetin is quantitatively a minor component of fruit juices, it is interesting to us for its biological activity, which has been widely studied. Quercetin inhibits 7,12-dimethylbenz[*a*]anthracene (DMBA)-induced skin tumor initiation in mice (10,11), possibly due to inhibition of epidermal aromatic hydrocarbon hydroxylase activity (28,29), modulation of epidermal metabolism of DMBA, and/or inhibition of the interaction of metabolites of DMBA with DNA (10,11,29). Quercetin also inhibits TPA-induced tumor promotion in the skin of mice (10–12). The inhibitory effect of quercetin on 12-*O*-tetradecanoylphorbol-13-acetate (TPA)-induced tumor promotion in skin of mice may be due to inhibition of epidermal protein kinase C activity (30), TPA-induced skin inflammation (31), TPA-induced epidermal ornithine decarboxylase activity (12), and/or TPA-induced phospholipid biosynthesis (11). In addition, quercetin has antioxidant activity, as well as the ability to scavenge active oxygen and electrophiles and to inhibit lipoxygenase activity (12,32).

Wall *et al.* (33) evaluated some 17 flavonoids as antimutagenic agents. They reported that flavonoid glycosides such as naringin, hesperidin, and rutin had only weak activity; tangeretin and nobiletin were somewhat more active.

Yasukawa, K. *et al.* (34) suggested that inhibitory effects of flavonol glycosides against tumor promotion may have been partly due to activation of immune responses against tumors.

Antitumor Activity of Citrus Flavonoids

Anti-proliferative Activity. Middleton and coworkers studied the effects of quercetin on squamous cell carcinoma of head and neck, and the antiproliferative effects of citrus flavonoids on human squamous cell carcinoma *in vitro* (21). In their quercetin paper, they investigated the effect of the flavonoid on the growth of two squamous cell carcinoma lines, both *in vitro* and *in vivo,* and a normal human lung fibroblast-like cell line. Concentrations used were 2, 20, 65, 110, 155, and 200 µM. Quercetin was found to inhibit growth in both squamous cell carcinoma lines in only 3 days, with the inhibitory effect being dose-dependent. The normal fibroblast-like human lung cells were only affected at the maximum concentrations suggesting that the carcinoma cell lines were differentially sensitive to the effect of the flavonoid.

Of greater interest to us as fruit juice producers is research by Middleton's group — Kandaswami *et al.* (20) — on citrus flavonoids in which they investigated the effects of tangeretin and nobiletin, as well as quercetin and taxifolin, on the *in vitro* growth of a human squamous cell carcinoma cell line. Cell cultures were treated with 2–8 µg/ml of each flavonoid for 3–7 days. It was found that the

Table I. Some Studies on the Anticarcinogenic and Antitumor Activities of Citrus Flavonoids

Site	Inhibitory action	References
Anticarcinogenic activity		
Mouse skin	Inhibit BP-, DMBA-, NMU-induced tumor initiation	(10,11)
Mouse skin	Inhibit TPA-induced tumor promotion	(10–12)
Mouse skin	Inhibit 3-MC complete carcinogenesis	(10)
Rat mammary gland	Inhibit DMBA- and NMU-induced carcinogenesis	(13)
Mouse colon	Inhibit AOM-induced colon carcinogenesis	(14)
Antitumor activity		
L1210 and P-388 leukemia and Ehrlich ascites tumor cells	Inhibit proliferative growth	(15)
NK/Ly tumor cells	Inhibit proliferative growth	(16)
Human squamous carcinoma cells	Inhibit proliferative growth	(17)
Mammary gland carcinoma cells	Inhibit proliferative growth	(18)
HeLa cells	Inhibit proliferative growth	(19)
Human squamous carcinoma cells	Inhibit proliferative growth	(20,21)
MO$_4$ cells into embryonic chick heart fragment	Inhibit invasion	(22)
Liver in tumor bearing mice	Inhibit invasion	(23)

BP, benzo[a]pyrene; DMBA, 7,12-dimethylbenz[a]anthracene; NMU, N-nitrosomethylurea; TPA, 12-O-tetradecanoylphorbol-13-acetate; AOM, azoxymethane; 3MC, 3-methylcholanthrene

polymethoxylated flavonoids, nobiletin and tangeretin, markedly inhibited cell growth at all concentrations tested on days 5 and 7. On day 3, the inhibition observed was 70–72% at 8 μg/ml, while on day 5 it ranged from 61–88% at 2-4 μg/ml. Quercetin and taxifolin exhibited no significant inhibition at any of the concentrations tested. The authors speculated that the difference in activity might be due to the relatively greater membrane uptake of the polymethoxylated flavonoids, since methoxylation of the phenolic groups decreases the polarity of the flavonoid.

Present and future work by Middleton's group (*21*) involves the role of the gastrointestinal (G.I.) tract in carcinogen metabolism. The G.I. tract is the route of entry of dietary flavonoids, which may modify the effect of enzyme systems involved in activation of chemical carcinogens. Research is being directed specifically toward a study of the effect of dietary administration of plant flavonoids on the metabolism of polynuclear aromatic hydrocarbon *in vivo*. BP will be used as a representative procarcinogen in an effort to determine possible inhibitory effects on tumors induced in mice.

Yoshida *et al.* (*35*) studied the effect of quercetin on human malignant cells from the gastrointestinal tract and found that it markedly inhibited the growth of these cells.

Hirano *et al.* (*18*) evaluated the effects of 21 synthetic and naturally occurring flavonoids on the *in vitro* growth of human breast carcinoma cells and concluded that flavonones of plant origin possess significant epidemiological properties either in their present form or as metabolites.

Anti-invasive Activity. The prognosis of cancer in man partially correlates with tumor growth, but is mainly determined by (a) the invasiveness of the tumor and (b) the metastatic capability of the tumor cell population (*36,37*). Growth of benign tumors will eventually lead to compression of surrounding tissues, but the tumor cell population respects the boundaries of its tissue of origin and the prognosis after surgical removal of such tumors is excellent. On the other hand, malignant tumors not only grow, but also invade into surrounding normal tissues. During invasion, these tumors cross membranes, and as invasion proceeds, gain access to the circulatory system and are transported via lymph or blood vessels into distant organs. These secondary tumor deposits, called metastases or invasive tumors, lead to the further origin of tertiary tumor deposits, causing the onset of an escalation, called the "metastatic cascade," which makes the disease much less responsive to treatment and can lead to death of the patient.

Viadana *et al.* (*38*) indicate that in almost all cases spread is not directly from the primary cancer throughout the human body, but requires previous seeding of key disseminating organs. For example, the lungs are the disseminating organ for primary skin melanomas, which will then spread from the lungs to the liver, the endocrine system and central nervous system. Liver metastases are also disseminated through the lung and then to the endocrine system and central nervous system. Primary gastrointestinal cancers, such as pancreatic, stomach, colon and rectal cancers disseminate to the liver which in turn spreads metastases throughout the body via the lung. In the case of cancer of the esophagus, bladder and female genitals, the lung and liver are usually seeded independently of each other.

The currently used techniques of chemotherapy and irradiation attempt to slow the rate of growth and proliferation, but these treatments have serious side effects and little anti-invasion and anti-metastasis effect specifically.

Although invasion appears to be the deciding factor in the survival of the patient, its mechanism is only beginning to be understood. Assays and tools to study invasion *in vitro*, and in some instances *in vivo*, are now available, however, and can generate data which may lead to our understanding of invasion mechanisms.

The Florida Department of Citrus furnished samples of hesperidin, naringin, nobiletin, and tangeretin to Bracke at Ghent in 1986 for screening by Mareel's *in vivo* embryonic chick heart assay. Tangeretin was the most potent of all citrus flavonoids tested for ability to inhibit the invasion of malignant mouse tumor cells (MO_4) into normal tissue fragments in organ culture (*22*). Certain catechins were also effective. The consequences of this were immediately significant, as inhibition of cell invasion would hinder or prevent the metastatic cascade effect by which cancer cells can spread rapidly from primary tumor to secondary sites throughout the body.

Following up this initial important discovery, Bracke *et al.* (*22,23*) carried out further research to determine the mechanism of action and whether flavonoids have anti-metastatic activity as well as anti-invasive activity. It was found that tangeretin did not inhibit invasion via the extracellular matrix, the cell surface glycoproteins or the cytoplasmic microtubule complex, so it was assumed that the activity found was brought about by inhibition of cell motility.

To test the possible anti-invasive activity in tumor-bearing mice, it was first necessary to study the distribution of tangeretin in the body organs after oral and parenteral administration to laboratory mice. The results of these tests showed that detectable levels of the compound could be achieved for a considerable length of time in the liver and the kidneys after oral administration via the drinking water, and in the lungs after intraperitoneal injection (*39*). This suggested that the effects of orally administered tangeretin be tested against invasion in the kidney and liver and the effects of intraperitoneal administration against lung metastasis.

Testing for liver invasion and metastasis was done by injecting MO_4 cells into the spleen where they were transported via the circulation to the liver, giving rise to invasive tumors. Invasion in the kidney was measured after introduction of MO_4 cells underneath the renal capsule, and spontaneous lung metastases were obtained after subcutaneous implantation in the tail. The initial experiment showed that oral administration of tangeretin could affect liver invasion in tumor-bearing mice.

In summary, research to this point has shown that the citrus flavonoid tangeretin inhibits the invasion of MO_4 mouse cells in chick heart fragments in organ culture and in the liver of syngeneic (nude) mice, and the group is now conducting studies to determine whether oral administration of tangeretin can reduce liver metastasis of one or more types of human tumors. To accomplish this, an assay was developed through which human tumor cells are injected into the spleen of syngeneic mice. For example, in the initial experiments, a suspension of 1 x 10^6 MCF-7 human mammary carcinoma cells was injected in the spleen of an estrogen-primed, 6 weeks old female nu/nu mouse. After 10 months, the moribund animal was killed, and tumors were found in both the spleen and the liver. Bracke's group (Bracke, M., Laboratory of Experimental Cancerology, Ghent University, personal communication, 1992) is now following a group of animals treated in this manner to determine whether tangeretin affects this phenomenon.

Summary

For more than 30 years, the Florida Department of Citrus has sponsored research related to the health effects of flavonoids which occur naturally in citrus fruit and juices. Many of them possess varied biological activities. Although it is not the most prevalent in citrus fruits, the hydroxylated flavonoid quercetin is the most widely distributed flavonoid in fruits and vegetables. Quercetin inhibits mouse skin tumor initiation by the polycyclic aromatic hydrocarbons BP and DMBA and other carcinogens, and skin tumor promotion by TPA. In addition, quercetin is able to selectively inhibit growth of a variety of tumor cells. Polymethoxylated flavonoids, such as tangeretin and nobiletin, are more potent inhibitors of tumor cell growth than hydroxylated flavonoids. More importantly, the polymethoxylated flavonoids tangeretin and nobiletin are capable of inhibiting tumor cell invasion into normal cells *in vitro* and *in vivo*. It is expected that polymethoxylated flavonoids may become important chemotherapeutic agents.

Literature Cited

1. National Research Council *Diet and Health*; National Academy Press: Washington, DC, 1989.
2. Szent-Gyorgyi, A. *Hoppe-Seyler's Z. Physiol. Chem.* **1938**, *255*, 126–131.
3. Szent-Gyorgyi, A. *Ann. N. Y. Acad. Sci.* **1955**, *61*, 732–735.
4. Horowitz, R. M.; Gentilli, B. In *Citrus Science and Technology*; Nagy, S., Shaw, P.; Veldhuis, M., Eds.; AVI Publishing Co.: Westport, CT, 1977; pp 397–426.
5. Robbins, R. C. In *Citrus Nutrition and Quality*; Nagy, S.; Attaway, J. A., Eds.; ACS Symposium Series No. 143; American Chemical Society: Washington, DC, 1980; pp 23–62.
6. Wattenberg, L. W.; Page, M. A.; Leong, J. L. *Cancer Res.* **1968**, *28*, 934–937.
7. Wattenberg, L. W.; Leong, J. L. *Cancer Res.* **1970**, *30*, 1922–1925.
8. Wattenberg, L. W. *Cancer Res.* **1975**, *35*, 3326–3330.
9. Buening, M. K.; Chang, R. L.; Huang, M.-T.; Fortner, J. G.; Wood, A. W.; Conney, A. H. *Cancer Res.* **1981**, *41*, 67–72.
10. Mukhtar, H.; Das, M.; Khan, W. A.; Wang, Z. Y.; Bik, D. P.; Bickers, D. R. *Cancer Res.* **1988**, *48*, 2361–2365.
11. Verma, A. K. In *Phenolic Compounds in Food and Their Effects on Health II: Antioxidants and Cancer Prevention*; Huang, M.-T., Ho, C.-T.; Lee, C. Y., Eds.; ACS Symposium Series 507; American Chemical Society: Washington, DC, 1992; pp 250–264.
12. Kato, R.; Nakadate, T.; Yamamoto, S.; Sugimura, T. *Carcinogenesis* **1983**, *4*(10), 1301–1305.
13. Verma, A. K.; Johnson, J. A.; Gould, M. N.; Tanner, M. A. *Cancer Res.* **1988**, *48*, 5754-5758.
14. Deschner, E. E.; Ruperto, J.; Wong, G.; Newmark, H. L. *Carcinogenesis* **1991**, *12*(7), 1193–1196.
15. Suolinna, E. M.; Buchsbaum, R. N.; Racker, E. *Cancer Res.* **1975**, *35*, 1865–1872.
16. Molnar, J.; Beladi, I.; Domonkos, K.; Foldeak, S.; Boda, K.; Veckenstedt, A. *Neoplasma* **1981**, *28*, 11–18.

17. Castillo, M. H.; Perkins, E.; Campbell, J. H.; Doerr, R.; Hassett, J. M.; Kandaswami, C.; Middleton, E., Jr. *Am. J. Surgery* **1989**, *158*, 351–355.
18. Hirano, T.; Oka, K.; Akiba, M. *Res. Commun. Chem. Pathol. Pharmacol.* **1989**, *64*(1), 69–78.
19. Mori, A.; Nishino, C.; Enoki, N.; Tawata, S. *Phytochem.* **1988**, *27*, 1017–1020.
20. Kandaswami, C.; Perkins, E.; Soloniuk, D. S.; Drzewiecki, G.; Middleton, E., Jr. *Cancer Lett.* **1991**, *56*, 147–152.
21. Middleton, E., Jr, The State University of New York at Buffalo, personal communication, 1992.
22. Bracke, M. E.; Vyncke, B. M.; Van Lanebeke, N. A.; Bruynell, E. A.; DeBruyne, G. K.; DePestel, G. H.; DeCoster, W. L.; Espeel, M. F.; Marred, M. M. *Clin. Experim. Metastasis* **1989**, *7*, 287–300.
23. Bracke, M. E. *Clin. Experim. Metastasis* **1991**, *9*, 13–25.
24. Fujiki, H. In *Plant Flavonoids in Biology and Medicine: Biochemical, Pharmacological and Structure-Activity Relationships*; Cody, V., Middleton, E., Jr.; Harborne, J. B., Eds.; Alan R. Liss, Inc.: New York, 1986; pp 429–440.
25. Glusker, J. P.; Russi, M. In *Plant Flavonoids in Biology and Medicine: Biochemical, Pharmacological and Structure-Activity Relationships*; Cody, V., Middleton, E., Jr.; Harborne, J. B., Eds.; Alan R. Liss, Inc.: New York, 1986; pp 395–410.
26. Bracke, M. E.; Van Cauwenberge, R. M.-L.; Mareel, M. M.; Castronovo, V.; Foidart, J. M. In *Plant Flavonoids in Biology and Medicine: Biochemical, Pharmacological and Structure-Activity Relationships*; Cody, V., Middleton, E., Jr.; Harborne, J. B., Eds.; Alan R. Liss, Inc.: New York, 1986; pp 441–444.
27. Bracke, M. E. In *Plant Flavonoids in Biology and Medicine II: Biochemical, Cellular and Medicinal Properties*; Cody, V., Middleton., E., Jr, Harborne, J. B.; Beretz, A., Eds.; Alan R. Liss, Inc.: New York, 1988; pp 219–233.
28. Das, M.; Mukhtar, H.; Bik, D. P.; Bickers, D. R. *Cancer Res,* **1987**, *47*, 760.
29. Das, M.; Khan, W. A.; Asokan, P.; Bickers, D. R.; Mukhtar, H. *Cancer Res.* **1987**, *47*, 767–773.
30. Horiuchi, T.; Fujiki, H.; Hakii, H.; Saganuma, M.; Yanashita, K.; Sugimura, T. *Jpn. J. Cancer Res. (Gann)* **1986**, *77*, 526–531.
31. Huang, M.-T.; Lysz, T.; Ferraro, T.; Abidi, T. F.; Laskin, J. D.; Conney, A. H. *Cancer Res.* **1991**, *51*, 813–819.
32. Huang, M.-T.; Ferraro, T. In *Phenolic Compounds in Food and Their Effects on Health II: Antioxidant and Cancer Prevention*; Huang, M.-T., Ho, C.-T.; Lee, C. Y., Eds.; ACS Symposium Series 507; American Chemical Society: Washington, DC, 1992; pp 8–34.
33. Wall, M. E.; Wan, M. C.; Manikumar, G.; Graham, P. A.; Taylor, H.; Hughes, T. J.; Walker, J.; McGivney, R. *J. Natural Prod.* **1988**, *51*(6), 1084–1091.
34. Yasukawa, K. *Chem. Pharm. Bull.* **1990**, *38*(3), 774–776.
35. Yoshida, M.; Sakai, T.; Hosokawa, N.; Marui, N.; Matsumoto, K.; Fujioka, A.; Nishino, H.; Aoike, A. *FEBS Lett.* **1990**, *260*(1), 10–13.
36. Poste, G.; Fidler, I. D. J. *Nature* **1980**, *283*, 139–146.
37. Mareel, M. M. *Int. Rec. Exper. Pathol.* **1988**, *22*, 65–129.
38. Viadana, E.; Bross, I. D. J.; Pickren, J. W. In *Pulmonary Metastasis*; Weiss, L.; Gilbert, H. A., Eds.; G. H. Hall: Boston, 1978; pp 143–167.
39. Bracke, M. E. In *Plant Flavonoids in Biology and Medicine III: Current Issues in Flavonoid Research*; Das, N. P., Ed.; National University of Singapore Press: Singapore, 1990; pp 279–292.

RECEIVED October 8, 1993

Chapter 20

Determination of Free and Glycosidically Bound Organic Compounds in an Umbelliferous Vegetable Drink

Robert T. Rosen, Tarik H. Roshdy, Thomas G. Hartman, and Chi-Tang Ho

Center for Advanced Food Technology and Food Science Department, Cook College, Rutgers, The State University of New Jersey, New Brunswick, NJ 08903

Gas chromatography and gas chromatography-mass spectrometry was used to identify and quantitate free and glycosidically bound phytochemicals in an Umbelliferous vegetable drink supplied by the National Cancer Institute (NCI). This drink was a blend of tomato, celery and carrot juices with added spices. Over twenty organics that were present as glycosides were determined as were over sixty "free" (not glycosidically bound) compounds. Methodology included XAD column chromatography to separate the free from the bound fractions, followed by enzymatic cleavage of the glycosides, with subsequent determination by GC and GC-MS

Phytochemicals in fruits and vegetables may exist as glycosides or aglycones. The enzymatically induced ripening of glycosides in vegetables and fruits is a major pathway leading to the formation of flavor compounds. Examples include the formation of the key flavor compounds 2,5-dimethyl-4-hydroxy-3-(2H)-furanone from pineapple, vanillin from the vanilla bean, and isobutyl 3-hydroxybutanone from the tropical fruit called the hog plum (*1,2,3*). Examples of beneficial compounds in vegetables include the indole alcohols and related compounds from the *Cruciferae* (*4*). The indoles are directly bound to sugars as glucosinolates.

The objective of this study was to determine two phthalides, two coumarins and two diacetylenics in an Umbelliferous vegetable drink supplied by the National Cancer Institute (the drink is used as part of NCI's designer food project under the directorship of Dr. Herbert Pierson). Also to be determined was if such compounds existed as glycosidically bound residues. These bound residues would, on digestion, liberate more of the target compounds of interest. Combined capillary column gas chromatography-mass spectrometry was used to identify the species of interest and gas chromatography was used for quantitation.

0097–6156/94/0546–0249$06.00/0

Materials and Methods

The drink was manufactured by Dr. Phillip Crandall of the University of Arkansas Food Science Department for the National Cancer Institute. Individual pint cans were received and stored in a 4°C refrigerator. The materials used are shown in Table I.

Table I. Materials Used

Standards

Standard	Purity	Supplier
Phthalide	98%	Aldrich
Psoralen	98%	Sigma
5-Methoxypsoralen	98%	Sigma
8-Methoxypsoralen	99.8%	Sigma
13-Octadecyn-1-ol	96%	Sigma
11-Octadecyn-1-ol	99.9%	Sigma
9,12,15-Octadecatrienol	98.6%	Sigma
Erucyl Alcohol	99%	Sigma
Methoxy-4-methylcoumarin	99%	Aldrich
7-Methoxycoumarin	98%	Aldrich
2-Methoxydibenzofuran	98%	Aldrich
2-Coumaranone	97%	Aldrich
Coumarin	unknown	Aldrich
2,3-Benzofuran	99.5%	Aldrich

Other
- β-Glucosidase, 50,000 units contained approx. 100% protein: Sigma Chemical
- Supelpak adsorbent, a purified form of Amberlite XAD-2 resin: Supelco Inc.
- Glass column for column chromatography, 1 cm I.D., 50 cm length: Fisher
- Dichloromethane, HPLC grade; methanol, HPLC grade; n-pentane, HPLC grade: Fisher
- Diethyl ether, HPLC grade: J. T. Baker

Sample preparation. One hundred ml of the Umbelliferous vegetable beverage was divided and placed into four 50 ml heavy-duty centrifuge glass tubes with screw caps. The tubes were then centrifuged for 20 min at 2500 RPM. Forty g dry Amberlite XAD-2 resin was placed into a glass column (50 cm x 1 cm). Glass wool was packed tightly at the top and bottom of the column to prevent the loss of the stationary phase and to prevent the column bed from rising with the water level and producing air bubbles. The XAD-2 column was sequentially conditioned (washed) with water, methylene chloride/n-pentane (1:1, v:v), methanol, and water. The second water elution was used to clear the methanol from the column prior to adding the sample. The supernatant from the centrifuge tubes was then loaded on the top of the glass column. HPLC grade distilled water was used to elute the water soluble (yellow-orange) fraction until the eluted water was clear. This fraction represents sugars and other very polar materials which eluted from the column. HPLC grade *n*-pentane/methylene chloride (1:1, v:v) was used to elute the second (yellow-red) fraction until the eluted solvent was clear of color. This fraction contains the free (not glycosidically bound) target compounds of interest. Finally, 100 ml HPLC grade methanol was used to elute the polar organic soluble compounds (yellow fraction). This fraction contained the glycosidically bound organics.

The water fraction was discarded. The free fraction (agycones) was dried over anhydrous sodium sulfate and was subsequently concentrated to 1 ml, followed by injection into either the GC or GC-MS. The GC and GC-MS conditions are shown in Table II. The internal standard used for the free fraction was the compound phthalide.

The methanol fraction (bound fraction) was concentrated to 1 ml using a rotary evaporator with the assistance of a water bath at 50°C. The concentrate was transferred into a 250 ml flask and 100 ml citrate-phosphate buffer (pH 5) was added. Nitrogen gas was bubbled into the solution at a slow rate to eliminate traces of solvent which would inhibit the enzyme. The sample was then hydrolyzed by adding 60 mg β-glucosidase (5.5 units/mg) and warmed to 37°C for 72 hrs in a shaking incubator. The liberated aglycones were extracted with three 150 ml portions of methylene chloride using a separatory funnel. The organic phase was dried over anhydrous sodium sulfate. This fraction was concentrated to 0.3 ml under a gentle stream of nitrogen gas prior to analysis by GC and GC-MS. The internal standard used for the bound fraction was triacontane.

Table II. Gas Chromatography and Mass Spectrometry Operating Conditions

GC Conditions:	
GC type:	Varian 3400
Column type:	DB-1, fused silica capillary (J&W Scientific)
Column film thickness:	0.25 μm
Column I.D.:	0.32 mm
Carrier gas:	40 psi high purity grade He, 1 ml/min (splitless mode)
Initial temp., hold:	50°C, 5 min
Rate:	2°/min
Attenuation	Range: 1×10^{-12}
Final temp., hold:	290°C, 90 min
Total time:	215 min
Data acquisition:	Output was split and recorded on a strip chart recorder as well as a VG Multichrom chromatography data acquisition system whose software resides on a MicroVax 3800.
MS Conditions:	
Instrument:	Finnigan MAT 8230 high performance gas chromatograph-mass spectrometer directly interfaced to a Varian 3400 gas chromatograph
Mode:	Electron ionization, 70 eV, 1 ma filament emission current
Accelerating voltage:	3 KV
Multiplier voltage:	1800 V corresponding to a gain of 10^6
Ion source temp.:	250°C.
Data acquisition:	Finnigan MAT SS300 data system.

Results and Discussion

The first step of this work was to choose six appropriate compounds which could be adequately identified and quantitated by GC and GC-MS. Free target compounds which were chosen by us were sedanenolide, sedanolide, falcarinone, 4-methoxypsoralen, falcarindiol and dimethoxypsoralen. The diacetylenics, especially falcarinol and falcarindiol from the added carrots, give rather nondescriptive mass spectra and were found in low concentration in the Umbelliferous drink. Identification of these in the drink was initially quite difficult. Therefore, it was decided to first analyze carrot juice, which was one of the ingredients (also supplied by NCI) of the Umbelliferous beverage. The levels in the carrot drink were naturally higher than that of the blended drink.

The diacetylenics or acetylenics are C17 straight chain ketones and alcohols with two triple bonds, found in a number of plants, including carrots (5). These diacetylenic analogs are very reactive compounds known to be produced by *Umbelliferae* and *Leguminosae*, and have the potential for modifying metabolism. A methodology for celery has been published (6) for the identification and quantitation of phthalides and coumarins, and this methodology was applied to the carrot concentrate for the diacetylenics.

Three diacetylenics were found in the carrot drink and the Umbelliferous vegetable drink. These included falcarinol, falcarindiol, and a falcarinone, which is a keto tautomer of falcarinol. The falcarinol itself was too low to quantitate (low ppb to sub-ppb level) and was hidden by an overlapping GC peak, and this meant that quantitation must be accomplished on the other two components. There is no available standard nor standard mass spectrum for falcarinone as there are for the other two, and thus, identification was accomplished on the basis of the electron impact fragmentation pattern and its similarity to the spectra of the other two analogs, and well as the chemical ionization (CI) mass spectrum.

The CI mass spectrum was obtained on the identical GC peak from the carrot concentrate, and clearly showed a strong protonated molecular ion (MH$^+$) at m/z 245. This proved the molecular weight of 244 which is correct for both falcarinol and falcarinone, and showed the molecule to be a good Brønsted base, that is, having a carbonyl moiety. As would be expected with long chain unsaturated alcohols, neither falcarinol nor falcarindiol gave strong MH$^+$ ions at m/z 245 and 261, respectively, as neither are good Brønsted bases. The structures of these compounds are shown in Figure 1.

The internal standard first tried for quantitative purposes was phthalide. A clear region of the chromatogram was picked and it was determined that phthalide was not present in the sample extract and that it would elute in the open window. Calibration curves were prepared for three psoralen analogs vs. phthalide, as these analogs can be commercially obtained.

Recovery Studies. The next step to be accomplished in the methods development portion of the study was recovery. As the exact compounds that were determined cannot be purchased as standards, model compounds of similar functionality were spiked into the Umbelliferous drink. Erucyl alcohol (cis-13-docosen-1-ol) was chosen as a surrogate standard as it elutes in a clear region on the chromatograph and is also representative of the species of interest. The level was determined versus the phthalide internal standard. Recovery was calculated as 105%. This was the average of two determinations. In a separate spiking study, erucyl alcohol had a recovery of 99% based on an average of two determinations using an external calibration method. This latter study also determined the recovery of 7-methoxy-coumarin, 2-methoxydibenzofuran, 7-methoxy-4-methylcoumarin and 13-octade-cyn-1-ol after passing the spiked drink through XAD-2 as per our method. The results for this experiment are shown in Table III.

Sedanenolide (M.W. 192)

Sedanolide (M.W. 194)

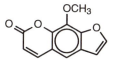

7H-Furo[3,2-g][1]benzopyran-7-one, methoxy
4-Methoxypsoralen (M.W. 216)

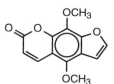

7H-Furo[3,2-g][1]benzopyran-7-one,
4,9-dimethoxy
Dimethoxypsoralen (M.W. 246)

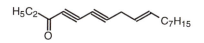

Falcarinone (M.W. 244)

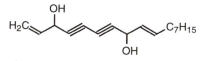

Falcarindiol (M.W. 260)

Figure 1. Structures of the free target compounds.

Table III. Recovery Experiment

Compound	Retention time (min)	recovery (%)
7-methoxycoumarin	77.91	103
2-methoxydibenzofuran	83.00	98
7-methoxy-4-methylcoumarin	87.31	109
13-octadecyn-1-ol	100.98	104
Erucyl alcohol	118.22	99

Reproducibility. The Umbelliferous drink extracts (2) then run to determine the reproducibility of the assay of the six target compounds. The two runs shown in Table IV summarize the results for the first "methods development" sample and were done about three months apart.

Table IV. Reproducibility Study on Methods Development

M.W.	Compound identified	RT	ppm	
			(Jan. 1991)	(March 1991)
134	Phthalide (internal standard)	51.29	—	—
192	Sedanenolide	77.47	3.88	3.88
194	Sedanolide	77.83	0.63	0.77
244	Falcarinone	99.85	0.04	0.02
256	4-Methoxypsoralen	112.6	0.04	0.02
260	Falcarindiol	117.60	0.03	0.02
246	4,9-Dimethoxypsoralen	128.64	0.16	0.09

Other "Free" and Bound Organics

The summary of the results for the bound fraction are given in Table V. It was noted that none of the six target compounds listed above were found in the bound fraction. Compounds that were indeed found as glycosides are listed. Table VI lists the organics found in the free fraction other than the target compounds of interest. These also were determined by GC-MS.

Conclusion

The identification and quantitation of conjugated species in fruits and vegetables is important in human biochemistry and in flavor chemistry. The human intestine has enzymes which liberate aglycones from glycosides, subsequently allowing adsorption of potentially important phytochemicals through the colon. Some of the phenolic and other compounds identified in this study have important antimutagenic activity. Ferulic acid has activity inhibiting formation of nitrosamines *in vivo* (7) and protecting DNA from electrophilic attack (8). Other phenolics inhibit formation of certain prostaglandins implicated in tumor growth (9). Indole carbinols and some homologs increase the rate of metabolism with subsequent reduction of estradiol, a compound thought important in induction of breast cancer (10).

Table V. Compounds Identified in the Bound Fraction of Umbelliferous Vegetable Juice

Compound	Spec #	Count	MW	μg/1.5 μL	ppm[1]	ppm[2]
Benzaldehyde	503	209200	106	0.213	141.8	0.97
Methionol	548	26539	108	0.027	18.0	0.12
Benzene methanol	678	378624	108	0.385	256.6	1.76
Maltol or 3-hydroxy-2-methyl-4*H*-pyran-4-one	839	202314	126	0.206	137.1	0.94
2,3-Dihydro-3,5-dihydroxy-6-methyl-4*H*-pyran-4-one	916	14768	144	0.015	10.0	0.07
Eugenol or 2-methoxy-4-(2-propenyl)-phenol	1402	110752	164	0.113	75.1	0.51
Vanillin or 3-hydroxy-4-methoxybenzaldehyde	1445	51280	152	0.052	34.8	0.24
4-Hydroxyacetophenone	1560	75456	136	0.077	51.1	0.35
Acetovanillone or 1-(4-hydroxy-3-methoxyphenyl)-ethanone	1629	191039	166	0.194	129.5	0.89
Phenylacetic acid	1752	134304	152	0.137	91.0	0.62
4-(5-Hydroxy-2,6,6-trimethyl-1-cyclohexen-1-yl)-3-buten-2-one	1908	18752	208	0.019	12.7	0.09
Homovanillic Acid or (4-hydroxy-3-methoxyphenyl)acetic acid	1959	158478	182	0.161	107.4	0.74
Eugenol isomer	2084	25424	164	0.026	17.2	0.12
Acetosyringone or 1-(4-hydroxy-3,5-dimethoxyphenyl)-ethanone	2100	131440	196	0.134	89.1	0.61
Coniferyl alcohol or γ-hydroxy isoeugenol or 3-(4-hydroxy-3-methoxyphenyl)-2-propen-1-ol	2117	131000	180	0.133	88.9	0.61
1*H*-Indole-3-ethanol	2129	183888	161	0.187	124.6	0.85
Homovanillic acid methyl ester	2141	147008	196	0.149	99.6	0.68
Hydroxycinnamic acid or 3-phenyl-2-propenoic acid	2237	19988	164	0.020	13.5	0.09
Ferulic acid or 3-(4-hydroxy-3-methoxyphenyl)-2-propenoic acid	2362	104608	194	0.106	70.9	0.49
2*H*-1-Benzopyran-2-one-7-hydroxy-6-methoxy	2467	132084	192	0.134	89.5	0.61
Palmitic acid	2572	95408	256	0.097	64.7	0.44
Linoleic acid	2828	130416	280	0.133	88.4	0.61
Stearic acid	2876	11040	284	0.011	7.5	0.05

[1]in concentrate [2]in Umbelliferous vegetable drink

**Table VI. Other Compounds Identified in the Free Fraction of
Umbelliferous Vegetable Juice**

Compound	Spec #	Count	MW	Formula	ppm
Butanal	140	510688	72	C_4H_8O	1.73
Methyl propanoate	157	344384	88	$C_3H_8N_2O$	1.17
Methylbutanone	172	524272	86	$C_5H_{10}O$	>2
Dimethylpentanol	283	524272	102	$C_6H_{14}O$	>2
Methyl methylbutanoate	307	413776	116	$C_6H_{12}O_2$	1.40
2-Furancarboxaldehyde	366	176432	96	$C_5H_4O_2$	0.60
3-Hexen-1-ol	449	140316	100	$C_6H_{12}O$	0.48
Ethylbenzene or dimethylbenzene	469	145504	106	C_8H_{10}	0.49
Benzaldehyde	636	51232	106	C_7H_6O	0.17
Methylheptanol	739	157904	126	$C_8H_{14}O$	0.53
β-Pinene	779	60976	136	$C_{10}H_{16}$	0.21
Benzene acetaldehyde	837	278855	120	C_8H_8O	0.94
Terpinene or isomer	862	145024	154	$C_{10}H_{18}O$	0.49
Acetophenone	902	26960	120	C_8H_8O	0.09
α-Pinene	938	53264	136	$C_{10}H_{16}$	0.18
Methyl benzoate	988	524272	136	$C_8H_8O_2$	>2
Terpinene or isomer	1011	23072	136	$C_{10}H_{16}$	0.08
Terpineol isomer	1047	522064	154	$C_{10}H_{18}O$	1.77
Camphor	1105	61646	152	$C_{10}H_{16}O$	0.21
Naphthalene	1130	31136	128	$C_{10}H_8$	0.11
Ethyl benzaldehyde	1138	37728	134	$C_9H_{10}O$	0.13
Methylene-1*H*-indene or isomer	1189	34000	128	$C_{10}H_8$	0.12
Terpineol isomer	1210	198102	154	$C_{10}H_{18}O$	0.67
Hydroxymethylbenzoate	1220	16464	152	$C_8H_8O_3$	0.06
Terpineol isomer	1234	131376	154	$C_{10}H_{18}O$	0.44
Estragole	1241	247728	148	$C_{10}H_{12}O$	0.84
Benzoic acid	1256	69680	122	$C_7H_6O_2$	0.24
Ethylphenylethanone	1375	41072	148	$C_{10}H_{12}O$	0.14
Methyldimethylbenzoate	1460	79327	164	$C_{10}H_{12}O_2$	0.27
Eugenol	1584	312528	164	$C_{10}H_{12}O_2$	1.06
Methylester 3-phenyl propenoic acid	1634	242928	162	$C_{10}H_{10}O_2$	0.82
Vanillin	1639	57882	152	$C_8H_8O_3$	0.20
Methyl eugenol	1687	524272	178	$C_{11}H_{14}O_2$	>2
Caryophylline or isomer	1759	108204	204	$C_{15}H_{24}$	0.37
Bergamotene or isomer	1800	41120	204	$C_{15}H_{24}$	0.14
Acetovanillinone	1823	29376	166	$C_9H_{10}O_3$	0.10
Unknown sesquiterpene	1839	16352	204	$C_{15}H_{24}$	0.06
Selinene or isomer	1890	97408	204	$C_{15}H_{24}$	0.33
Dihydroactinidiolide	1899	50544	180	$C_{11}H_{16}O_2$	0.17
Gerjunene or isomer	1910	31343	204	$C_{15}H_{24}$	0.11
BHT	1935	50835	220	$C_{15}H_{24}O$	0.17
Patchoulene or isomer	1983	155307	204	$C_{15}H_{24}$	0.53
Humulene or isomer	2006	89280	204	$C_{15}H_{24}$	0.30
Dodecanoic acid	2057	10960	200	$C_{12}H_{24}O$	0.04

Continued on next page

Table VI. Continued

Compound	Spec #	Count	MW	Formula	ppm
Dimethoxypropenylphenol	2068	34000	194	$C_{11}H_{14}O_3$	0.12
Butyl phthalide	2144	524272	194	$C_{12}H_{14}O_2$	>2
Ligustilide	2304	25680	190	$C_{12}H_{14}O_2$	0.09
Tetradecanoic acid	2423	32064	228	$C_{14}H_{28}O_2$	0.11
Psoralen	2428	13280	186	$C_{11}H_6O_3$	0.04
Isocoumarin or isomer	2452	177810	208	$C_{11}H_{12}O_4$	0.60
Scopoletin	2648	48304	192	$C_{10}H_8O_4$	0.16
Scoprone	2665	10058	206	$C_{11}H_{10}O_4$	0.03
Methyl palmitate	2692	62224	270	$C_{17}H_{34}O_2$	0.21
Palmitic acid	2763	219488	256	$C_{16}H_{32}O_2$	0.74
Piperanine	3589	305778	287	$C_{17}H_{21}NO_3$	1.04
Piperylin	3874	35056	271	$C_{16}H_{17}NO_3$	0.12
Piperine	3911	190510	285	$C_{17}H_{19}NO_3$	0.65
γ- or β-Tocopherol	4140	32800	416	$C_{28}H_{48}O_2$	0.11
Methyl tocopherol	4233	519456	430	$C_{29}H_{50}O_2$	1.76
Ergostenol	4347	33000	400	$C_{28}H_{48}O$	0.11
Stigmastadienol	4397	126976	412	$C_{29}H_{48}O$	0.43
Stigmastenol	4496	166111	414	$C_{29}H_{50}O$	0.56

Acknowledgments

The authors thank Dr. Herbert Pierson, formerly of NCI, for his comments and encouragement during the period in which the work was done and the NCI for the funding of the project. The Center For Advanced Food Technology is a New Jersey Commission on Science and Technology Center. This is Agricultural Experiment Station paper number F10569-1-93.

Literature Cited

1. Wu, P.; Kuo, M.-C.; Hartman, T. G.; Rosen, R. T.; Ho, C.-T. *J. Agric. Food Chem.* 1991, *39*,170–172.
2. Wu, P.; Kuo, M.-C.; Hartman, T. G.; Rosen, R. T.; Ho, C.-T. *Perfumer and Flavorist* 1990, *15*, 51–54.
3. Adedeji, J.; Hartman, T. G.; Rosen, R. T.; Ho, C.-T. *J. Agric. Food Chem.* 1991, *39*, 1494–1497.
4. Wall, W. E.; Taylor, H.; Perera, P.; Wani, M. C. *J. Natural Prod.* 1988, *51* 129–135.
5. Jones, E. *Chemistry in Britain* 1966, 3, 6–13.
6. Tang, J.; Zhang, Y.; Hartman, T. G.; Rosen, R. T.; Ho, C.-T. *J. Agric. Food Chem.* 1990, 38, 1937–1940.
7. Kuenzig, W.; Chan, J.; Norkus, E.; Holowaschenko, H.; Newmark, H.; Mergens, W.; Conney, A.H. *Carcinogenisis* 1984, 5, 309–313.
8. Newmark, H. L. *Nutr. Cancer* 1984, 6, 58–70.
9. Dehirst, F. E. *Prostaglandin* 1980, 20, 209–214
10. Bradlow, L. H.; Michnovicz, J. J. *J. Natl. Cancer Inst.* 1990, 82, 613–615.

RECEIVED June 7, 1993

Chapter 21

Effects of Consumption of an Umbelliferous Vegetable Beverage on Constituents in Human Sera

H. E. Sauberlich, D. S. Weinberg, L. E. Freeberg, W.-Y. Juan, T. R. Sullivan, T. Tamura, and C. B. Craig

University of Alabama at Birmingham and Southern Research Institute, Birmingham, AL 35294

A study was conducted to identify and quantify carotenoids and other phytochemicals in human sera following the consumption of a vegetable beverage prepared from Umbelliferous vegetables (carrot, celery, parsley) and tomato. Methodology was developed to quantify 5-methoxypsoralen, 8-methoxypsoralen, 5,8-dimethoxypsoralen, butylidene phthalide, 3-n-butyl phthalide, β-caryophyllene, and α-humulene in the beverage and in human sera. Fifteen adults were divided into three groups that received either 8, 16, or 24 oz of the beverage daily for three months. Blood samples were obtained regularly throughout this period. Concentrations of the indicated phytochemicals in the beverage were low and could not be detected in the serum of any of the subjects. In contrast, marked increases in serum levels of carotenoids occurred in all subjects. The degree of increase was related to the amount of beverage consumed. The higher levels of intakes, however, resulted in the development of a "bronzing-effect" on the palms of the hands and portions of the face. This effect was associated with high levels of carotenoids in the serum, particularly of α-carotene, β-carotene, γ-carotene, ζ-carotene, and lycopene. The carotenodermia disappered rapidly following the completion of the study.

The *Umbelliferae* plant family consists of over 2,000 species. The family includes edible items such as celery, parsley, dill, celeriac, lovage, parsnips, carrots, angelica root, anise, cumin, chervil, coriander, fennel, and caraway (*1*). These vegetables are a rich source of the following phytochemicals: polyacetylenes, phthalides, coumarins, alkenylbenzenes, phenolic acids, flavonoids, terpenes, and carotenoids. *Compositae* plants, such as artichoke, Jerusalem artichoke, chicory, dandelion, and sunflower (seeds) may also contain these phytochemicals. On the other hand, non-edible members of the *Umbelliferae* family are rich in potent photosensitizing substances and various toxic constituents. Edible members of the family, however, may also contain toxic constituents at low levels. Carrots and celery, for example,

0097–6156/94/0546–0258$06.00/0

contain the neurotoxin, falcarinol. Parnips root may contain photocarcinogenic furo-coumarins. Celery, parsnips, and parsley commonly contain psoralens, a group of furocoumarins. These constituents have been associated with contact dermititis and other allergic responses in the human.

Celery, celeriac, parsley, and coriander contain small amounts of phthalides. Phenolic acids, such as caffeic acid, are common constituents of *Umbelliferae*. The main flavor compounds in the *Umbelliferae* are flavonoids, terpenes, phthalides, and to some extent, coumarins (*1*). Coumarins are precursors of the furocoumarins, which are widely distributed in the *Umbelliferae*. Investigations in mice and rats have found coumarins to have anticarcinogenic effects (*1*).

The terpenoids include sesquiterpenols, such as β-caryophyllene and α-humulene. Both are present in most Umbelliferous plants. Several polyacetylenes, such as falcarinol and falcarindiol, have been found in carrots. Carrot root is a rich source of carotenoids, particularly β-carotene. It is estimated that approximately 14% of our vitamin A intake is provided by carrot root.

Numerous animal investigations and epidemiologic studies indicate that carotenoids inhibit carcinogenesis in mice and rats and may have anticarcinogenic properties for the human (*1–9*). Although studies have focused on the protective association of β-carotene with cancer, recently consideration has been given to the potential anticarcinogenic properties of other carotenoids, including those without provitamin A activity, such as lycopene and canthaxanthin (*8–11*). The studies of Khachik *et al.* (*20*) have shown that over 40 carotenoids may be available to the human from common fruits and vegetables in the diet.

Although the studies are not conclusive, there is growing evidence of an inverse association between the intake of Umbelliferous and green/yellow vegetables and the incidence of cancer (*1–4,7,8,10,11,32*). The reports have considered oral, gastric, colorectal, lung, endometrial, pancreatic, prostate, and bladder cancers.

This report is concerned with the effect of the consumption of an Umbelliferous vegetable beverage on the constituents found in human sera.

Methods

Investigations were conducted on an Umbelliferous vegetable beverage that was fed to human volunteer subjects. In the initial phase, phytochemicals were identified in the beverage and methodology developed for their quantitation. Although a considerable number of phytochemicals are present in Umbelliferous plants, few are present in amounts that can be readily measured.

The Umbelliferous Beverage.   The beverage was composed of 50% carrot juice, 10% celery juice, and 15% tomato juice from a high lycopene variety. Additional water was added. Spices that were added included black pepper, basil, paprika, all-spice, cilantro, sweet cicely, garlic, and rosemary. The beverage was provided by the National Cancer Institute in 8 oz and 16 oz containers.

Beverage Composition.   Based on evidence in the literature of their presence, certain phytochemicals were selected for studies on their identification and quantitation in the Umbelliferous vegetable beverage used. The *Umbelliferae* plants contain a number of groups of chemical constituents including coumarins, polyacetylenes, furocoumarins, alkenylbenzenes, psoralens, phenolic acids, flavonoids, terpenes, phthalides and carotenoids. This investigation was limited to the compounds listed in Table I. Their structures are presented in Figure 1.

Measurement of Beverage Components. Methyl-4-chlorobenzoate, 4-chlorobenzophenone, 1-chlorodecane, 1,4-dichlorobenzen-D4, naphthalene-D8, and phenanthrene-D10 were used as surrogates or internal standards in the identification and quantitation of the phytochemicals. The surrogates and internal standards were added to the beverage. Extracts were prepared from the beverage and aliquots were analyzed by gas chromatography/mass spectrometry. The measurements were made using a Hewlett-Packard Model 5890 gas chromatograph equipped with split/splitless and "on-column" capillary injectors. A packed-column injector was coupled to a VG 7OS high-resolution mass spectrometer by means of a direct inlet for capillary-column gas chromatography and a jet-separator for packed-column gas chromatography.

Table I. Concentration Of Identified Components In The Umbelliferous Vegetable Beverage Used

Component	Concentration (ng/g)
1. Coumarins/Furocoumarins	
5-Methoxypsoralen	194
8-Methoxypsoralen	86
5,8-Dimethoxypsoralen	240
2. Phthalides	
Butylidene phthalide	16
3-n-Butyl phthalide	820
3. Polyacetylenes	
Falcarinol	885
4. Terpenoids	
β-Caryophyllene	1560
α-Humulene	146

Concentration and Recovery of Beverage Components

The concentrations of the beverage components were quite low, ranging from 15.8 ng/g for butylidene phthalide to 1560 ng/g for β-caryophylene (Table I). Table II shows the recovery results when the Umbelliferous vegetable was spiked with known amounts of the phytochemicals, surrogates, and internal standards. Recoveries of over 80% were attained for each component. The average recovery of all the components in the spikes with repeated samples ranged from 95–99%. The high recovery and good precision obtained supported the validity of the analytical procedures. Data were analyzed with the Statistical Analysis System (SAS Institute, Inc., Cary, NC).

Human Study

After the development of methods for the measurement of selected phytochemicals, the Umbelliferous vegetable beverage was fed to adult volunteer subjects for 90 days. The study received approval from the University Institutional Review Board and informed consent was obtained from each subject.

Subjects. Fifteen healthy subjects, males and females from several ethnic groups who were non-smoking and non-obese, were selected. All subjects received physical examinations and clinical laboratory evaluations to ensure their disease-free state.

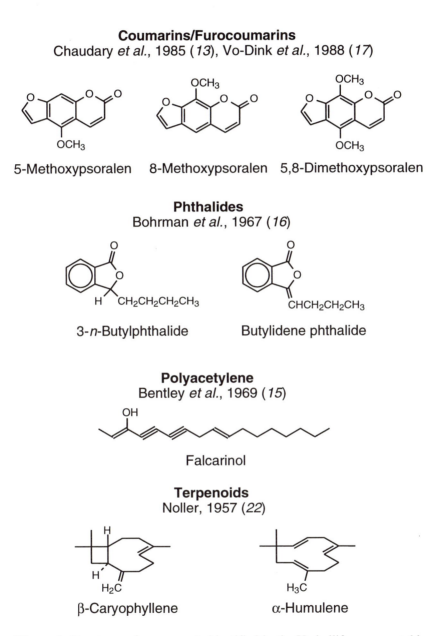

Coumarins/Furocoumarins
Chaudary *et al.*, 1985 (*13*), Vo-Dink *et al.*, 1988 (*17*)

5-Methoxypsoralen   8-Methoxypsoralen   5,8-Dimethoxypsoralen

Phthalides
Bohrman *et al.*, 1967 (*16*)

3-*n*-Butylphthalide   Butylidene phthalide

Polyacetylene
Bentley *et al.*, 1969 (*15*)

Falcarinol

Terpenoids
Noller, 1957 (*22*)

β-Caryophyllene   α-Humulene

Figure 1. Structures of components identified in the Umbelliferous vegetable beverage.

Information on the subjects and their assignments is summarized in Table III. The subjects were divided into three groups. Group A consumed 8 oz of the beverage daily at breakfast. Group B consumed 8 oz of the beverage at breakfast and 8 oz again with the evening meal. Group C consumed 8 oz of the beverage with each meal, a daily total of 24 oz.

The subjects were given a list of Umbelliferous vegetables to avoid throughout the study. The subjects were interviewed to determine the likelihood of compliance and availability for the duration of the study. The Health Habits and Diet Questionnaire developed and validated by the National Cancer Institute was used to provide additional information on personal health and dietary habits. Registered dietitians conducted 24-hour recalls periodically throughout the study and reminded the subjects as to foods to avoid during the study.

Blood samples were collected in the fasting state at 8 AM before the start of the study, after 1 day, 2 weeks, and 1, 2 and 3 months of daily beverage consumption and 1 and 7 days after stopping beverage consumption. The serum was collected and held at -72°C until analyzed.

Tolerance and Compliance. All subjects but one appeared to tolerate the consumption of the beverage with minimal difficulty. In the one subject the consumption of the 24 oz of the beverage per day produced a severe gastric upset and abdominal pains. The severity of the condition was such that the subject was required to withdraw from the study.

Subjects consuming the beverage three times a day for over two months showed a "bronzing-effect" on the palms of the hands and portions of the face. Bronzing of a lesser degree occurred in several of the subjects receiving the beverage twice each day. The coloration disappeared rapidly with the completion of the consumption of the beverage.

Subjects enrolled in the study were very cooperative with little difficulty expressed in complying with the requirements of the investigation. Compliance was considered excellent.

Results of the Human Study. The sera collected during the feeding study were analyzed for the phytochemicals observed in the Umbelliferous vegetable beverage. The methods used were the same as for the analysis of the components in the beverage. The exception was that the sera extracts were concentrated 83-fold to permit the detection of low concentrations of the components.

Target Compounds in Sera. None of the target compounds was detected in sera samples taken before a human volunteer consumed any of the Umbelliferous vegetable beverage, in sera samples taken 1 hour or 3 hours after the volunteer had consumed 8 oz of the beverage, or 24 hours, 2 weeks, 1 month, or 3 months after volunteers had consumed the vegetable beverage at a rate of 24 oz every day.

Recovery of Surrogates From Sera. The recoveries of the surrogates spiked into sera were generally in the range of 75 to 110% (Table IV). The sensitivity of the analytical methods is indicated in Table V. For example, 8-methoxypsoralen at a level of 1.25 ng/g was 98% recovered. This would suggest that when the beverage was consumed, the target phytochemical compounds were either poorly absorbed, diluted too much in the body, or were cleared too rapidly from the sera to be detected by the methodology used. Consequently, their accumulation in

Table II. Recovery Of Components Spiked Into
the Umbelliferous Vegetable Beverage

Component	Sample (ng/g)	Spiking mix (ng/g)	Found in spiked sample (ng/g)	Recovery of spike components (%)
5-Methoxypsoralen	194	250	419	90
8-Methoxypsoralen	86	80	151	81
5,8-Dimethoxypsoralen	240	147	445	119
Butylidene phthalide	16	61	83	111
3-*n*-Butyl phthalide	820	780	1631	104
Falcarinol	885	630	1607	114
β-Caryophyllen	1560	2000	3505	97
α-Humulene	146	206	346	97
4-Chlorobenzophenone	0	808	791	98
Methyl 4-chlorobenzoate	0	790	1327	168
1-Chlorodecane	0	851	796	94
Naphthalene-D8	0	790	791	100

Table III. Information On Subjects And Their Assignments

Subject No.	Age (years)	Weight (lbs/kg)	Race	Sex
Group A				
1	34	175/79.5	Caucasian	Male
2	51	145/65.8	Asian	Male
3	28	130/59.0	Caucasian	Female
4	27	142/64.5	Caucasian	Female
5	26	133/60.4	Caucasian	Female
Group B				
6	28	107/48.6	Asian	Female
7	28	170/77.2	Black	Female
8	39	218/99.0	Caucasian	Male
9	27	143/64.9	Caucasian	Male
10	32	101/45.9	Caucasian	Female
Group C				
11	39	150/68.1	Caucasian	Male
12	26	130/59.0	Caucasian	Female
13	39	143/64.9	Black	Female
14	34	190/86.3	Caucasian	Female
15	31	125/56.8	Caucasian	Female

Table IV. Recovery Of Surrogates Spiked Into Sera

Component	Recovery of surrogates (%)			
	Sample 1[a]	Sample 2[b]	Sample 3[b]	Sample 4[b]
4-Chlorobenzophenone	104	102	93.5	101
Methyl 4-chlorobenzoate	74.1	98.2	91.3	103
1-Chlorodecane	64.0	90.9	84.5	96.1
Naphthalene-D8	59.0	86.2	79.0	85.0

[a] Sample 1 was serum from a subject 3 hrs after consuming 8 oz of beverage.
[b] Samples 2–4 were serum samples from 3 different subjects after consuming 24 oz of the beverage daily for 3 months.

Table V. Recovery Of Components Spiked Into Sera

Component	Concentration in sample (ng/g)	Spiking mix (ng/g)	Spike components found (ng/g)	Recovery of spike components (%)
5-Methoxypsoralen	0	3.90	4.42	113
8-Methoxypsoralen	0	1.25	1.23	98
5,8-Dimethoxypsoralen	0	2.30	2.31	101
Butylidene phthalide	0	0.945	1.11	117
3-*n*-Butyl phthalide	0	12.2	12.3	101
β-Caryophyllene	0	31.2	25.9	83
α-Humulene	0	3.22	2.79	87
4-Chlorobenzophenone	0	40.3	42.0	104
Methyl 4-chlorobenzoate	0	39.4	39.4	100
1-Chlorodecane	0	42.5	39.2	92
Naphthalene-D8	0	40.3	36.1	90

the blood did not occur. It is known that orally administered 8-methoxypsoralen is metabolized rapidly, with over 95% excreted within 24 hours (*1*).

Vitamin Levels in Sera. The consumption of the beverage had little effect on vitamin A and vitamin E levels in the sera (Figure 2). The slight increase in the average serum vitamin A levels with time on the beverage was not significant. One subject in Group A started the study with a low serum vitamin A level (0.32 µg/ml). The level remained unchanged throughout the study despite the high intake of provitamin A carotenoids provided by the daily intake of 8 oz (237 ml) of the beverage. One subject in Group B also had a low serum vitamin A level (0.31 µg/ml) that increased slightly throughout the study. One subject in Group A had unusually high levels of vitamin E in the plasma (21.0 µg/ml), which distorted the average vitamin E levels for this group.

Plasma cholesterol and triglyceride levels were determined for each subject. This information was utilized to evaluate the vitamin E status of each subject according to the procedures of Horwitt (*19*) and Thurnham *et al.* (*18*). The vitamin E status was considered acceptable for all subjects. The subject in Group A with the high vitamin E levels was observed to have moderately elevated serum levels of triglycerides and cholesterol (575 mg/dl and 276 mg/dl, respectively).

Carotenoids. Prolonged consumption of high amounts of carotinoids can induce carotenemia and carotenodermia in the human. As a result of the bronzing observed in a number of the subjects, the Umbelliferous vegetable beverage was analyzed for carotenoids. Upon high performance liquid chromatography (HPLC) analysis (*20,21,24–26*), appreciable amounts of α-carotene, β-carotene, and lycopene were found in the beverage (Table VI). Addi-

Table VI. Level of Carotenoids Present in the Umbelliferous Beverage

Component	Concentration (µg/ml)	Amount in 8 oz. (mg)
β-Carotene	62.1	14.7
α-Carotene	28.0	6.6
Lycopene	21.9	5.2
Lutein	2.5	0.6
Zeaxanthin	0.25	0.06

tional carotenoids were observed but were present in lesser amounts. Consequently, each 8 ounces of the beverage consumed provided approximately 15 mg of β-carotene and 7 mg of α-carotene. Subjects on the high intake of the beverage would have ingested daily approximmately 45 mg of β-carotene and 20 mg of α-carotene.

The carotenoids that have been commonly observed in human sera with pro-vitamin A activity are β-carotene, α-carotene, γ-carotene, cryptoxanthin, β-apo-8′-carotenal (C_{30}), and those without pro-vitamin A activity are lutein (xanthophyll), zeaxanthin, canthaxanthin, lycopene, neurosporene, ζ-carotene, phytofluene, phytoene (*20,21,24–29*). Echinenone, which has pro-vitamin A activity, is rare in nature (it is found in sea urchin), therefore, its presence in sera is unlikely. Consequently, echinenone may be used as an internal standard in the HPLC analyses for carotenoids (*26*). The structures of the more common carotenoids are depicted in Figures 3A and B.

The serum carotenoid levels before and after the ingestion of the Umbelliferous vegetable beverage for two months are summarized in Table VII. Intake of the beverage resulted in an increased concentration in sera of all of the carotenoids measured, except of cryptoxanthin, which remained relatively

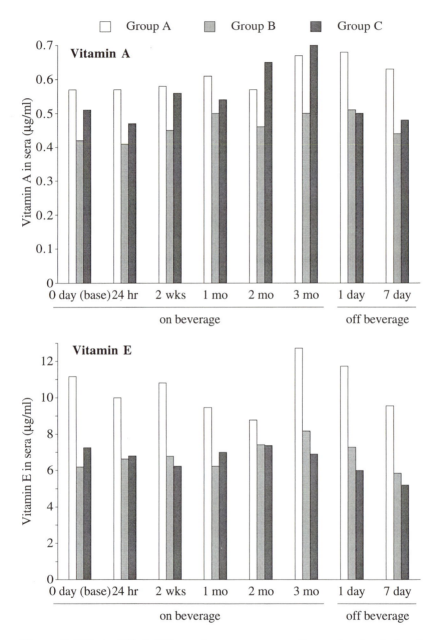

Figure 2. Effect of Umbelliferous beverage consumption on levels of vitamins A and E in sera.

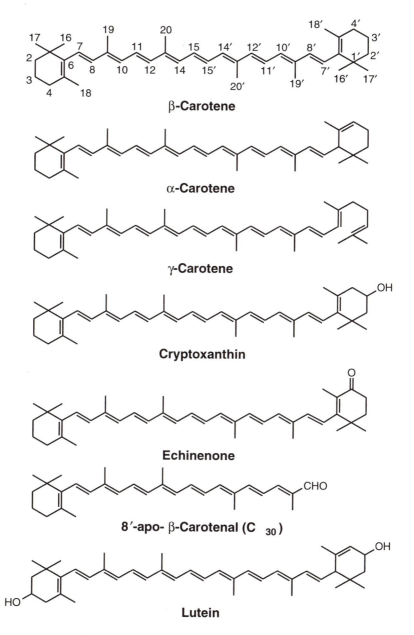

Figure 3A. Structures of common carotenoids.

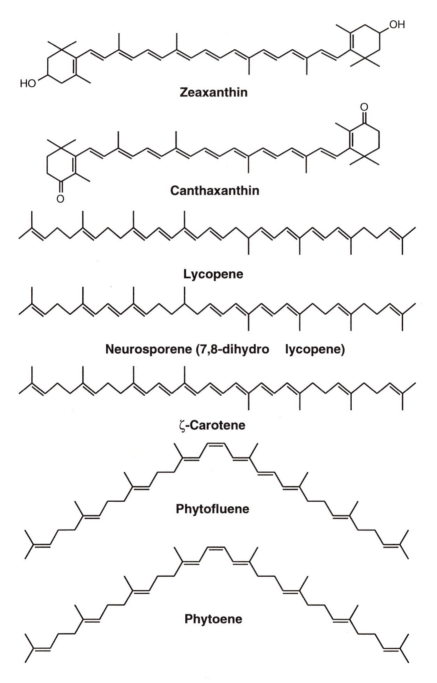

Figure 3B. Structures of common carotenoids.

unchanged. The highest percentage increase (1171%) occurred in α-carotene in Group C, which received the beverage with each meal. Although the beverage provided amounts of lycopene nearly equal to that of α-carotene, the percentage increase in serum concentrations of lycopene was considerable less than that for α-carotene. This may be a reflection of the relatively high initial plasma concentrations of lycopene. Maximum absoption for lycopene may have been reached.

Table VII. Serum Carotenoid Levels Before and After Intake of a Umbelliferous Beverage

Carotenoid	Group A Initial	Group A 2 mos	Group B Initial	Group B 2 mos	Group C Initial	Group C 2 mos
α-Carotene	5.5	29.5 (436)	25.0	73.3 (193)	4.2	53.4 (1171)
β-Carotene	22.6	53.9 (138)	34.0	79.6 (134)	16.9	81.8 (384)
γ-Carotene	10.6	38.9 (267)	16.3	47.5 (191)	11.0	53.3 (384)
ζ-Carotene	10.7	32.1 (300)	18.1	52.6 (190)	11.4	57.7 (406)
Lycopene	47.6	49.5 (40)	23.0	39.8 (73)	38.0	63.4 (67)
Lutein/Zeaxanthin	26.1	37.9 (45)	23.0	43.9 (90)	27.5	38.8 (41)
Phytofluene	18.3	64.8 (254)	15.6	95.5 (512)	14.5	110.0 (658)
Phytoene	8.6	25.0 (191)	8.1	27.0 (133)	8.1	29.6 (265)
β-Cryptoxanthin	9.8	11.2 (14)	9.4	8.0 (-15)	9.3	12.2 (31)
α-Cryptoxanthin	4.5	3.5 (-22)	4.4	3.5 (-20)	4.4	3.8 (-13)
Anhydrolutein	8.4	13.2 (57)	6.5	17.6 (171)	11.0	17.8 (61)
Neurosporene	2.6	7.4 (185)	5.2	9.8 (88)	3.0	6.7 (123)

[a]μg/dl, value in parentheses is the percent change from intial time to two months

Serum levels of the carotenoids for Group B and Group C were quite comparable (Table VII). Since bronzing appeared in subjects of Group C after 2 months on the beverage, these serum carotenoid levels could be considered associated with the induction of the carotenodermia. The carotenodermia also occurred in the women in Group B after 3 months of consumption of the beverage at the intermediate level (16 oz/day). The consumption for 3 months of only 8 oz per day of the beverage, however, produced marked increases in most of the carotenoids producing a carotenemia state. For Group A, the serum levels of β-carotene and α-carotene after 3 months of consumption of the beverage had not reached the levels of Group B and C. The α-carotene and β-carotene levels for both Groups B and C had largely plateaued after one month on the beverage. Thus, the intake of about 22 mg per day of α-carotene and β-carotene combined for 3 months was tolerated without evidence of carotenodermia. Presumably α-carotene, β-carotene, and γ-carotene were the major contributors in the production of the carotenodermia. Within one week after discontinuence of the study, the carotenemia had largely disappeared.

Phytofluene appears to be readily absorbed. The significance of a high serum level of phytofluene metabolically or as a potential anti-carcinogen remains unknown.

A number of other investigators have studied the effects of β-carotene supplements on serum β-carotene concentrations (*12,23,27,29,33*). Other factors, such as smoking, may effect the serum concentrations of β-carotene (*23-25,29,30*). Only a few reports have been published on the effects of dietary intake of carotenoid-containing foods on the serum concentrations of the various carotinoids (*12,20,21,29*). If it becomes established that the consumption of carotinoids is associated with a reduced risk of certain cancers, a beverage of the type used in this study could serve as a convenient source of these compounds.

Acknowledgements

We appreciate the cooperation and interest of the volunteer subjects who made this study possible. We are indebted to Gary R. Beecher of the Beltsville Human Nutrition Research Center, USDA, Beltsville, MD, for assistance in the identification and quantitation of carotinoids in serum. We wish to thank Neal Craft of the National Institute of Standards and Technology for corroboration of carotenoid analyses of the Umbelliferous vegetable beverage. We also wish to thank Hemmige Bhagavan of Hoffmann-La Roche, Nutley, NJ, for providing us with various reference carotinoid samples. The assistance of Harry Vaughn in the preparation of the figures is greatly appreciated. This research was supported in part by National Cancer Institute Contract No 1-CN-05278.

Literature Cited

1. Diet, Nutrition, and Cancer Program *Umbelliferae: Biomedical Literature Review and Analysis*; Division of Cancer Prevention and Control, National Cancer Institute: Bethesda, MD, 1987.
2. Ross, A.C. Evaluation of publicly available scientific evidence regarding certain nutrient-disease relationships: 8A. Vitamin A and Cancer Report of the Life Sciences Research Office, FASEB, Bethesda, MD, December, 1991.
3. Garewal, H. S.; Meyskens, F., Jr. *Cancer Epidemiology, Biomarkers & Prevention* **1992**, *1*, 155–159.
4. Ziegler, R. G. *J. Nutr.* **1989**, *119*, 116–122.
5. Schwartz, J.; Suda, D.; Shklar, G. *Biochem. Biophys. Res. Commun.* **1986**, *136*, 1130–1135.
6. Shklar, G; Schwartz, J.; Trickler, D.; Reid, S. *Nutr. Cancer* **1989**, *12*, 321–325.
7. Willett, W. C. *Nutr. Revs.* **1990**, *48*, 201–211.
8. VanEenwyk, J; Davis, F. G.; Bowen, P. E. *Int. J. Cancer* **1991**, *48*, 34–38.
9. Grubbs, C. J.; Eto, I.; Juliana, M. M.; Whitaker, L. M. *Oncology* **1991**, *48*, 239–245.
10. Micozzi, M. S.; Beecher, G. R.; Taylor, P. R.; Khachik, F. *J. Natl. Cancer Inst.* **1990**, *82*, 282–285.
11. Colditz, G. A.; Branch, L. G.; Lipnick, R. J.; Willet, W. C.; Rosner, B.; Posner, B. M.; Henneckens, C. H. *Am. J. Clin. Nutr.* **1985**, *41*, 32–36.
12. Micozzi, M. S.; Brown, E. D.; Edwards, B. K.; Bieri, J. G.; Taylor, P. R.; Khachik, F.; Beecher, G. R.; Smith, J. C., Jr. *Am. J. Clin Nutr.* **1992**, *55*, 1120–1125.
13. Chaudhary, S. K.; Ceska, O.; Warrington, P. J.; Ashwood-Smith, M. J. *J. Agric. Food Chem.* **1985**, *33*, 1153–1157.
14. Bjeldanes, L. F.; Kim, I-S. *J. Food Sci.* **1978**, *43*, 143.

15. Bentley, R. K.; Bhattacharjee, D.; Jones, E. R. H.; Thaller, V. *J. Chem. Soc.* **1969**, (*C*), 685–688.
16. Bohrmann, H; Stahl, E.; Mitsubashi, H. *Chem. Pharm. Bull.* **1967**, *15*(10), 1606.
17. Vo-Dink, T.; White, D. A.; O'Malley, M. A.; Seliginann, P. J.; Beier, R. *J. Agric. Food Chem.* **1988**, *36*, 333.
18. Thurnham, D. I.; Davies, J. A.; Crump, B. J.; Situnayake, R. D.; Davis, M. *Am. Clin. Chem.* **1986**, *23*, 514–520.
19. Horwitt, M. K.; Harvey, C. C.; Dahm, C. H.; Searcy, M. T. *Ann. NY Acad. Sci.* **1972**, *203*, 223–236.
20. Khachik, F.; Beecher, G. R.; Goli, M. B. *Pure Applied Chem.* **1991**, *63*, 71–80.
21. Khachik, F; Beecher, G. R.; Goli, M. B.; Lusby, W. R.; Smith, J. C., Jr. *Anal. Chem.* **1992**, *64*, 2111–2122.
22. Noller, C.R. *Chemistry of Organic Compounds*, 2nd ed.; W. B. Saunders Company: Philadelphia, 1957.
23. Nierenberg, D. W.; Stukel, T. A.; Baron, J. A.; Dain, B. J.; Greenberg, E. R. *Am. J. Epidemiol.* **1989**, *30*, 511–521.
24. Stacewicz-Sapuntzakis, M.; Bowen, P. E.; Kikendall, J. W.; Burgess, M. *J. Micronutrient Analysis* **1987**, *3*, 27–45.
25. Cantilena, L. R.; Stukel, T. A.; Greenberg, E. R.; Nann, S.; Nierenberg, D. W. *Am. J. Clin. Nutr.* **1992**, *55*, 659–663.
26. Bieri, J. G.; Brown, E. D.; Smith, J. C., Jr. *J. Liquid Chromatog.* **1985**, *8*, 473–484.
27. Krinsky, N. I.; Russett, M. D.; Handelman, G. J.; Max, D. *J. Nutr.* **1990**, *120*, 1654–1662.
28. Nierenberg, D. W.; Nann, S. L. *Am. J. Clin. Nutr.* **1992**, *56*, 417–426.
29. Micozzi, M. S.; Brown, E. D.; Taylor, P. R.; Wolfe, E. *Am. J. Clin. Nutr.* **1988**, *48*, 1061–1064.
30. Dimitrov, N. V.; Meyer, C.; Ullrey, D. E.; Chenoweth, W.; Michelakis, A; Malone, W.; Boone, C.; Fink, G. *Am. J. Clin. Nutr.* **1988**, *48*, 298–304.
31. Costantino, J. P.; Kuller, L. H.; Begg, L.; Redmond, C. K.; Bates, M. W. *Am. J. Clin. Nutr.* **1988**, *48*, 1277–1283.
32. Menkes, M. S.; Comstock, G. W.; Vuilleumier, J. P.; Heising, K. J.; Rider, A. A.; Brookmeyer, R. *N. Engl. J. Med.* **1986**, *315*, 1250–1254.
33. Nierenberg, D. W.; Stukel, T. A.; Baron, J. A.; Dain, B. J.; Greenberg, E. R. *Am. J. Clin. Nutr.* **1991**, *53*, 1443–1449.

RECEIVED July 6, 1993

Chapter 22

Chlorophyllin: An Antigenotoxic Agent

T. Ong[1], H. E. Brockman[2], and W-Z. Whong[1]

[1]Division of Respiratory Disease Studies, National Institute for Occupational Safety and Health, Morgantown, WV 26505
[2]Department of Biological Sciences, Illinois State University, Normal, IL 61761

Chlorophyllin, an aqueous soluble chemical derived from chlorophyll, has been shown to inhibit the mutagenic activity of certain chemicals. Studies have been performed in our laboratories to determine the antimutagenic activity of chlorophyllin against complex mixtures by using bacterial mutagenesis assays. The results indicate that chlorophyllin, at non-toxic concentrations, totally or almost totally inhibits the mutagenic activity of the extracts of 10 dietary and environmental complex mixtures. The antimutagenic activity appears to be heat stable. Results of a comparative study show that chlorophyllin is a more effective antimutagen against complex mixtures than retinol, β-carotene, vitamin C, or vitamin E. With the morphological transformation assay in cultured mammalian cells, chlorophyllin also was found to inhibit the transforming activity of several carcinogens and complex mixtures tested in our laboratories. It appears, based on our results and those from other studies, that chlorophyllin is a potent antigenotoxic agent.

The human population is exposed to various genotoxic agents of both man-made and natural origin. For example, genotoxic agents are found in airborne particles (1,2), diesel engine emission (3), cigarette smoke (4), tobacco (5), beverages, and foods (6). Through the effort of regulatory agencies and the scientific community, the level of mutagens and carcinogens may be reduced in our environment, but it is unlikely that exposure of humans to these agents can be completely eliminated. The consequence to humans of exposure to mutagens is not known in most cases. Nevertheless, damage to or alteration of DNA by environmental genotoxic agents is likely to be a major cause of cancer and genetic disorders, and may contribute to heart disease, aging, and developmental birth defects (7). As there is no known safe concentration or threshold limit, there should be increased efforts to identify agents

0097–6156/94/0546–0272$06.00/0

that can counteract or eliminate the activity of genotoxic agents in our diet and environment.

Various agents, including vitamins (such as A, C, and E), glutathione, propyl gallate, retinyl acetate, germanium oxide, β-carotene, selenium, uric acid, phenol, cinnamaldehyde, cobaltous chloride and coumarin, have been shown to inhibit the mutagenic and/or carcinogenic activity of certain chemicals (6,8–10). Chlorophyll, the green pigment of vegetables and other plants, has been shown to be responsible for most of the antimutagenic activity of certain vegetable extracts (11–14). A sample of commercial chlorophyll containing a and b is known to inhibit the genotoxicity of 3-amino-1-methyl-5H-pyrido[4,3-b]indole, a carcinogenic heterocyclic amine found in certain foods (15). Both chlorophylls a and b are also known to inhibit the mutagenicity of 4-nitro-o-phenylenediamine (16) and cigarette-smoke condensate (17).

Unlike chlorophyll, chlorophyllin (Figure 1), a sodium and copper derivative of chlorophyll, is water soluble. It has been shown to inhibit the mutagenic activity of several known mutagens (12–21), meat extract (22), amino acid pyrolysis products (23), and cigarette smoke condensate (17). Since humans are exposed mainly to environmental and dietary complex mixtures rather than single chemicals, studies have been conducted in our laboratories to: (l) determine the antimutagenic activity of chlorophyllin against a variety of such mixtures; (2) compare the antimutagenic activity of chlorophyllin, vitamins, and related compounds against mutagenic complex mixtures; and (3) determine the antitransforming activity of chlorophyllin against selected carcinogens and complex mixtures. The results are summarized in this report. The data related to antimutagenicity and comparative antimutagenicity studies have been published previously (24–26).

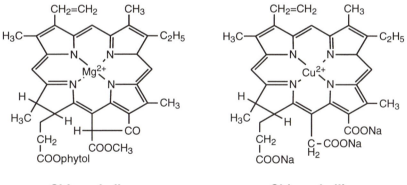

Chlorophyll a   **Chlorophyllin**

Figure 1. Chemical structures of chlorophyll and chlorophyllin.

Antimutagenic Activity Against Environmental and Dietary Complex Mixtures

The Ames Salmonella reverse mutation (27) and the arabinose-resistant forward mutation (28) assays were used as the test systems. The Ames assay is the most commonly used test system for the detection of mutagens and antigenotoxic agents. Both bacterial test systems are sensitive and easy to perform. The complex mixtures used for the study included diesel emission particles (DEP), airborne particles (AP), coal dust (CD), cigarette smoke (CS), tobacco snuff (TS), chewing tobacco (CT), fried shredded pork (FSP), fried beef (FB), red grape juice (RGJ), red wine (RW), and black pepper (BP). These complex mixtures were selected for the study because they are closely related to our daily life. The solvent extract of each test sample was prepared according to reported procedures (24–26). The CD and CT extracts were nitrosated by sodium nitrite in an acidic environment and the TS extract was acidified before use in the antimutagenesis assay. Tester strain TA98 (27) or SV50 (29) of Salmonella typhimurium was treated with a predetermined mutagenic concentration of each extract and different concentrations of chlorophyllin. The standard plate incorporation assay with and without in vitro metabolic activation as recommended by Ames et al. (27) was used for the assay procedure.

The results from the Ames assay (Figure 2) indicate that chlorophyllin inhibited the mutagenic activity of each of the complex mixtures in a dose-dependent manner. At the concentration of 1.25 mg per plate, it completely inhibited the mutagenic activity of extracts of AP, CS, FSP, and FB. Chlorophyllin also strongly inhibited (92-98%) the mutagenic activity of extracts of DEP, CD, RGJ, and RW, but at a concentration greater than 1.25 mg per plate. The only complex mixtures whose mutagenicities were inhibited less than 90% by chlorophyllin were extracts of TS and CT, which were inhibited 75-80% by 10 mg of chlorophyllin per plate. In the forward-mutation assay, chlorophyllin at the concentration of 2.5 mg per plate inhibited 60 to 100% of the mutagenic activity of extracts of AP, CD, DEP, TS, and BP (Table I). Chlorophyllin, at 10 mg per plate, the highest concentration tested, was neither mutagenic nor toxic to S. typhimurium.

Table I. Antimutagenic Activity of Chlorophyllin Against Extracts of Selected Complex Mixtures in the Ara^r Assay System with Tester Strain SV50 of Salmonella typhimurium

Complex Mixture (amount/plate)	Amount of chlorophyllin per plate (mg)					
	0	0.31	0.63	1.25	2.5	5.0
Solvent control	96[a]	103	106	93	95	72
AP (20.9 mg)	719	370 (54)	261 (73)	231 (76)	189 (84)	204 (77)
CD (0.167 mg)	459	NT[b]	114 (98)	91 (100)	84 (100)	74 (99)
DEP (0.30 mg)	387	206 (64)	146 (86)	123 (90)	102 (98)	NT
TS (0.05 mg)	449	NT	NT	218 (65)	185 (75)	128 (84)
BP (740.0 mg)	864	576 (35)	473 (49)	396 (58)	372 (62)	298 (69)

SOURCE: Adapted from ref. 26.
[a] Number of mutants per plate (percent inhibition).
[b] NT, not tested.

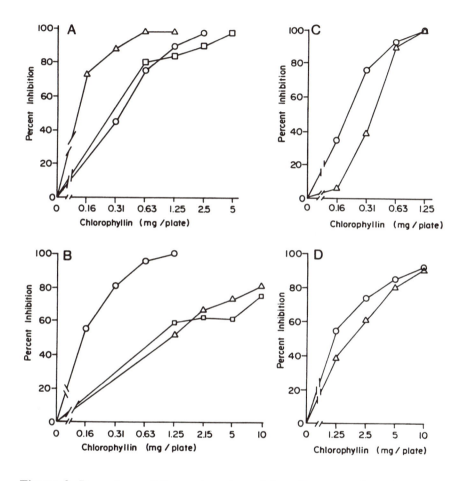

Figure 2. Percentage of the mutagenic activity of ten complex mixtures inhibited by chlorophyllin. (A) Diesel emission particles, O; airborne particles, Δ; coal dust, ◻. (B) Cigarette smoke, O; tobacco snuff, Δ; chewing tobacco, ◻. (C) Fried shredded pork, O; fried beef, Δ. (D) Red grape juice, O; red wine, Δ. (Adapted from ref. 24).

To determine whether the antimutagenic property of chlorophyllin is heat resistant, a sample of chlorophyllin was heated for 10 min at 100°C. The antimutagenic activities of the heated and unheated chlorophyllin were similar when tested with the extract of DEP (Table II).

Table II. Antimutagenic Activity of Heated and Non-heated Chlorophyllin Against the Extract of Diesel Emission Particles in *Salmonella typhimurium* TA98

Chlorophyllin concentration (mg/plate)	DEP (mg/plate)	Revertants per plate	Percent inhibition
Without heat treatment			
0.00	0.0	9	
0.00	0.5	415	
1.25	0.5	115	72.3
2.50	0.5	69	83.4
5.00	0.5	22	94.7
5.00	0.0	6	
With heat treatment[a]			
0.00	0.5	415	
1.25	0.5	107	74.2
2.50	0.5	28	86.0
5.00	0.5	28	93.3
5.00	0.0	5	

[a]Chlorophyllin was heated at 100°C for 10 min.

A reconstruction experiment also was performed to determine whether the reduction in the number of revertants is due to inhibition of chlorophyllin against the mutagenic activity of complex mixtures and not against the growth of TA98 revertants. Different numbers of cells from DEP induced revertants were plated along with the usual 0.1 ml of an overnight culture of TA98 in the absence or presence of 5 mg chlorophyllin per plate. The revertant colonies were counted after 2 days of incubation. The results showed that the recovery of revertants was close to 100% at the different numbers of revertants plated in the absence or presence of chlorophyllin (Table III). These results clearly indicate that the decreased number of revertants observed in the previous experiments was due to inhibition of the mutagenic activity of the complex mixture and not to inhibition of cell growth by chlorophyllin.

Comparative Antimutagenicity Studies

Several vitamins and their related compounds have been shown to possess anticarcinogenic and/or antigenotoxic activities (*6,9,10,30*). β-carotene and vitamins C and E are known to be antioxidants and radical scavengers. They are capable of inhibiting mutations caused by oxygen radicals. Vitamins C and E are known to inhibit the formation of genotoxic nitroso compounds. Inhibition of mutagenicity by vitamin A and its natural and synthetic analogues, the retinoids, may be due to alteration of the metabolism of chemicals by these compounds.

Table III. Effect of Chlorophyllin on the Growth of
***Salmonella typhimurium* TA98 Revertants**

Number of revertants plated	Revertant colonies recovered	
	without chlorophyllin	with 5 mg chlorophyllin per plate
200	192	202
300	304	333
600	556	601
1300	1306	1340

Studies have been performed in our laboratories, using the Ames plate incorporation assay, to compare the inhibitory activity of chlorophyllin, vitamins C and E, retinol, and β-carotene against the mutagenicity of solvent extracts of CD, DEP, AP, FB, and TS. Tester strain TA98 was treated with a predetermined mutagenic concentration of each extract in the presence of different concentrations of each antigenotoxic agent.

As summarized in Figure 3, the maximum antimutagenic activity of chlorophyllin against the 5 complex mixtures studied is much higher (69-94% inhibition with the concentrations tested) than that of retinol, β-carotene, or vitamin C or E. Retinol is the only other compound that showed a significant inhibition against all 5 complex mixtures. Based on the same concentration (1.72 μmol/plate), however, the percentage of inhibition caused by retinol, in all cases, was only about one-half of that caused by chlorophyllin. β-Carotene and vitamin E inhibited less than 20%, if any, of the mutagenicity of the complex mixtures tested, with the exception of AP, which was inhibited 39% by 3.45 μmol β-carotene/plate. None of the mutagenic activities of the 5 complex mixtures was inhibited by vitamin C. In the case of AP, vitamin C enhanced the mutagenic activity by as much as 60%. The reason for this enhancement is not known. With the same mutagenesis assay system, vitamin C has been reported to increase the mutagenic activities of several genotoxic chemicals (see ref. 25 for references).

Antitransformation Studies

Several established mammalian cell lines, such as BALB/3T3 cells, can grow as a monolayer in a culture flask or dish. When the culture reaches confluence, the cells cease dividing. Due to genetic and/or epigenetic changes, however, some cells lose their contact inhibition property and continue to divide, resulting in cells piling up and forming foci. Such a change, referred to as morphological transformation, has been used as an endpoint for the detection of potential carcinogens. Many genotoxic agents and carcinogens have been shown to induce morphological transformation in BALB/3T3 cells (*31*). Studies have been performed in our laboratories to determine whether chlorophyllin can inhibit the transforming

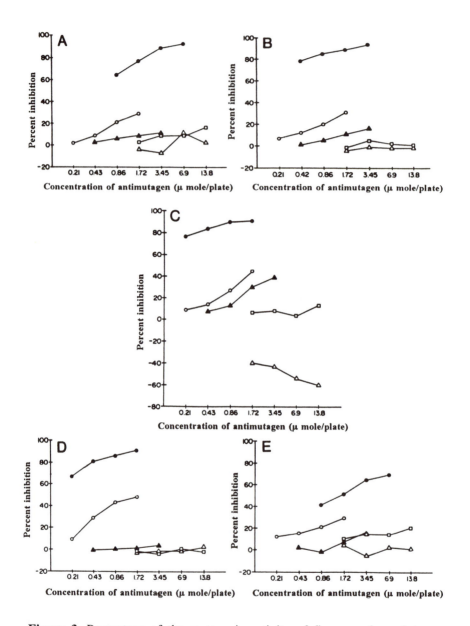

Figure 3. Percentage of the mutagenic activity of five complex mixtures inhibited by chlorophyllin and four other antigenotoxic agents. (A) coal dust, (B) diesel emission particles, (C) airborne particles, (D) fried beef, (E) tobacco snuff. Chlorophyllin, ●; retinol, O; β-carotene, ▲; vitamin C, △; vitamin E, □. (Reproduced with permission from ref. 25. Copyright 1989 Elsevier.)

activity of benzo[*a*]pyrene (BaP), *N*-methyl-*N'*-nitro-*N*-nitrosoguanidine (MNNG), 3-methylcholanthrene (MCA), 7,12-dimethylbenz[*a*]anthracene (DMBA), aflatoxin B_1 (AFB_1), CD, CT, and TS.

The procedure for performing the morphological transformation assay followed that reported in the literature (*31*). Two-day old cultures were exposed to a predetermined concentration of each selected carcinogen or complex mixture with or without chlorophyllin (150 µg/ml) for 72 hr. After washing, cells were grown in a low-serum medium with medium change every 3 days. Four weeks following treatment, the cultures (20 flasks/treatment) were fixed with methanol, stained with crystal violet, and scored for the presence of transformed foci. The detailed information regarding the antitransformation studies of chlorophyllin against morphological transformation induced by selected carcinogens and complex mixtures will be published elsewhere.

Chlorophyllin at the concentration of 150 µg/ml showed no significant influence on either plating efficiency or cell growth of BALB/3T3 cells at the density (2 x 10^4 cells/flask) used for the transformation assay. Results of the antitransformation studies are summarized in Tables IV and V. Chlorophyllin (150

Table IV. Antitransforming Activities of Chlorophyllin Against Selected Carcinogens in the BALB/3T3 Transformation Assay

Compound	Concentration (µg/ml)	Foci/flask mean ± SD	Percent inhibition[a]
Solvent control	0	0.07 ± 0.09	
Chlorophyllin	150	0.32 ± 0.15	
BaP	0.2	7.73 ± 0.13	
BaP + chlorophyllin	0.2 + 150	3.24 ± 0.28	61
MNNG	2.5	4.03 ± 0.19	
MNNG + chlorophyllin	2.5 + 150	0.35 ± 0.22	99
MCA	5.0	3.21 ± 0.24	
MCA + chlorophyllin	5.0 + 150	0.89 ± 0.25	80
AFB_1	0.1	1.72 ± 0.20	
AFB_1 + chlorophyllin	0.1 + 150	0.56 ± 0.19	83
DMBA	0.25	2.96 ± 0.10	
DMBA + chlorophyllin	0.25 + 150	1.05 ± 0.24	72

[a] $100 - \left[\dfrac{\text{foci/flask in the presence of chlorophyllin}}{\text{foci/flask in the absence of chlorophyllin}} \right] \times 100$

The number of background foci/flask (0.32) was subtracted from the numerator and denominator.

**Table V. Antitransforming Activities of Chlorophyllin Against
Three Complex Mixtures in the BALB/3T3 Transformation Assay**

Complex mixture	Concentration (mg/ml)	Foci/flask mean ± SD	Percent inhibition
Distilled water	0	0.04 ± 0.07	
DMSO	5 μl/ml	0.23 ± 0.14	
Chlorophyllin	0.15	0.32 ± 0.15	
CD	4.16	3.57 ± 0.29	
CD + chlorophyllin	4.16 + 0.15	1.01 ± 0.29	79
CT	1.76	1.97 ± 0.33	
CT + chlorophyllin	1.76 + 0.15	0.80 ± 0.23	71
TS	1.4	3.65 ± 0.23	
TS + chlorophyllin	1.4 + 0.15	0.68 ± 0.22	89

μg/ml) inhibited the transforming activity of all the carcinogens tested. The percent of inhibition ranged from 61% for BaP to 99% for MNNG. Chlorophyllin, at the same concentration, also inhibited (>70%) the transforming activity of the three complex mixtures tested. These results indicate that chlorophyllin is also an effective antitransforming agent.

Conclusion

Chlorophyllin, a derivative of chlorophyll, is highly effective against the mutagenicity of various environmental and dietary complex mixtures. It also is effective against the transforming activity of the selected carcinogens and complex mixtures. The antigenotoxic activity of chlorophyllin appears to be heat stable. Based on the five complex mixtures studied, the antigenotoxic activity of chloro-phyllin is higher than that of retinol, β-carotene, vitamin C, and vitamin E. The mechanism(s) by which chlorophyllin inhibits the genotoxicity of carcinogens and complex mixtures is not known. It may scavenge radicals and/or interact with the genotoxic agent to form a complex. Further studies are needed to understand the mechanisms of chlorophyllin's antigenotoxicity and to determine whether chloro-phyllin is also an effective anticarcinogen.

Literature Cited

1. Chrisp, C. E.; Fisher, G. L. *Mutation Res.* **1980**, *76*, 143–164.
2. Hughes, T. J.; Pellizzari, E.; Little, L.; Sparacino, C.; Kolber, A. *Mutation Res.* **1980**, *76*, 51–83.
3. *Health Effect of Diesel Engine Emissions: Proceedings of an International Symposium;* Pepelko, W. E.; Danner, R. M.; Clarke, N. A., Eds; EPA: Washington, DC, 1980; Vol. 1.
4. DeMarini, D. M. *Mutation Res.* **1983**, *114*, 59–89.
5. Whong, W-Z.; Ames, R. G.; Ong, T. *J. Toxicol. Environ. Health* **1984**, *14*, 491–496.
6. Ames, B. N. *Science* **1983**, *221*, 1256–1264.
7. Ames, B. N. *Science* **1979**, *204*, 587–593.
8. Kada, T.; Kaneko, K.; Matsuzaki, S.; Matsuzaki, T.; Hara, Y. *Mutation Res.* **1985**, *150*, 127–132.
9. Hayatsu, H.; Arimoto, S.; Negishi, T. *Mutation Res.* **1988**, *202*, 429–446.
10. Hartman, P. E.; Shankel, D. M. *Environ. Mol. Mutagen.* **1990**, *15*, 145–182.
11. Barale, R.; Zucconi, D.; Bertani, R.; Loprieno, N. *Mutation Res.* **1983**, *120*, 145–150.
12. Kimm, S.; Park, S.; Kang, S. *Korean J. Biochem.* **1982**, *14*, 1–8.
13. Lai, C. *Nutrition and Cancer* **1979**, *1*, 19–21.
14. Lai, C., Butler, M. A.; Matney, T. S. *Mutation Res.* **1980**, *77*, 245–250.
15. Negishi, T.; Arimoto, S.; Nishizaki, C.; Hayatsu, H. *Carcinogenesis* **1989**, *10*, 145–149.
16. Gentile, J. M.; Gentile, G. J. *Mutation Res.* **1991**, *250*, 79–86.
17. Terwel, L.; van der Hoeven, J. C. M. *Mutation Res.* **1985**, *152*, 1–4.
18. Arimoto, S.; Negishi, T.; Hayatsu, H. *Cancer Lett.* **1980**, *11*, 29–33.
19. Bronzetti, G.; Galli, A.; Croce, D. M. In *Antimutagenesis and Anticarcinogenesis Mechanisms II*; Kuroda, Y.; Shankel, D. M.; Waters, M. D., Eds.; Plenum Press: New York, 1990; pp 463–468.
20. Dashwood, R. H.; Breinholt, V.; Bailey, G. S. *Carcinogenesis* **1991**, *12*, 939–942.
21. Katoh, Y.; Nemoto, N.; Tanaka, M.; Takayama, S. *Mutation Res.* **1983**, *121*, 153–157.
22. Münzner, R. *Fleischwirtschaft* **1981**, *61*, 1586–1588.
23. Arimoto, S.; Ohara, Y.; Namba, T.; Negishi, T.; Hayatsu, H. *Biochem. Biophys. Res. Commun.* **1980**, *92*, 662–668.
24. Ong, T.; Whong, W-Z.; Stewart, J. D.; Brockman, H. E. *Mutation Res.* **1986**, *173*, 111–115.
25. Ong, T.; Whong, W-Z.; Stewart, J. D.; Brockman, H. E. *Mutation Res.* **1989**, *222*, 19–25.
26. Warner, J. R.; Nath, J.; Ong, T. *Mutation Res.* **1991**, *262*, 25–30.
27. Ames, B. N.; McCann, J.; Yamasaki, E. *Mutation Res.* **1975**, *31*, 347-363.
28. Pueyo, C. *Mutation Res.* **1978**, *64*, 249–258.
29. Whong, W-Z.; Stewart, J. D.; Ong, T. *Environ. Mutagen.* **1981**, *3*, 95–99.
30. Ramel, C.; Alekperov, U. K.; Ames, B. N.; Kada, T.; Wattenberg, L. W. *Mutation Res.* **1986**, *168*, 47–65.
31. Matthews, E. J. *J. Tissue Cult. Methods* **1986**, *3*, 157–164.

RECEIVED July 27, 1993

Chapter 23

Dietary Cytochrome P-450 Modifiers in the Control of Estrogen Metabolism

Jon J. Michnovicz[1] and H. Leon Bradlow[2]

[1]The Institute for Hormone Research, 145 East 32nd Street, 10th Floor, New York, NY 10016
[2]Strang Cancer Prevention Center, 428 East 72nd Street, New York, NY 10021

Extensive studies of the cytochrome P450 enzymes, and of the related system of Phase II drug-detoxifying enzymes, have pointed the way to understanding how dietary components affect the metabolism of drugs and hormones in humans. Estrogens are metabolized by P450 enzymes to produce both 2-hydroxyestrogens and 16α-hydroxyestrogens. Many phytochemicals in our diets are capable of altering the relative amounts of these metabolites formed in a given individual. Since formation of different estrogen metabolites is linked to breast and uterine cancer risk in women, several natural products may exert their beneficial effects through a modification of hormone metabolism. Our studies have focused on the structure-activity relationship between dietary indoles derived from cabbage-type vegetables and estrogen P450-dependent metabolism. We propose that phytochemicals active in these P450 pathways may be useful as chemopreventive agents in humans.

A battery of inducible enzymes is known to exist in humans and other animals, the general purpose of which is to metabolize and/or detoxify both endogenous and exogenous chemical compounds. These metabolizing enzymes fall into two broad groups. The first consists of the superfamily of "Phase I" monooxygenase enzymes, collectively called cytochrome P450, which all act similarly in oxidizing a substrate compound (1–3). Several hundred isozymes of P450 are known to exist in mammals. Endogenous chemical substrates, such as steroids and prostaglandins, are metabolized by P450 enzymes, the former including corticosteroids and sex steroids, such as estradiol. Exogenous chemicals metabolized by P450 enzymes include toxic foreign substances, e.g., dioxin or cigarette smoke, and pharmacologic and dietary drugs and chemicals.

The "Phase II" system refers to a more loosely knit collection of enzymes, which generally function to produce less reactive, more water-soluble compounds from those initially metabolized by the Phase I system. This family of enzymes includes, among others, glutathione S-transferase, UDP-glucuronyl transferase,

0097–6156/94/0546–0282$06.00/0
© 1994 American Chemical Society

NADPH:quinone oxidoreductase, sulfotransferase, and epoxide hydrolase (*4–7*). Nebert has shown that these two enzyme families, Phase I and Phase II, often work together to reduce the toxicity of foreign chemicals (*8*).

Cytochrome P450 Induction

Numerous drugs or foreign chemicals, when ingested (or inhaled in the case of cigarette smoke), have been observed to increase, or "induce," the activity of these drug-detoxifying enzymes, thereby leading to an enhanced capacity of the animal to inactivate and excrete the inducing drug or chemical. While some chemicals elicit a widespread enzyme induction of both Phase I and Phase II enzymes (*8–10*), others elicit predominantly a Phase II enzyme pattern following exposure (*11–13*).

Three enzyme families within the cytochrome P450 superfamily are particularly inducible in humans: P450I (prototype inducer: polycyclic hydrocarbon compounds, e.g., 3-methylcholanthrene), P450II (prototype inducer: phenobarbital), and P450III (prototype inducer: synthetic corticosteroids, such as dexamethasone). P450I enzymes are induced following the interaction of a specific cellular receptor protein (the Ah receptor) with the inducing chemical. Thus, agents binding the Ah receptor are usually good inducers of P450I. Specific receptor proteins for the P450II and P450III systems have not yet been identified (*1,2,14*).

Estradiol 2- and 16α-Hydroxylases are P450-Dependent Enzymes

Two pivotal metabolic steps involving the sex steroid estradiol are C-2 and C-16α hydroxylation, both of which are performed by isozymes of P450 (Figure 1). It is now known that each of the three inducible P450 families described above (i.e., P450I, P450II, and P450III) is capable of performing estradiol C-2 hydroxylation (*15–19*). The relative activities of these three enzyme groups are largely unknown for a given individual. The P450 isozyme(s) performing 16α-hydroxylation in humans are not well understood at this time.

The monooxygenase enzymes catalyzing 2-hydroxylation and 16α-hydroxylation are often capable of metabolizing the same substrates, estradiol (E_2) and estrone (E_1). Therefore, competition exists between the two enzyme pathways for a limited pool of estrogens within the cells where these enzymes are located. This fact becomes particularly important when only one of the two P450 enzyme pathways is selectively induced, following exposure to the inducers discussed above. Increased activity of one enzyme pathway will "steal" the available estrogen substrate from the other pathway. The practical consequence is that greater activity in only one of the two enzymes (e.g., C-2) will be reflected by an apparent decrease in the activity of the other pathway (C-16α).

C-2/C-16α Estradiol Metabolism is Linked to Hormone-Dependent Diseases

Research over the past decade has shown that activity of the P450 estrogen metabolizing pathways is critically linked to hormone-dependent disease states in women. Studies indicate C-16α hydroxylation is higher in women with breast cancer and endometrial cancer (*20,21*). Some preliminary evidence exists that it is also higher in women whose mothers had breast cancer, and are themselves at higher risk (*22*). The enzyme product of this pathway, 16α-hydroxyestrone (16-OHE_1), has the ability to bind covalently to proteins and DNA (*23–26*), thereby

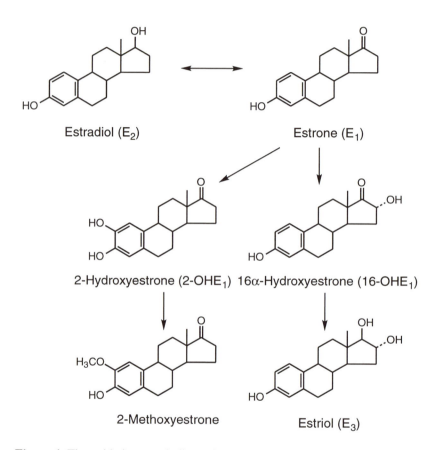

Figure 1. The oxidative metabolism of estrogens. The two competing pathways of C-2 (left) and C-16α (right) hydroxylation are shown.

augmenting or prolonging its action as a hormone (27). 16-OHE$_1$ has also been found to be mutagenic in some assays (28).

The other principal enzyme metabolite, 2-hydroxyestrone (2-OHE$_1$), has properties opposite to 16-OHE$_1$ (29,30). 2-OHE$_1$ behaves in breast and uterine tissues as an inactive metabolite, and is actually anti-estrogenic in cell culture experiments (31,32).

It is now clear that C-16α hydroxylation leads to metabolites that promote estrogen action, while C-2 hydroxylation produces metabolites opposing estrogen action. Increases or decreases in either pathway generally have a consistent effect upon disease risk. Higher 16α-hydroxylation or lower 2-hydroxylation appears to elevate risk, while lower 16α-hydroxylation or higher 2-hydroxylation seems to reduce risk.

Several examples of this principle can be cited. Elevated C-16α activity is associated with increased mouse mammary tumors (33), and with higher rates of breast and endometrial cancer in women (see above). The C-16α pathway is elevated by high-fat diets and decreased by low-fat diets (34–36), consistent with the observed association of dietary fat and breast cancer. C-16α hydroxylation is associated with ras oncogene activity, as well as human papilloma virus activity (37,38).

In contrast, C-2 hydroxylation activity is greater in many settings where breast and uterine cancer risk is lower. Thinness elevates C-2 activity, while obesity suppresses it (39,40). In postmenopausal women, excessive body weight is linked to breast cancer risk (41). Aerobic exercise elevates the C-2 pathway, as does cigarette smoking (42,43), and in the former case, exercise is clearly linked to lower incidence of breast cancers in women (44,45). Smoking has been repeatedly observed to reduce the risk of uterine carcinoma, an estrogen-dependent disease (46,47). In addition, smoking is generally observed to be anti-estrogenic in other tissues, such as bone (48).

The effect of smoking on breast cancer risk, however, is less clear and merits further discussion. Some reports have documented a protective effect of cigarette smoking against breast cancer, while others have not (47,49). Two possible reasons for this discrepancy in the effect of smoking on breast and endometrial cancer risk have been put forward. First, because breast cancer has a long latency period, researchers may not be able to optimally evaluate the protective effects of cigarette smoke in their subjects.

Second, it is known several exogenous chemicals accumulate in high concentration in breast fluids (50). Thousands of chemical substances are produced in cigarette smoke (51), many of which are absorbed and potentially able to accumulate in the fluids bathing breast epithelial cells. Since many smoke compounds are known carcinogens, their presence in the breast may counteract the anti-estrogenic (protective) effect of elevated 2-hydroxylation. In fact, one such compound in smoke, benzo[a]pyrene, has been shown to elevate ras oncogene protein expression in cultured breast cell organelles (52).

The C-2 pathway has also been shown to be very sensitive to pharmacologic control. It is easily suppressed by the P450 inhibitor cimetidine (53), and increased by drugs such as thyroxine (54).

The concept has gradually emerged that relative activities of the C-2 to C-16α pathways help to determine the risk for hormone-dependent cancers and other diseases (20,55,56). The greater the C-2/C-16α ratio, the lower the risk for breast and uterine cancer, as well as endometriosis and fibrocystic breast disease.

We have discovered that ratios of urinary estrogen metabolites representing these two pathways can serve as a biomarker for the relative activity of the two pathways. The preferred urinary metabolite ratio used to date is 2-hydroxyestrone relative to estriol, or $(2\text{-}OHE_1)/(E_3)$, as published in several studies (57–60). We are currently refining our ability to measure other relevant metabolites, and thus determine the utility of other ratios.

Because the ratio of C-2/C-16α activities may be important physiologically, efforts have been focused on elevating this ratio, either by increasing C-2 activity, or by inhibiting C-16α enzyme activity. The former approach has so far proved more successful (56), because of the involvement of multiple inducible P450 pathways (e.g., P450I, P450II, P450III) in this enzyme step, as discussed above.

Cruciferous Vegetables as Sources of Natural Enzyme Inducers

Early researchers interested in identifying inducers of the cytochrome P450 system focused their attention not only on exogenous drugs and chemicals, but also on the many components of our diets. Studies performed in the early 1970's showed that cruciferous vegetables, a family of common vegetables comprising cabbage, broccoli, brussels sprouts, and others, were capable of inducing P450 activity in animals when added to an experimental diet ($61,62$). This proved to be true also in human clinical studies, in which drug-metabolizing activity was elevated following a brief period of dietary supplementation with cruciferous vegetables ($63,64$).

In detailed cancer prevention studies, it was discovered that adding cruciferous vegetables to the diet produced a marked decrease in the incidence of breast tumors in animals. This was shown for dimethyl[*a*]benzanthracene (DMBA)-induced tumors ($65,66$), as well as MCF-7 breast tumor metastases in athymic mice (67). The added cruciferous vegetables were also found to be protective against the induction of tumors in animals under various experimental conditions, including aflatoxin exposure (68–71).

Similarly, studies of cruciferous vegetable consumption by humans showed that a greater amount of these vegetables in the diet was very strongly associated with protection against cancers. Steinmetz and Potter recently reviewed this problem, and showed that cruciferous vegetables were the one family of vegetables whose elevated consumption was most frequently associated with reduced cancer rates in humans ($72,73$). The epidemiologic data are particularly strong for protection against colon cancer and lung cancer, and, to a lesser extent, breast cancer.

Cruciferous vegetables were soon demonstrated to contain three groups of chemicals likely to be associated with the cancer preventive properties. These included indole-3-carbinol (I3C) and related indole compounds; phenylethyl isothiocyanate and related isothiocyanates; and dithiolthione and other thiol-containing compounds ($74,75$). These substances are derived from precursor compounds in the vegetable termed glucosinolates, one of which, glucobrassicin (Figure 2), leads to the formation of indoles and other sulfur-containing compounds (76).

The indoles themselves were found to be effective inducers of both cytochrome P450 and Phase II enzymes (77–86). An important indole in cruciferous vegetables, I3C, was demonstrated to produce much of the same protection against cancer in animals previously ascribed to the whole cruciferous vegetable. This included reduction of DMBA-induced mammary tumors and protection against a host of different tumors produced experimentally in other animals.

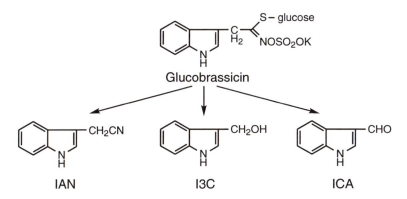

Figure 2. Formation of indole-3-carbinol (I3C) in cruciferous vegetables. I3C and other indole-containing compounds are formed from the enzymatic breakdown of glucobrassicin.

Indole-3-Carbinol is an Estradiol 2-Hydroxylase Inducer

Because of the strong positive association of I3C with cancer reduction in animals, as well as the important protective effects of cruciferous vegetables in human populations, we undertook a series of novel experiments to determine the effects of pure I3C on estrogen metabolism in healthy men and women, particularly as an inducer of estrogen 2-hydroxylase.

The initial experiment evaluated the use of I3C for a short period of 7–10 days at a moderate dose (6 mg/kg/day), using an *in vivo* radiometric procedure as evidence of altered 2-hydroxylation. We discovered that I3C is indeed a powerful inducer of the C-2 pathway in both men and women, affecting nearly every research subject studied (Figure 3) (*60,87*). In animal testing involving mice with a normally high rate of spontaneous breast tumors, I3C was found both to elevate C-2 hydroxylation and to lead to reduced breast tumor formation (*88*).

Subsequent studies showed that I3C increased P450I activity in breast cells in cell culture (Figure 4). In addition, the I3C compound was shown to compete with other polycyclic hydrocarbons for binding to the Ah receptor (unpublished studies). The chemistry of I3C in aqueous solutions is quite complex, with various condensation products formed under acidic conditions such as those found in gastric secretions (*89–92*). In fact, in addition to I3C, the dimer diindolylmethane and a product termed indolo(3,2-b)carbazole also bound avidly to the Ah receptor (*82,93*).

In further studies, I3C was demonstrated to elevate the ratio of C-2/C-16α estrogen metabolites in urine samples of men and women (*60*). This discovery suggested that the biological effects of I3C and related compounds could be monitored directly by studies of urinary estrogen measurements. We documented this fact in a larger extended clinical trial, utilizing a three-way placebo-controlled study of I3C in groups of 20 healthy young women studied for three consecutive months. We found, using a radioimmunoassay technique, that an elevation of the $2OHE_1/E_3$ ratio was produced by I3C; there was no change in the placebo group. The effect of I3C lasted throughout the trial (Table I).

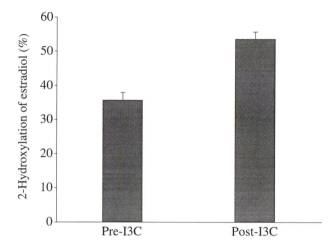

Figure 3. Elevation of 2-hydroxylation by I3C. 2-Hydroxylation tested in 12 men and women rose from an average of $34.3 \pm 4.8\%$ to $52.8 \pm 6.3\%$ following a 7-day course of treatment with 6–7 mg/kg/day.

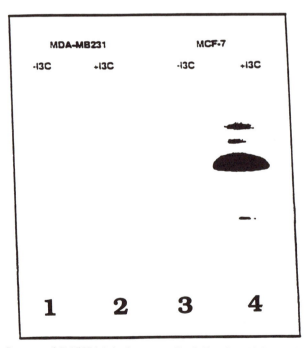

Figure 4. Increased P450IA1 in breast cells following I3C. In the estrogen-receptor positive (ER+) MCF-7 cell line following I3C, Western immunoblots showed marked increase in a band corresponding to P450IA1. The ER- cell line, MDA-MB231 showed no increase with I3C.

Table I. Elevated 2-Hydroxylation in Women Following Exposure to I3C

Group	[2OHE$_1$]/[E$_3$] Baseline	After 3 months	p value
Indole-3-carbinol	0.72 ± 0.31	1.19 ± 0.57	<0.001
Placebo	0.73 ± 0.34	0.76 ± 0.34	NS

Urine samples from this 3-month study were subsequently analyzed by gas chromatography/mass spectrometry. This technique measures all urinary estrogen metabolites in human urine samples. The results clearly indicated that I3C not only significantly elevated 2-hydroxylated estrogens, but also severely reduced the levels of other active estrogens in the urine, including 16OHE$_1$, estrone, estradiol, and estriol (Table II).

Most of our studies to date have been carried out using a dose of 350–500 mg/day of the I3C (usually around 6–7 mg/kg/day). All available data thus far indicate oral consumption of I3C is generally well tolerated, without side effects in women and men, in doses up to 500–1000 mg/day.

Table II. Changes in Various Urinary Estrogens in Women Following I3C

Metabolite	Pre	Post	p value
2-Hydroxyestrone	3.33 ± 1.97	6.62 ± 3.63	0.03
2-Hydroxyestradiol	0.82 ± 0.46	1.26 ± 0.49	0.05
Estradiol	0.76 ± 0.64	0.46 ± 0.21	NS
Estrone	2.09 ± 0.93	1.46 ± 0.52	0.02
16α-Hydroxyestrone	0.57 ± 0.26	0.42 ± 0.36	0.04
Estriol	1.05 ± 0.44	0.77 ± 0.37	0.05

All metabolites expressed as nmol steroid/mmol creatinine.

Other Natural Modifiers of P450-Dependent Metabolism of Estrogens in Humans

We have explored other natural sources of phytochemicals that affect the P450 enzymes metabolizing estradiol. For example, we studied the effects of adding a polyphenolic extract derived from green tea (GTE) to the diets of BALB/c mice on hepatic estradiol 2-hydroxylation. This extract has been studied extensively by other workers as a source of chemopreventive polyphenolic compounds (*94,95*). We found that it was capable of increasing the activity of mouse liver 2-hydroxylase (Figure 5).

Further studies are under way of the effects of this and other complex phytochemical mixtures, such as licorice root extract, on 2-hydroxylase and other P450 isozymes. The possibility exists that these complex mixtures contain phytochemicals with unique effects on steroid metabolizing enzymes. Further research will be needed to fully understand the role played by these phytochemicals in human metabolism and their effects on risk for breast and other cancers.

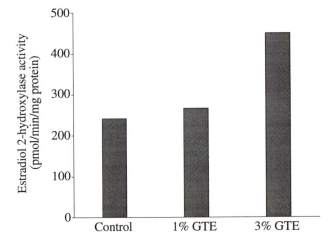

Figure 5. Metabolic effects of green tea extract (GTE). Estrogen metabolism was measured radiometrically with [2-^{3}H]-estradiol in mouse hepatic microsomes following two concentrations of GTE. A moderate increase was observed at the higher concentration (3% GTE).

Literature Cited

1. Gonzalez, F. J. *Pharmacol. Rev.* **1989**, *40*, 243–288.
2. Okey, A. B. *Pharmacol. Ther.* **1990**, *45*, 241–298.
3. Guengerich, F. P. *Crit. Rev. Biochem. Mol. Biol.* **1990**, *25*, 97–153.
4. Bock, K. W. *Crit. Rev. Biochem. Mol. Biol.* **1991**, *26*, 129–150.
5. Falany, C. N. *Trends Pharmacol. Sci.* **1991**, *12*, 255–259.
6. Boyer, T. D. *Hepatology* **1989**, *9*, 486–496.
7. De Long, M. J.; Prochaska H. J.; Talalay, P. *Proc. Natl. Acad. Sci. USA* **1986**, *83*, 787–791.
8. Nebert, D. W.; Petersen D. D.; Fornace, A. J., Jr.*Environ. Health Perspect.* **1990**, *88*, 13–25.
9. Bock, K. W.; Wiltfang, J.; Blume, R.; Ullrich, D.; Bircher, J. *Eur. J. Clin. Pharmacol.* **1987**, *31*, 677–683.
10. Jaiswal, A. K.; McBride, O. W.; Adesnik, M.; Nebert, D. W. *J. Biol. Chem.* **1988**, *263*, 13572–13578.
11. Chen, L. H.; Shiau, C. A. *Anticancer Res.* **1989**, *9*, 1069–1072.
12. Spencer, S. R.; Wilczak, C. A.; Talalay, P. *Cancer Res.* **1990**, *50*, 7871–7875.
13. Prochaska, H. J.; Talalay, P. *Cancer Res.* **1988**, *48*, 4776–4782.

14. Chung, F-L.; Morse, M. A.; Eklind, K. I.; Lewis, J. *Cancer Epidemiol. Biomarkers Prev.* **1992**, *1*, 383–388.
15. Gierthy, J. F.; Lincoln, D. W., II; Kampcik, S. J.; Dickerman, H. W.; Bradlow, H. L.; Niwa, T; Swaneck, G. E. *Biochem. Biophys. Res. Commun.* **1988**, *157*, 515–520.
16. Graham, M. J.; Lucier, G. W.; Linko, P.; Maronpot, R. R.; Goldstein, J. A. *Carcinogenesis* **1988**, *9*, 1935–1941.
17. Yang, C. S.; Yoo, J-S. H. *Pharmacol. Ther.* **1988**, *38*, 53–72.
18. Ball, S. E.; Forrester, L. M.; Wolf, C. R.; Back, D. J. *Biochem. J.* **1990**, *267*, 221–226.
19. Waxman, D. J. *Biochem. Pharmacol.* **1988**, *37*, 71–84.
20. Schneider, J.; Kinne, D.; Fracchia, A.; Pierce, V.; Anderson, K. E.; Bradlow, H. L.; Fishman, J, *Proc. Natl. Acad. Sci. USA* **1982**, *79*, 3047–3051.
21. Fishman, J.; Schneider, J.; Hershcopf, R. J.; Bradlow, H. L. *J. Steroid Biochem. Mol. Biol.* **1984**, *20*, 1077–1081.
22. Osborne, M. P.; Karmali, R. A.; Hershcopf, R. J.; Bradlow, H. L.; Kourides, I. A.; Williams, W. R.; Rosen, P. P.; Fishman, J. *Cancer Invest* **1988**, *8*, 629–631.
23. Bucala, R.; Fishman, J.; Cerami, A. *Proc. Natl. Acad. Sci. USA* **1982**, *79*, 3320–3324.
24. Swaneck, G. E.; Fishman, J. *Proc. Natl. Acad. Sci. USA* **1988**, *85*, 7831–7835.
25. Yu, S. C.; Fishman, J. *Biochem.* **1985**, *24*, 8017–8021.
26. Miyairi, S.; Ichikawa, T.; Nambara, T. *Steroids* **1991**, *56*, 361–366.
27. Lustig, R. H.; Mobbs, C.V.; Pfaff, D. W.; Fishman, J. *J Steroid Biochem Mol. Biol.* **1989**, *33*, 417–421.
28. Telang, N. T.; Suto, A.; Wong, G. Y.; Osborne, M. P.; Bradlow, H. L. *J. Natl. Cancer Inst.* **1992**, *82*.
29. Martucci, C. P.; Fishman, J. *Endocrinol.* **1979**, *105*, 1288–1292.
30. Fishman, J.; Martucci, C. *J. Clin. Endocrinol. Metab.* **1980**, *51*, 611–615.
31. Schneider, J.; Huh, M. M.; Bradlow, H. L.; Fishman, J. *J. Biol. Chem.* **1984**, *259*, 4840–4845.
32. Vandewalle, B.; Lefebvre, J. *Mol. Cell Endocrinol.* **1989**, *61*, 239–246.
33. Bradlow, H. L.; Hershcopf, R. J.; Martucci, C. P.; Fishman, J. *Proc. Natl. Acad. Sci. USA* **1985**, *82*, 6295–6299.
34. Longcope, C.; Gorbach, S.; Goldin, B.; Woods, M.; Dwyer, J.; Morrill, A; Warram, J. *J. Clin. Endocrinol. Metab.* **1987**, *64*, 1246–1249.
35. Musey, P. I.; Collins, D. C.; Bradlow, H. L.; Gould, K. G.; Preedy. J. R. K. *J. Clin. Endocrinol. Metab.* **1987**, *65*, 792–795.
36. Moller, H.; Lindvig, K.; Klefter, R.; Mosbech, J.; Moller Jensen, O. *Gut* **1989**, *30*, 1558–1562.
37. Telang, N. T.; Narayanan, R.; Bradlow, H. L.; Osborne, M. P. *Br. Cancer Res. Treat.* **1991**, *18*, 155–163.
38. Auborn, K. J.; Woodworth, C.; DiPaolo, J. A.; Bradlow, H. L. *Int. J. Cancer* **1991**, *49*, 867–869.
39. Fishman, J.; Boyar, R. M.; Hellman, L. *J. Clin. Endocrinol. Metab.* **1975**, *41*, 989–991.
40. Schneider, J.; Bradlow, H. L.; Strain, G.; Levin, J.; Anderson, K.; Fishman, J. *J. Clin. Endocrinol. Metab.* **1983**, *56*, 973–978.
41. Helmrich, S. P.; Shapiro, S.; Rosenberg, L. *Am. J. Epidemiol.* **1983**, *117*, 35–45.

42. Snow, R. C.; Barbieri, R. L.; Frisch, R. E. *J. Clin. Endocrinol. Metab.* **1989**, *69*, 369–376.
43. Michnovicz, J. J.; Hershcopf, R. J.; Naganuma, H.; Bradlow, H. L.; Fishman, J. *N. Engl. J. Med.* **1986**, *315*, 1305–1309.
44. Frisch, R. E.; Wyshak, G.; Albright, N. L.; Albright, T. E.; Schiff, I.; Jones, K. P.; Witschi, J.; Shiang, E.; Koff, E.; Marguglio, M. *Cancer* **1985**, *52*, 885–891.
45. Frisch, R. E.; Wyshak, G.; Witschi, J.; Albright, N. L.; Albright, T. E.; Schiff, I. *Int. J. Fertil.* **1987**, *32*, 217–225.
46. Lesko, S. M.; Rosenberg, L.; Kaufman, D. W.; Helmrich, S. P.; Miller, D. R.; Strom, B.; Schottenfeld, D.; Rosenshein, N. B.; Knapp, R. C.; Lewis, J.; Shapiro, S. *N. Engl. J. Med.* **1985**, *313*, 593–596.
47. Baron, J. A.; La Vecchia, C.; Levi, F. *Am. J. Obst. Gynecol.* **1990**, *162*, 502–514.
48. Kiel, D. P.; Baron, J. A.; Anderson, J. J.; Hannan, M. T.; Felson, D. T. *Ann. Intern. Med.* **1992**, *116*, 716–721.
49. Baron, J. A. *Am. J. Epidemiol.* **1984**, *119*, 9–22.
50. Wrensch, M. R.; Petrakis, N. L.; Gruenke, L. D.; Miike, R.; Ernster, V. L.; King, E. B.; Hauck, W. W.; Craig, J. C.; Goodson, W. H., III *Cancer Res.* **1989**, *49*, 2168–2174.
51. Stedman, R. L. *Chemical Reviews* **1968**, *68*, 153–207.
52. Telang, N. T.; Axelrod, D. M.; Bradlow, H. L.; Osborne, M. P. *Ann. N. Y. Acad. Sci.* **1990**, *586*, 70–78.
53. Galbraith, R. A.; Michnovicz, J. *N. Engl. J. Med.* **1989**, *321*, 269–274.
54. Michnovicz, J. J.; Galbraith, R. A. *Steroids* **1990**, *55*, 22–26.
55. Hershcopf, R. J.; Bradlow, H. L. *Am. J. Clin. Nutr.* **1987**, *45*, 283–289.
56. Bradlow, H. L.; Michnovicz, J. J. *Proc. R. Soc. Edinburgh* **1989**, *95B*, 77–86.
57. Michnovicz, J. J.; Naganuma, H.; Hershcopf, R. J.; Bradlow, H. L.; Fishman, J. *Steroids* **1988**, *52*, 69–83.
58. Michnovicz, J. J., Hershcopf, R. J.; Haley, N. J.; Bradlow, H. L.; Fishman, J. *Metabolism* **1989**, *38*, 537–541.
59. Michnovicz, J. J.; Galbraith, R. A. *Metabolism* **1991**, *40*, 170–174.
60. Michnovicz, J. J.; Bradlow, H. L. *Nutr. Cancer* **1991**, *16*, 59–66.
61. McDanell, R.; McLean, A. E. M.; Hanley, A. B.; Heaney, R. K.; Fenwick, G. R. *Food Chem. Toxicol.* **1987**, *25*, 363–368.
62. Hendrich, S.; Bjeldanes, L. F. *Food Chem. Toxicol.* **1983**, *21*, 479–486.
63. Pantuck, E. J.; Pantuck, C. B.; Garland, W. A.; Min, B. H.; Wattenberg, L. W.; Anderson, K. E.; Kappas, A.; Conney, A. H. *Clin. Pharmacol. Ther.* **1979**, *25*, 88–95.
64. Pantuck, E. J.; Pantuck, C. B.; Anderson, K. E.; Wattenberg, L. W.; Conney, A. H.; Kappas, A. *Clin. Pharmacol. Ther.* **1984**, *35*, 161–169.
65. Stoewsand, G.; Anderson, J. L.; Munson, L. *Cancer Lett.* **1988**, *39*, 199–207.
66. Stoewsand G. S.; Anderson, J. L.; Munson, L.; Lisk, D. J. *Cancer Lett.* **1989**, *45*, 43–48.
67. Scholar, E. M.; Wolterman, K.; Birt, D. F.; Bresnick, E. *Nutr. Cancer* **1989**, *12*, 121–126.
68. Bresnick, E.; Birt, D. F.; Wolterman, K.; Wheeler, M.; Markin, R. *Carcinogenesis* **1990**, *11*, 1159–1163.
69. Dashwood, R. H.; Arbogast, D. N.; Fong, A. T.; Pereira, C.; Hendricks, J. D.; Bailey, G. S. *Carcinogenesis* **1989**, *10*, 175–181.

70. Boyd, J. N.; Babish, J. G.; Stoewsand, G. S. *Food Chem. Toxicol.* **1982**, *20*, 47–52.
71. Fong, A. T.; Hendricks, J.D.; Dashwood, R.H.; Van Winkle, S.; Lee, B. C.; Bailey, G. S. *Toxicol. Appl. Pharmacol.* **1988**, *96*, 93–100.
72. Steinmetz, K. A.; Potter, J. D. *Cancer Causes Cont.* **1991**, *2*, 325–357.
73. Steinmetz, K. A.; Potter, J. D. *Cancer Causes Cont.* **1991**, *2*, 427–442.
74. Virtanen, A. *Angew. Chem. Int. Ed. Engl.* **1962**, *1*, 299–301.
75. Wattenberg, L. W. *J. Natl. Can. Inst.* **1974**, *52*, 1583–1586.
76. Sones, K.; Heaney, R. K.; Fenwick, G. R. *J. Sci. Food Agric.* **1984**, *35*, 712–720.
77. Loub, W. D.; Wattenberg, L. W.; Davis, D. W. *J. Natl. Can. Inst.* **1975**, *54*, 985– 988.
78. Bradfield, C. A.; Bjeldanes, L. *Food Chem. Toxicol.* **1984**, *22*, 977–982.
79. Arcos, J. C.; Myers, S. C.; Neuburger, J.; Argus, M. F. *Cancer Lett.* **1980**, *9*, 161–167.
80. Shertzer, H. G. *Toxicol. Appl. Pharmacol.* **1982**, *64*, 353–361.
81. Babish, J. G.; Stoewsand, G. S. *Fd. Cosmet. Toxicol.* **1978**, *16*, 151–155.
82. Gillner, M.; Bergman, J.; Cambillau, C.; Fernstrom, B.; Gustafsson, J. *Molec. Pharmacol.* **1985**, *28*, 357.
83. Salbe, A. D.; Bjeldanes, L. F. *Food Chem. Toxicol.* **1985**, *23*, 57–65.
84. Sparnins, V. L.; Venegas, P. L.; Wattenberg, L. W. *J. Natl. Can. Inst.* **1982**, *68*, 493–496.
85. Godlewski, C. E.; Boyd, J. N.; Sherman, W. K.; Anderson, J. L.; Stoewsand, G. S. *Cancer Lett.* **1985**, *28*, 151–157.
86. Ramsdell, H. S.; Eaton, D. L. *J. Toxicol. Env. Health* **1988**, *25*, 269–284.
87. Michnovicz, J. J.; Bradlow, H. *J. Natl. Can. Inst.* **1990**, *82*, 947–949.
88. Bradlow, H. L.; Michnovicz, J. J.; Telang, N. T.; Osborne, M. P. *Carcinogenesis* **1991**, *12*, 1571–1574.
89. Bradfield, C. A.; Bjeldanes, L. *J. Toxicol. Env. Health* **1987**, *21*, 311–323.
90. Bergman, J.; Norrby, P-O.; Tilstam, U.; Venemalm, L. *Tetrahedron* **1989**, *45*, 5549–5564.
91. Grose, K. R.; Bjeldanes, L. F. *Chem. Res. Toxicol.* **1992**, *5*, 188–193.
92. Wortelboer, H. M.; de Kruif, C. A.; van Iersel, A. A. J.; Falke, H. E.; Noordhoek, J.; Blaauboer, B. J. *Biochem. Pharmacol.* **1992**, *43*, 1439–1447.
93. Perdew, G. H.; Babbs, C. F. *Nutr. Cancer* **1991**, *16*, 209–218.
94. Wang, Z. Y.; Cheng, S. J.; Zhou, Z. C.; Athar, M.; Khan, W. A., Bickers, D. R.; Mukhtar, H. *Mutat. Res.* **1989**, *223*, 273–285.
95. Wang, Z. Y.; Khan, W. A.; Bickers, D. R.; Mukhtar, H. *Carcinogenesis* **1989**, *10*, 411–415.

RECEIVED May 28, 1993

Chapter 24

Comparative Study of Ellagic Acid and Its Analogues as Chemopreventive Agents against Lung Tumorigenesis

Andre Castonguay, Mohamed Boukharta, and Guylaine Jalbert

Laboratory of Cancer Etiology and Chemoprevention, School of Pharmacy, Laval University, Quebec G1K 7P4, Canada

The polyphenol ellagic acid inhibits lung tumorigenesis induced by a nicotine-derived nitrosamine in A/J mice. This inhibition was related to the logarithm of the dose of ellagic acid added to the diet. The biodistribution of ellagic acid was studied in mice gavaged with ellagic acid. Pulmonary levels of ellagic acid reach a maximum 30 min after gavage and were directly proportional to the dose between 0.2 and 2.0 mmol EA/kg b.w. Ellagitannins extracted from pomegranate are hydrolyzed extensively in mice leading to the excretion of ellagic acid in the feces and urine. Feeding mice pomegranate ellagitannins (10 g/kg diet) did not inhibit lung tumorigenesis.

Nicotine accounts for 90% of the alkaloids present in tobacco. Levels of nicotine in mainstream smoke of cigarette range from 0.1 to 3 mg/cigarette (*1*). Cigarettes also deliver about 0.3 mg of minor alkaloids. Among those is nornicotine. Combustion of cigar and pipe tobacco also yields nicotine (*2*). Thus nicotine is an essential component in tobacco products and there are many lines of evidence that it is the pharmacological reinforcer in tobacco smoke (*3*).

Freshly harvested tobacco leaves contain between 0.6 and 14 mg nitrate per cigarette. During the curing and processing of tobacco leaves, reaction of nicotine and nornicotine with nitrate produces *N*-nitrosamines (*4*), one of which is 4-(methyl-nitrosamino)-1-(3-pyridyl)-1-butanone (NNK). Levels of NNK in mainstream smoke range from 4 to 1700 ng per cigarette (*5*). Tumorigenic assays of NNK in laboratory animals have revealed a remarkable specificity of NNK for lung tissues (*6*). These observations led Hoffmann and Hecht to conclude that NNK is important in the etiology of tobacco smoke-induced lung cancer (*7*).

It is estimated that humans consume as much as 1 g of plant phenols per day (*8*). Ellagic acid is a polyphenol generated from ellagitannins present in grapes, strawberries raspberries, and pomegranate, which are normally consumed by humans (*9*). This phenol is one of most promising chemopreventive agent likely to reduce the risk of human cancer and which could be introduced in human intervention trials (*10*). The aim of this study was to characterize the chemopreventive

0097–6156/94/0546–0294$06.00/0

efficacy of ellagic acid in lung tumorigenesis and to compare the efficacies of EA and ellagitannins, the precursors of EA.

Materials and Methods

Chemicals. NNK was synthesized as previously described (*11*). Its purity was higher than 98.5% as determined by reverse phase HPLC (*12*). EA was purified by crystallization in pyridine and drying at 40°C for 3 days.

Lung Adenoma Assay. In a first bioassay, tumor response was studied as a function of the dose of EA. Female A/J mice (6 to 7 weeks old) were housed in groups of 5 in cages with wire bottoms. They were fed *ad libitum* a powdered AIN-76A diet (Teklad, Madison, WI) or AIN-76A diet with EA, starting 2 weeks before carcinogen treatment and throughout the experiment (25 weeks). The initial concentration of NNK was 62.4 µg/ml and was adjusted thereafter for each cage according to water consumption. The cumulative dose of NNK was 9.1 mg per mouse. Five groups received EA at doses of 0, 0.06, 0.25, 1.00 or 4.00 g/kg diet. All the mice were necropsied 16 weeks after the end of the carcinogen treatment.

In a second bioassay, the efficacies of EA and pomegranate ellagitannin were compared. The first group of 10 mice was fed control diet only. Three groups of 25 mice were given a cumulative dose of 9.1 mg of NNK per mouse. Two of these groups were fed EA (4 g/kg diet) or pomegranate ellagitannins (10 g/kg diet) starting 2 weeks before carcinogen treatment and throughout the bioassay.

Biodistribution of EA. The biodistribution of EA was studied as a function of the dose and time after gavage with a suspension of EA. Six groups of 3 mice were fasted for 24 hours. They were gavaged with 0.5 ml of a suspension containing 0, 0.2, 0.5, 1.0, 1.5 or 2 mmol EA/kg b.w. in distilled water and sacrificed 30 min later. Seven groups of 3 mice were fasted for 24 hours and gavaged with 0.5 ml of a suspension containing 2 mmol EA/kg b.w. They were sacrificed 10, 15, 30, 45, 60, 120 or 240 min later. Liver (300 mg) or lung tissues (100 mg) were homogenized in 2 ml 1 M NaH_2PO_4 (pH 3) for 15 sec. After the addition of 20 ml of a mixture of acetone and ethyl acetate (1:2, v:v) the homogenization was continued for 2 min. The homogenate was centrifuged (12,000 g, 15 min., 4°C) and a 5 ml aliquot of the supernatant was evaporated to dryness under nitrogen at room temperature and dissolved in 100 µl methanol. A 20 µl aliquot of this solution was injected on a µBondapak C_{18} column (10 µm, 0.4 x 30 cm). The EA was eluted with 2% formic acid in methanol:water (2:3, v:v). The column was washed with methanol for 12 min before each assay. The flow rate was 1 ml/min and EA was detected at 254 nm.

Pomegranate Extraction. Pomegranate peels were extracted 3 times with 80% aqueous acetone. Concentration of the extracts under reduced pressure left a brown precipitate, which was removed by filtration. The filtrate was applied to a Sephadex LH-20 column. After washing with water, the column was eluted with 80% aqueous acetone. The solvent was evaporated and the residue was dried over P_2O_5. An aliquot of the residue was analyzed by HPLC on µBondapak C_{18} 10 µm. Solvent A was a 2% aqueous solution of formic acid and solvent B was 2% solution of formic acid in methanol:water (3:7; v:v). The solvent gradient was as follows: 100% solvent A for 3 minutes, then linear to 100% solvent B in 72 minutes.

A 1 mg aliquot of the residue was dissolved in 2 ml methanol/water (1:1, v:v), mixed with 0.75 ml concentrated HCl and heated at 100°C for two hours. The hydrolysate was analyzed by HPLC as described for pomegranate extract.

Analysis of Feces and Urine. Two groups of five A/J mice were fasted for 6 hours. Group 1 received 0.5 ml of a pomegranate ellagitannins solution in 10% Acacia (588 mg/kg b.w.). Group 2 received EA (147 mg/kg b.w.). The feces and urine were collected for 24 hours. Feces were extracted with methanol and analyzed by HPLC as described above. Levels of EA and ellagitannins in feces were determined by interpolation from calibration curves which were constructed by extracting feces from untreated mice to which were added known amounts of EA or ellagitannins. The extraction of free EA from urine was carried out as described for lung tissues.

Results

The inhibition of lung tumorigenesis was studied at four dose levels of EA. Feeding mice 4 g EA/kg diet reduced the lung tumor multiplicity by 54% (Figure 1). In mice fed a diet containing 0.06–4.00 g EA/kg diet, the lung multiplicity was inversely proportional to the logarithm of the dose of EA ($r^2 = 0.992$). On the other hand, pomegranate ellagitannins had no effect on tumor multiplicity (Table I).

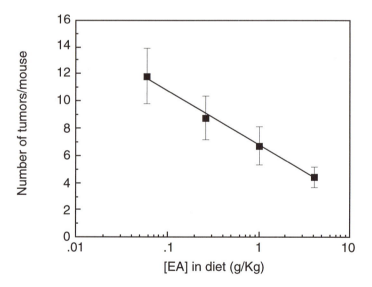

Figure 1. Inhibition of NNK-induced lung tumors by EA at various concentrations. Data are the means ± SE from 23 to 25 mice. The curve was fit by the least square method and the data analyzed by linear regression ($r^2 = 0.992$).

A HPLC assay of EA in EA-treated mice was developed. Tissues from mice gavaged with EA were extracted and the extract fractionated by HPLC. HPLC profiles of lung tissues extracts from EA-treated and untreated mice are compared in Figure 2. In EA treated mice, we observed a peak which co-eluted with EA and was base separated. The limit of detection of this assay was 0.4 nmol/g tissue.

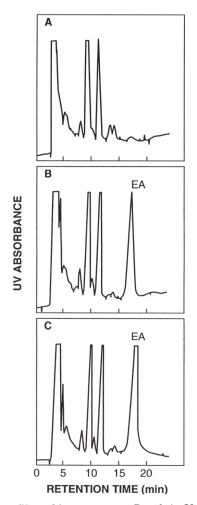

Figure 2. HPLC profiles of lung extracts. Panel A: Untreated mice; panel B: EA added to lung tissues from untreated mice before extraction; panel C: mice gavaged with 2 mmol EA/kg and sacrificed 30 min later.

**Table I. Effects of Ellagic Acid and Pomegranate Ellagitannins on
NNK-induced Lung Tumors in A/J Mice**

Total dose of NNK (mg/mouse)	Chemopreventive agent	Number of tumors per mouse[a]
None	None	0.6 ± 0.2
9.1	None	6.8 ± 0.8
9.1	Ellagic acid 4 g/kg diet	3.9 ± 0.4*
9.1	Pomegranate ellagitannin 10 g/kg diet	9.1 ± 2.0

[a]mean ± SE, n = 25, *p<0.005

The tissue distribution of EA was investigated as a function of time, after gavage with 2 mmol EA/kg b.w. The results are illustrated in Figure 3. In the lung, a maximum level of EA, corresponding to 21.3 nmol/g, was observed 30 min after EA administration. This level corresponds to only 70 ppm of the administered dose. Levels in liver tissues were 10-fold lower and reached a maximum 30 min after gavage. At this time, the blood level of EA was 1 nmol/ml. As shown in Figure 4, the levels of EA in the lung were directly proportional to the dose of EA (r^2 = 0.993), between 0.2 and 2.0 mmol EA/kg b.w.

Acid hydrolysis of ellagitannins from pomegranate yielded 25% EA on a weight basis. HPLC profile of hydrolyzed ellagitannins are indicated in Figure 5.

Elimination of ellagitannins and EA in urine and feces were compared. Two groups of mice were gavaged with ellagitannins or EA. The results summarized in Table II indicate that ellagitannins were hydrolyzed to EA *in vivo*. The analysis of feces and urine showed that 4.3% of the dose of ellagitannin was excreted unchanged in the feces and 0.04 of the dose was excreted in the urine as EA.

**Table II. Recovery of Ellagitannins and Ellagic Acid (EA)
in Feces and Urine of Mice Gavaged With Pomegranate Ellagitannins or EA**

Treatment	Recovery (% of dose)			
	Pomegranate ellagitannins		EA	
	Feces	Urine	Feces	Urine
Pomegranate ellagitannins	4.3	—	1.5	0.04
EA	—	—	9.4	0.13

Discussion

It is estimated that 85% of lung cancer could be prevented by the eradication of tobacco smoking (*13*). In spite of extensive laboratory investigations on mechanism of action of tobacco smoke, our better understanding has led to only a modest decrease in tobacco smoking in North American countries (*14*). Chemoprevention has been proposed as a second approach to lung cancer control (*15*). Cancer chemo-prevention is defined as the administration of substances inhibiting the overall

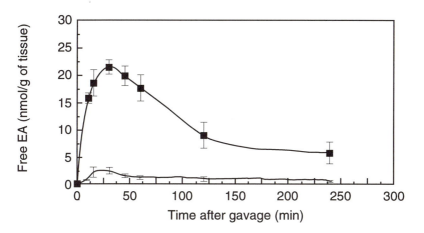

Figure 3. Levels of EA in lung (black squares) and liver (dots) tissues following gavage of EA (2 mmol/kg body weight). Data are the means ± SE from 3 mice.

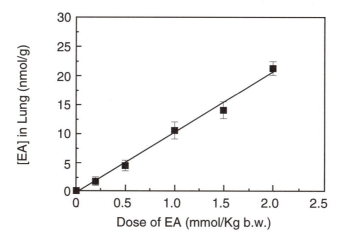

Figure 4. Pulmonary levels of EA as a function of the dose of EA given by gavage. Data are the means ± SE from 3 mice. The curve was fit by the least square method and the data analyzed by linear regression ($r^2 = 0.993$).

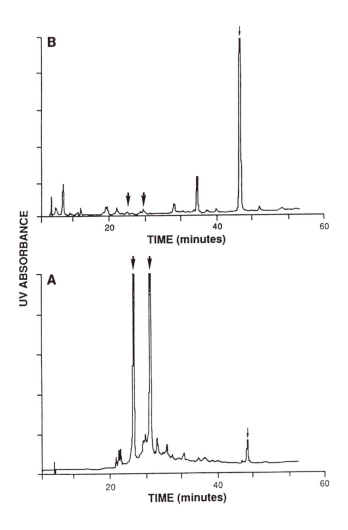

Figure 5. Panel A: HPLC profile of pomegranate extract. Panel B: HPLC profile of a pomegranate extract hydrolyzed with concentrated HCl.
Bold arrows: ellagittannins, plain arrows: ellagic acid.

process of carcinogenesis. Chemopreventive agents can be divided into two major classes: synthetic compounds (including pharmaceuticals) and naturally occurring compounds. The latter class has the advantages of being more acceptable to the public and more likely to be introduced quickly in clinical trial. It is noteworthy that most on-going intervention trials in chemoprevention of lung cancer involve vitamin B_{12}, E, A and β-carotene (provitamin A) found in food and beverages (*15*).

Ellagitannin is a family of chemically complex substances characterized by the presence of one or more hexahydroxydiphenoyl groups. This group is cleaved during hydrolysis and undergoes intramolecular esterification. The resulting dilactone, ellagic acid, has been investigated for its efficacies to inhibit chemically induced carcinogenesis in laboratory animals. At high concentrations, it is generally effective against a wide variety of carcinogens in various tumor models (*16–20*). In this study we confirmed that EA is effective in preventing lung tumorigenesis. The efficacy of EA was dose related.

Studies of the mechanism of action of EA have concluded that it could inhibit phase I enzymes involved in the activation of procarcinogens (*21–22*). According to Das *et al.* (*23*), ellagic acid induces hepatic glutathione *S*-transferase, a phase II enzyme responsible for the detoxification of some carcinogen-generated electrophiles. Ellagic acid may also scavenge oxygen radicals involved in oxidative destruction of membrane lipids and involved in tumor promotion (*24*). The relative importance of various hypotheses on the mechanism of action of EA remains to be determined. In this study, we have demonstrated that localization of EA in lung tissues is dose related. Our results demonstrated that localization of EA or its metabolites in lung tissues coincide with the inhibition of tumorigenesis. We have recently reported that pulmonary localization of EA could be increased by including EA in a β-cyclodextrin polymer (*25*).

The chemopreventive properties of EA have been investigated extensively in spite of the fact that EA is not present as such in living plants (*26*). In contrast, the ellagitannins that are the precursors of EA have received little attention as chemopreventive agents. Recently, Gali *et al.* (*27*) showed that two ellagitannins, castagalagin and vescalagin, reduced TPA-induced ODC activity and hydroperoxide production in mouse epidermis. Geraniin and other ellagitannins exhibit strong antimutagenic properties (*28*). The major ellagitannins present in pomegranate are granatins A and B (*29*). The results of this study demonstrated for the first time that pomegranate ellagitannins are metabolized extensively in A/J mice. This metabolism leads to the hydrolysis of the hexahydroxyphenoic group and its cyclization to EA. These results suggest that pomegranate is a natural source of EA, a non-toxic lung tumorigenesis chemopreventive agent. A relatively high dietary dose might be necessary, however, to achieve significant inhibition of lung tumorigenesis.

Acknowledgments

This study was supported by a grant from the National Cancer Institute of Canada.

Literature Cited

1. Hoffmann, D.; Hecht, S. S. In *Handbook of Experimental Pharmacology*; Cooper, C.; Grover, L. P., Eds.; Springer–Verlag: Berlin, 1990, Vol 94/I; pp 63–102.

2. Brunnemann, K. D., Hoffmann, D., Wynder, E. L.; Gori, G. B. *Proceedings of 3rd World Conference on Smoking and Health*; US Department of Health, Education and Welfare: Washington DC [DHEW Publication No. (NIH) 76-1221], 1976; pp 441–449

3. Jaffe, J. H. In *Nicotine Psychopharmacology, Molecular, Cellular, and Behavioral aspects*; Wonnacott, S.; Russell, M. A. H.; Stolerman, I. P.; Oxford University Press: New York, 1990; pp 1–37

4. Fischer, S.; Spiegelhalder, B.; Eisenbarth, J.; Preussmann, R. *Carcinogenesis* **1990**, *11*, 723–730.

5. Fischer, S.; Spiegelhalder, B.; Preussmann, R. *Arch. Geschwulstforsch.* **1990**, *3*, 169–177.

6. Hecht, S. S.; Hoffmann, D. *Carcinogenesis* **1988**, *9*, 875–884.

7. Hecht, S. S.; Hoffmann, D. *Cancer Surv.* **1989**, *8*, 273–294.

8. Kühnau, J. *World Rev. Nutr. Diet* **1976**, *24*, 117–191.

9. Bate-Smith, E. C. In *The Pharmacology of Plant Phenolics* Fairbairn, J. W., Ed.; Academic Press: New York, 1959; pp 133–147.

10. Kelloff, G. J.; Malone, W. F.; Boone, C. W.; Sigman, C. C.; Fay, J. R. *Semin. Oncol.* **1990**, *17*, 438–455.

11. Hecht, S. S.; Chen, C. B.; Dong, M.; Ornaf, R. M.; Hoffmann, D.; Tso, T. C. *Beitr. Tabakforsch.* **1977**, *9*, 1–6.

12. Castonguay, A.; Lin, D.; Stoner, G. D.; Radok, P.; Furuya, K.; Hecht, S. S.; Schut, H. A. G.; Klaunig, J. E. *Cancer Res.* **1983**, *43*, 1223–1229.

13. U. S. Surgeon General *The Health Consequences of Smoking: Cancer* U. S. Department of Health and Human Services: Washington, D.C. (NIH Publ. No. 82–50179), 1982.

14. Ramström, L. M. In *Tobacco, a Major International Health Hazard* Zaridze, D.; Peto, R., Eds. IARC Sci. Publi., 1986, Vol. 74; 135–142.

15. Castonguay, A. *Cancer Res.* **1992**, *52*; 2641s–2651s.

16. Tanaka, T.; Iwata, H.; Niwa, K.; Mori, Y.; Mori, H. *Jpn. J. Cancer Res.* **1988**, *79*, 1297–1303.

17. Mukhtar, H.; Das, M.; Bickers, D. R. *Cancer Res.* **1986**, *46*, 2262–2265.

18. Mukhtar, H.; Das, M.; Del Tito, B., Jr; Bickers, D. R. *Biochem. Biophys. Res. Commun.* **1984**, *119*, 751–757.

19. Lesca, P. *Carcinogenesis* **1983**, *4*, 1651–1653.

20. Mandal, S.; Stoner, G. D. *Carcinogenesis* **1990**, *11*, 55–61.

21. Teel, R. W.; Dixit, R.; Stoner, G. D. *Carcinogenesis,* **1985**, *6*, 391–395.

22. Mandal, S.; Shivapurkar, N. M.; Galati, A. J.; Stoner, G. D. *Carcinogenesis* **1988**, *9*, 1313–1316.

23. Das, M.; Bickers, D. R.; Mukhtar, H. *Carcinogenesis* **1985**, *6*, 1409–1413.

24. Osawa, T.; Ide, A.; Su, J-D.; Namiki, M. *J. Agric. Food Chem.* **1987**, *35*, 808–812.

25. Boukharta, M.; Jalbert, G.; Castonguay, A. *Nutr. Cancer* **1992**, *18*, 181–189.

26. Okuda, T.; Mori, K., Hatano; T. *Phytochemistry* **1980**, *19*, 547–551.

27. Gali, H. U.; Perchellet, E. M.; Bottari, V.; Hemingway, R. W.; Scalbert, A.; Perchellet, J. P. *Proc. Am. Assoc. Cancer Res.* **1992**, *33,* 162.

28. Okuda, Y.; Mori, K.; Hayatsu, H. *Chem. Pharm. Bull.* **1984**, *32*, 3755–3758.

29. Tanaka, T.; Nonaka, G.; Nishioka, I., *Chem. Pharm. Bull.* **1990**, *38*, 2424–2428.

RECEIVED September 7, 1993

Chapter 25

Antitumor-Promoting Effects of Gallotannins, Ellagitannins, and Flavonoids in Mouse Skin In Vivo

J. P. Perchellet[1], H. U. Gali[1], E. M. Perchellet[1], P. E. Laks[2], V. Bottari[3], R. W. Hemingway[4], and A. Scalbert[5]

[1]Anti-Cancer Drug Laboratory, Division of Biology, Kansas State University, Ackert Hall, Manhattan, KS 66506–4901
[2]Institute of Wood Research, Michigan Technological University, Houghton, MI 49931–1295
[3]Silva S.r.l., 12080 San Michele Mondovi (Cuneo), Italy
[4]Southern Forest Experiment Station, U.S. Department of Agriculture, Forest Service, Pineville, LA 71360–5500
[5]Laboratoire de Chimie Biologique, Institut National de la Recherche Agronomique, Institut National Agronomique, Paris-Grignon, 78850 Thiverval-Grignon, France

Hydrolyzable (HTs) and condensed tannins (CTs) were tested topically for their ability to inhibit the biochemical and biological effects of 12-*O*-tetradecanoylphorbol-13-acetate (TPA) in mouse epidermis *in vivo*. Overall, commercial tannic acid (TA), ellagic acid (EA), and *n*-propyl gallate (PG) inhibit the promotion of skin papillomas and carcinomas by TPA in relation with their ability to inhibit TPA-induced epidermal ornithine decarboxylase (ODC) activity, hydroperoxide (HPx) production, and DNA synthesis. Pure pentagalloylglucose, castalagin, vescalagin, catechin dialkyl ketals, and epicatechin-4-alkylsulphides or heterogenous sumac leaf TA, Aleppo gall TA, tara pod TA, loblolly pine bark CT, guamuchil bark CT, and southern red oak bark CT also inhibit these biochemical markers of TPA promotion to various degrees. When applied to initiated skin 20 min before each promotion treatment, the different TA samples all remarkably inhibit complete tumor promotion by TPA. Sumac leaf TA is the most effective. The antitumor-promoting activity of a TA pretreatment can be further enhanced by the application of TA 24 h after each promotion treatment with TPA. Commercial TA and Aleppo gall TA inhibit the second stage of tumor promotion by mezerein but not the first stage of tumor promotion by TPA. Therefore, tannins in general might be valuable to prevent and/or inhibit tumor propagation, the only reversible stage of tumorigenesis.

0097–6156/94/0546–0303$07.25/0
© 1994 American Chemical Society

Tannins are found in high concentrations throughout the plant kingdom and are the sources of reactive polyphenols with great structural diversity (*1–3*). The HTs are esters of gallic acid (GA) or hexahydroxydiphenic acid (HHDP) and the CTs derive from the condensation of flavan-3,4-diol (Figure 1). Tannin extracts are often collections of HTs or CTs with undefined degree of polymerization, molecular weight (M.W.), monomer unit composition, and type of linkage.

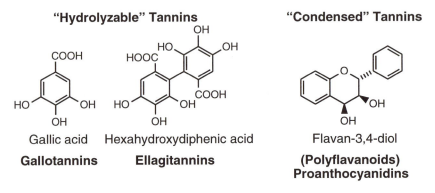

Figure 1. Classes and monomer units of tannins (adapted from ref. 3).

Structures of HTs and CTs

The HTs, or more correctly gallotannins and ellagitannins, are easily split into sugars and phenolic carboxylic acids. Upon acid hydrolysis, gallotannins and ellagitannins yield GA and EA, respectively. HTs, therefore, have a sugar core with pendant esterified GA or HHDP constituents and may possess a variable number of depsidically linked galloyl units in a polygalloyl chain (Figure 2).

The CTs, or more correctly proanthocyanidins or polyflavonoids, have no sugars, do not readily break down, and almost invariably contain one of or both of the flavan-3-ols, (+)-catechin or (-)-epicatechin. CT monomer units are flavonoids consisting of 2 aromatic rings A and B joined through a pyran ring C. Ten classes of CTs are distinguished on the basis of the hydroxylation patterns of the A- and B-rings. In oligomeric and polymeric CTs, a variable number of these skeletons are bound together by interflavonoid linkages occurring at one or more sites (Figure 2).

Tannins and Multistage Carcinogenesis

Experimental skin carcinogenesis has been divided into tumor initiation, promotion, and progression (*4–6*). This sequence is also called the multistage model of carcinogenesis. A subcarcinogenic dose of 7,12-dimethylbenz[*a*]anthracene (DMBA) produces no tumors during the lifespan of the animal but, after metabolic activation, the electrophilic ultimate carcinogen interacts covalently with, and mutates, epidermal DNA to initiate skin carcinogenesis. Tumor initiation is generally regarded as a permanent alteration of the cell genotype with as yet no neoplastic phenotype. The correlation between tumor initiation and resistance of epidermal cells to signals for terminal differentiation suggests that the initiating event in skin carcinogenesis causes a genetic alteration in the program of terminal differentiation.

Repetitive applications of the most potent tumor promoter TPA are required to trigger molecular events leading the immediate progeny of the initiated epidermal cells to the formation of growing skin tumors and achieve complete tumor promotion. It is theorized that TPA stimulates the expression of the abnormal genetic information within the initiated cells which, because of their altered program of differentiation, acquire a neoplastic phenotype (conversion) and a proliferative advantage over their normal neighbors (propagation).

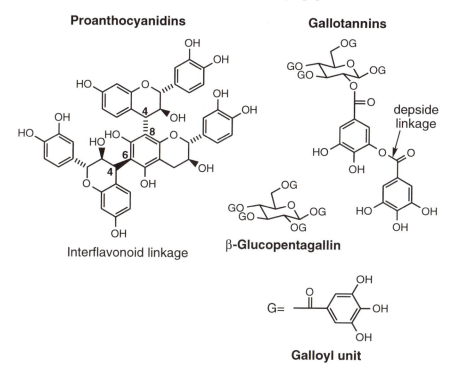

Figure 2. Examples of linkages between monomer units of HTs and CTs (adapted from ref. 3).

One to four applications of TPA are sufficient to trigger the first, partially irreversible, stage of promotion called "conversion." Multiple applications of the ineffective promoter mezerein are then required to achieve the second stage of promotion called "propagation" and complete the promotion process. Neither treatment alone is sufficient in CF-1 mice. The long-lasting effects of TPA that are essential for stage 1 promotion persist for almost 2 months before declining whereas those of mezerein in stage 2 promotion are rapidly reversible and require a certain frequency of application in order to induce tumors. The molecular mechanism by which tumor promoters select the mutation-bearing initiated epidermal cells and induce their neoplastic transformation and clonal expansion into skin tumors is not fully understood.

Finally, tumor progression requires a number of genetic alterations and is correlated to the level of DNA damage, chromosomal aberration, and aneuploidy.

Benign neoplastic cells that have accumulated additional sublethal genetic alterations, besides those associated with tumor initiation, may be selected to develop malignant characteristics.

EA and HTs, such as commercial TA, have already been shown to inhibit mutation, tumor initiation, and complete carcinogenesis, which are irreversible events (7–24). Therefore, it is crucial to determine if HTs can also inhibit the reversible phase of tumor promotion. When the tumor promoter TPA is applied topically to mouse skin at time 0, it triggers 3 major biochemical markers of tumor promotion: the induction of ODC activity at 5 h and the stimulation of HPx production and DNA synthesis at 16 h (25). Much of the biochemical significance of tannins may be linked to macromolecule complexation, mineral chelation, and antioxidation (1–3). Recently, we found that commercial TA, EA, and several GA derivatives applied topically to mouse skin can inhibit ODC, a marker of TPA promotion, by up to 85% (26).

Objectives of Current Investigation

New studies were designed to determine if the antioxidant activities of commercial HTs would enable them to inhibit TPA-stimulated HPx production and DNA synthesis in mouse epidermis *in vivo,* and the promotion of papillomas and carcinomas by TPA in initiated skin (27–29). Moreover, the antitumor-promoting effects of different TA extracts were compared. Vegetable tannins are common dietary components but CTs are far more abundant in food than are HTs. Therefore, the present study was also undertaken to determine if pure or heterogenous HTs and CTs prepared from various sources share the ability to inhibit the biochemical events linked to skin tumor promotion by TPA.

Methods of Study

Biochemical Markers of Skin Tumor Promotion. All tumor promoters were delivered to the shaved backs of female CF-1 mice in 0.2 ml acetone. HTs and CTs were also applied topically in 0.4 ml of the same solvent at various times before or after, and to the same area of skin as, each application of tumor promoter. EA and chestnut wood tannin were administered similarly in 0.4 ml methanol:acetone (55:45) and water:acetone (35:65), respectively. Doses of all TA samples and chestnut wood tannin were expressed in μmol based on average M.W.s of 1701 and 1100, respectively. At the appropriate times after TPA treatment, the mice were killed, their skins were excised, the epidermis was separated from the dermis, and homogenates were prepared by pooling the epidermises from 2 skins (26–29).

Epidermal ODC activity was determined 5 h after a single TPA treatment by measuring the release of $^{14}CO_2$ from L-[1-^{14}C]ornithine-HCl (30). The HPx-producing activity of the epidermis was assayed by a modification of the ferrithiocyanate method 16 h after the last of 2 applications of TPA at a 48 h interval (25,31). DNA synthesis was determined 16 h after a single TPA treatment. The mice were injected i.p. with 30 μCi ^{3}H-thymidine and the rate of incorporation of this radiolabeled precursor into epidermal DNA was measured by liquid scintillation counting after a 40 min period of pulse-labeling (25,32).

Tumor Induction Experiments. In the initiation-complete tumor promotion protocol (4–6), skin tumors were initiated in all female CF-1 mice by a single topical application of 100 nmol DMBA in 0.2 ml acetone. Two weeks later, all mice were promoted twice a week (on days 1 and 4) with 8.5 nmol TPA for the rest of the experiments (20 or 40 weeks). In the initiation/two-stage promotion protocol (*4–6*), skin tumors were initiated in all female SENCAR mice by a single application of 25 nmol DMBA. Two weeks later, all mice were promoted twice a week for 2 weeks with 4.25 nmol TPA (stage 1) and then twice a week for 18 weeks with 4.25 nmol mezerein (stage 2). Except where otherwise specified, HTs were applied 20 min before, and to the same area of skin as, each TPA treatment in complete and stage 1 tumor promotion or each mezerein treatment in stage 2 tumor promotion. Initially, there were 36 mice in each treatment group. The incidence and yield of skin tumors were recorded weekly and once every 2 weeks, respectively. The graphs of tumor data include 2 or 4 panels. (A) Average number of papillomas/ survivor. (B) Percentage of survivors with papillomas. (C) Average number of carcinomas/mouse and (D) percentage of mice with carcinomas based on the number of survivors in each group at the time of appearance of the first carcinoma in the experiment. Statistics for the differences between the means of papillomas/ mouse or carcinomas/mouse were performed using Student's *t*-test. Differences between papilloma and carcinoma incidences were compared using the χ^2 statistic. The level of significance was set in both cases at $p \leq 0.05$.

Antitumor-Promoting Effects of Commercial HTs

Commercial PG, EA, and TA (Figure 3) were from Sigma Chemical Co. Commercial EA, the dilactone form of HHDP, was purified from chestnut bark but the source of commercial TA was unknown. Commercial TA, usually Chinese gallotannin, is a mixture of molecules with a core of β-penta-*O*-galloyl-D-glucose to which other (approximately 2) galloyl residues are linked. Thus, TA samples are heterogenous because of the variation in the number of depsidically linked galloyl groups in the polygalloyl chain (Figure 3).

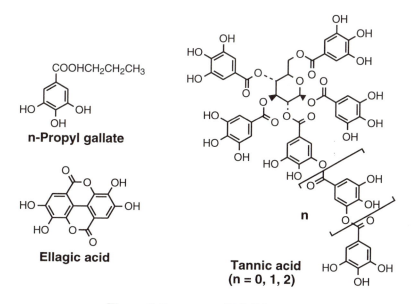

Figure 3. Structures of PG, EA and TA.

Inhibition of Complete Tumor Promotion. Commercial TA, EA, and PG — the treatments that most inhibit the ODC and HPx responses to TPA (*26,29*) — also inhibit remarkably the formation of papillomas and carcinomas when applied topically to DMBA-initiated skin 20 min before each promotion treatment with TPA (*27,28*).

Antitumor-promoting Activity of Commercial TA. In control mice promoted with TPA only, the first papillomas appear at 7 weeks, the incidence and yield of papillomas reach a plateau at 16 weeks, and maximal tumor promotion is observed at 22 weeks with 97% of the mice bearing papillomas and an average of 12 papillomas/mouse (Figures 4 and 5). That commercial TA is an outstanding inhibitor of TPA promotion is illustrated by the fact that, at this stage of 22 weeks, 2.5, 5, 10, or 20 µmol of TA inhibit the tumor incidence by 50–100% and the tumor yield by 80–100% (Figure 4). Incredibly, 20 µmol of TA can prolong the latency period for papilloma development by 20 weeks so that the first tumors appear at 27 weeks instead of 7 weeks in the control group. After 22 weeks, some of the mice promoted in the presence of TA develop new papillomas while the control mice treated only with TPA have decreasing numbers of papillomas. But the TA treatments still inhibit the incidence of papillomas by 40–80% and the yield of papillomas by 70–95% at 40 weeks, suggesting that this HT decreases the tumor-promoting activity of TPA and does not simply delay or slow down the rate of tumor development.

Less than 10% of the benign skin papillomas promoted by TPA progress to malignant skin carcinomas starting at week 26 and this event is also delayed by 2–11 weeks in the presence of increasing doses of TA (Figure 4). The treatments with 2.5, 5, and 10 µmol of TA inhibit both the incidence and yield of carcinomas at 40 weeks by 70–90%. Obviously, so few papillomas are promoted in the presence of 20 µmol of TA that none can progress to carcinomas before 40 weeks. Therefore, commercial TA clearly inhibits the promotion of both skin papillomas and carcinomas by TPA.

These chronic topical applications of TA are not toxic because the body weights and rates of survival are identical in all groups up to 32 weeks, whether or not they receive TA. Moreover, the protective effect of TA is illustrated by the fact that, between weeks 32 and 40, the rate of survival drops from 100 to 50% in the control mice promoted only with TPA but it remains at 100% in all the groups promoted in the presence of TA.

There are several explanations for the fluctuations and apparent declines in the number of papillomas after 22 weeks, especially in the control mice promoted with TPA only (Figure 4). The local and systemic toxicity of chronic TPA treatment and the high tumor burden decrease the rate of survival and the dead mice lost, especially those bearing carcinomas, are likely to have more papillomas than the group average. Furthermore, with increasing time there is coalescence of several small neighboring papillomas that combine together to form a single big papilloma and the continued conversion of papillomas to carcinomas results in tumors disappearing from Chart A to appear in Chart C (Figure 4).

Antitumor-promoting Activities of EA and PG. From this tumor experiment, it appears that TA is the most effective anti-tumor promoter among the commercial HTs tested. Topical application of 5 µmol EA or 20 µmol PG 20 min before each TPA treatment also inhibits the incidence and yield of skin papillomas

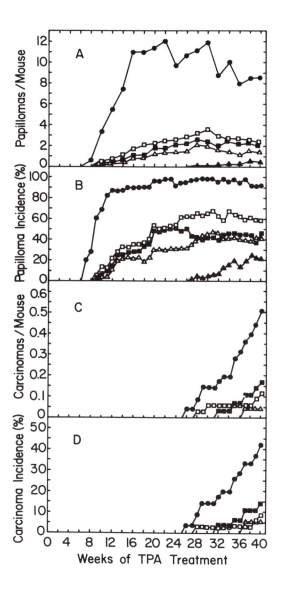

Figure 4. Inhibition of TPA promotion (●) by 2.5 (□), 5 (■), 10 (△), and 20 (▲) μmol TA. (Reproduced with permission from ref. 28. Copyright 1992 IRL Press.)

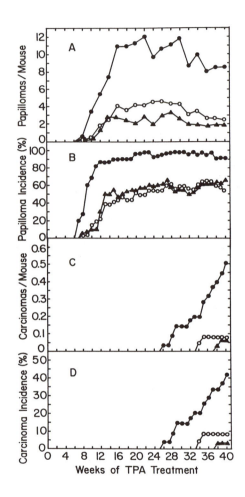

Figure 5. Inhibition of TPA promotion (●) by 5 μmol EA (○) or 20 μmol PG (▲). (Adapted from ref. 28.)

and carcinomas promoted by this compound but EA and PG are somewhat less potent than equal doses of TA (Figure 5). Overall, the antitumor-promoting activities of commercial TA, EA, and PG in the initiation-complete tumor promotion protocol correlate well with their combined inhibitory effects on the 3 major biochemical markers of TPA promotion.

Inhibition of the Biochemical Markers of Tumor Promotion. In Table I, the magnitudes of the inhibitory effects of commercial TA, EA, and PG on the biochemical and biological effects of TPA have been summarized. TPA-induced ODC activity and HPx production are 2 events that have been postulated to complement each other to trigger enough DNA synthesis and compensatory cell proliferation to achieve the prolonged hyperplastic response required for tumor propagation (25). On an equal dose basis, TA is as good an antioxidant as PG and less effective than EA (29) but it is a much more potent inhibitor of ODC induction than either of these two compounds (26). Consequently, TA also inhibits the stimulation of epidermal DNA synthesis and the promotion of skin tumors by TPA to a greater degree than similar doses of EA or PG. Indeed, the mean inhibitions of the ODC, HPx, and DNA responses to TPA by TA, EA, and PG match closely the means by which these HTs inhibit the incidence and yield of skin tumors promoted by TPA (Table I).

Table I. Comparison of the Abilities of TA, EA and PG to Inhibit the Biochemical and Biological Effects of TPA in Mouse Skin *In Vivo*

Treatment	Biochemical markers of tumor promotion[a]			Skin tumor promotion at plateau[a]	
	ODC induction	HPx production	DNA synthesis	Tumor incidence	Tumor yield
TPA (8.5 nmol)	100[b]	100[c]	100[d]	100[e]	100[f]
+ EA (5 μmol)	90	0	69	63	40
+ TA (5 μmol)	41	27	50	48	21
+ PG (20 μmol)	48	1	82	62	25
+ TA (20 μmol)	14	1	41	25	4

[a]Percent of TPA effect. [b]1957% of control; [c]283% of control; [d]486% of control; [e]95.7% of mice with papillomas; [f]11.01 papillomas/mouse.

Antitumor-Promoting Effects of Pure or Heterogenous HTs

Several samples of pure or heterogenous HTs and CTs prepared from various sources were screened for their antitumor-promoting effects in mouse epidermis *in vivo*.

Heterogenous TA Samples from Various Sources. Table II shows that topical applications of sumac leaf TA (from *Rhus coriaria*), Aleppo gall TA (from *Quercus infectoria*), and tara pod TA (from *Caesalpinia spinosa*) all mimic or even surpass the inhibitory effects of commercial TA on TPA-induced ODC activity and HPx production in mouse epidermis *in vivo*. But chestnut wood tannin (from *Castanea sativa*) inhibits only the HPx response to TPA, and to a lesser degree than the other TA samples tested (Table II). At a higher dose, sumac leaf TA, Aleppo gall TA, and tara pod TA also mimic more or less the inhibition of TPA-stimulated DNA synthesis caused by commercial TA (Table III). Aleppo gall TA, however, which inhibits ODC induction and HPx production the most (Table II), is not the best inhibitor of epidermal DNA synthesis (Table III) or skin tumor promotion (Figure 6) during TPA treatment, suggesting that one cannot always predict accurately the antitumor-promoting potential of HT extracts based solely on the magnitudes of their inhibitions of the ODC and HPx responses to TPA. Other inhibitory effects still unknown must play a role in the mechanism of anti-tumor promotion by TA.

Inhibition of Complete Tumor Promotion. The effectiveness of Aleppo gall TA, tara pod TA, and sumac leaf TA as inhibitors of complete tumor promotion by TPA was compared at 10 μmol (Figure 6), a dose at which it is possible to determine if these TA extracts are more potent than commercial TA (Figure 4). In the control mice promoted only with TPA, skin tumor promotion is maximal as early as 12 weeks with about 84% of the mice bearing papillomas and 12 papillomas/mouse. At this stage, topical applications of Aleppo gall TA, commercial TA, tara pod TA, and sumac leaf TA 20 min before each TPA treatment inhibit the tumor incidence by 29–55% and the tumor yield by 80–88% (Figure 6). After 20 weeks of TPA promotion, the same TA samples are still able to inhibit significantly the tumor incidence by 16–46% and the tumor yield by 53–78%. Interestingly, sumac leaf TA, which is the best inhibitor of TPA-stimulated DNA synthesis (Table III), is also the most effective against skin tumor promotion by TPA. At 20 weeks, the inhibitory effects of sumac leaf TA are significantly greater than those of Aleppo gall TA and commercial TA on the tumor incidence and those of Aleppo gall TA on the tumor yield. Body weights and rates of survival were identical in all treatment groups up to the end of the experiment. Thus, all TA extracts tested so far are anti-tumor promoters but their efficacy may vary considerably depending on their origin.

Effects of TA Post-Treatments. Since commercial TA can inhibit the HPx response to TPA even when it is administered 24 h after the tumor promoter (*29*), it is of interest to determine if such TA post-treatment can also inhibit complete tumor promotion by TPA. As shown in Figure 7, 10 μmol commercial TA applied 24 h after each promotion treatment with TPA do not inhibit skin tumor promotion, suggesting that the ability of TA to decrease the HPx-producing activity of the epidermis previously treated with TPA is not sufficient in itself to prevent the promotion of skin tumors.

Even though TA may exert potent antioxidant effects in the epidermis at any time (*29*), it must be applied at a time when it can prevent TPA from triggering the other biochemical events linked to skin tumor promotion in order to successfully decrease the tumor-promoting activity of this agent. Indeed, TA applied 12–15 h after TPA fails to alter the peak stimulation of DNA synthesis observed 16 h after the tumor promoter (data not shown). TA post-treatment at +24 h, however, enhances significantly the ability of TA pretreatment at -20 min to

Table II. Inhibitory Effects of Heterogenous HT Samples from Various Sources on TPA-induced Ornithine Decarboxylase Activity (ODC) and Hydroperoxide (HPx) Production in Mouse Epidermis *In Vivo*

Treatment[a] (Dose/application)	ODC activity[b]			HPx production[c]		
	nmol CO_2/h/ mg protein	% of control	% of TPA	nmol H_2O_2/4 h/ mg protein	% of control	% of TPA
Control	0.41 ± 0.05	100	—	10.8 ± 0.5	100	—
TPA (8.5 nmol)	11.70 ± 0.99	2854	100	28.9 ± 1.1	268	100
+ commercial TA (5 μmol)	4.63 ± 0.35	1129	37	15.0 ± 0.7	139	23
+ sumac leaf TA (5 μmol)	3.76 ± 0.44*	917	30	20.2 ± 1.6	187	52
+ Aleppo pod TA (5 μmol)	3.27 ± 0.33**	798	25	10.8 ± 2.8***	100	0
+ tara pod TA (5 μmol)	6.65 ± 0.53	1622	55	18.3 ± 4.0	169	42
+ chestnut wood tannin (5 μmol)	11.68 ± 1.18	2849	100	23.3 ± 2.0****	216	69

[a]TAs applied 20 min before each TPA treatment; [b]Five h after a single TPA treatment, mean ± S.D. (n=6); [c]Sixteen h after 2 TPA treatments 48 h apart, mean ± S.D. (n=4).
*$p < 0.005$, compared to TPA + commercial TA; **$p < 0.05$, compared to TPA + sumac leaf TA; ***$p < 0.025$, compared to TPA + commercial TA; ****< 0.05, compared to TPA.

Table III. Comparison of the Inhibitory Effects of Various TA Samples on TPA-induced DNA Synthesis in Mouse Epidermis *In Vivo*

Treatment[a]	^{3}H-Thymidine incorporation into DNA		
	cpm/μg DNA ± SD (n = 6)	% of control	% of TPA
Control	42.2 ± 7.6	100	—
TPA (8.5 nmol)	231.8 ± 30.1	549	100
+ commercial TA (10 μmol)	126.8 ± 21.5	300	45
+ sumac leaf TA (10 μmol)	100.4 ± 19.0*	238	31
+ Aleppo gall TA (10 μmol)	142.3 ± 15.6	337	53
+ tara pod TA (10 μmol)	159.6 ± 20.7**	378	62

[a] TAs applied 20 min before a single TPA treatment.
[b] Sixteen h after TPA treament.
* Significantly different than TPA + commercial TPA, $p < 0.05$.
** Significantly different than TPA + commercial TPA, $p < 0.025$.

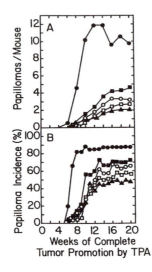

Figure 6. Effects of 10 μmol commercial TA (O), Aleppo gall TA (■), tara pod TA (□), or sumac leaf TA (▲) on complete tumor promotion by TPA (●).

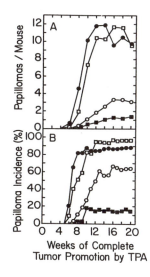

Figure 7. Effects of 10 μmol commercial TA applied either 20 min before (○), 24 h after (□), or 20 min before and 24 h after (■) each promotion treatment on complete tumor promotion by TPA (●).

inhibit the incidence and yield of skin tumors promoted by TPA (Figure 7). For instance, a single application of TA at -20 min before each promotion treatment with TPA inhibits the tumor incidence by 42 and 25% and the tumor yield by 85 and 68% at 12 and 20 weeks, respectively. But 2 consecutive applications of TA at -20 min before and +24 h after each promotion treatment with TPA inhibit the tumor incidence by 83 and 84% and the tumor yield by 94 and 86% at 12 and 20 weeks, respectively. The same phenomenon was observed with Aleppo gall TA pre- and post-treatment (data not shown).

These results suggest that once the early ODC, HPx, and DNA responses to TPA have been reduced by a TA pretreatment, a prolonged antioxidant protection might be required to maximally inhibit the tumor-promoting activity of TPA. The frequency of TPA promotion being twice a week (on Mondays and Thursdays), it is unclear whether a TA post-treatment applied at +24 h on Tuesday enhances the antitumor-promoting effects of either the TA pretreatment at -20 min given on Monday, the one given on Thursday, or both. Incidentally, the tumor incidence is significantly greater when each TPA treatment is followed by TA at +24 h perhaps because TA at this time can still inhibit some of the local toxic effects, which normally impair the tumor-promoting activity of TPA.

Effects in Two-Stage Tumor Promotion. Fewer skin tumors develop in SENCAR mice using the initiation/two-stage promotion protocol (Figures 8 and 9). In the control group promoted 4 times with TPA (stage 1) followed by mezerein up to 20 weeks (stage 2), about 89% of the mice bear papillomas and there are 6 papillomas/mouse. When applied 20 min before each of the 4 applications of TPA in the first stage, 10 μmol commercial TA or Aleppo gall TA and 20 μmol commercial TA do not inhibit significantly the incidence and yield of skin tumors promoted after 20 weeks (Figure 8). In contrast, when applied 20 min before each promotion treatment with mezerein in the second stage, these 3 TA pretreatments inhibit the tumor incidence by 28, 60, and 72% and the tumor yield by 75, 80, and 93%, respectively, at 20 weeks (Figure 9). Under these conditions, therefore, commercial TA and Aleppo gall TA inhibit the tumor-promoting activity of mezerein in stage 2 but not that of TPA in stage 1.

These results suggest that the effectiveness of HTs against complete skin tumor promotion in the preceding experiments (Figures 4, 5, and 6) may be linked to their ability to inhibit the stage 2 rather than the stage 1 tumor-promoting effects of TPA. This hypothesis is substantiated by the fact that HTs inhibit the effects of various tumor promoters on ODC induction and HPx production (26,29), two responses that have been associated with the second stage of tumor promotion (4,25,31). It should be noted that the protease inhibitor N-tosyl-L-phenylalanine chloromethyl ketone inhibits stage 1 promotion by TPA but neither stage 2 promotion by mezerein nor the ODC, xanthine oxidase (XO), HPx, or hyperplastic responses to TPA (4,33,34). Even though TA can inhibit proteolytic reactions (35), it does not mimic the effects of protease inhibitors on TPA promotion. The mechanism by which HTs inhibit skin tumor promotion, therefore, is unlikely to involve the inhibition of epidermal proteolytic enzymes.

Pure Ellagi- and Gallotannins. Since the TA preparations used in the above studies are mixtures of various polygalloylglucoses, it is important to confirm the antitumor-promoting effects of HTs with pure, chemically defined compounds. Pentagalloylglucose (Figure 2) is the core of Chinese and sumac gallotannins (1).

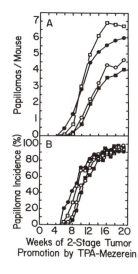

Figure 8. Promotion of skin tumors by the two-stage TPA-mezerein protocol (●): effects of 10 μmol Aleppo gall TA (O), and 10 (□) or 20 (■) μmol commercial TA applied 20 min before each promotion treatment with TPA in stage 1.

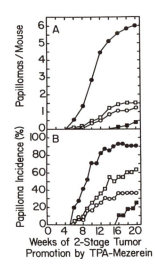

Figure 9. Promotion of skin tumors by the two-stage TPA-mezerein protocol (●): effects of 10 μmol Aleppo gall TA (O), and 10 (□) or 20 (■) μmol commercial TA applied 20 min before each promotion treatment with mezerein in stage 2.

Castalagin and vescalagin (Figure 10) are the 2 main esters of HHDP purified from pedonculate oak heartwood (*36*). These ellagitannins are unique because they contain the D-glucose residue in its open-chain form and are C-glycosides (*1*). Castalagin is isomerized to vescalagin by an epimerization at the benzylic C-1 position of the sugar residue. When applied at a dose of 5 μmol 20 min before TPA, these pure ellagi- and gallotannins can inhibit TPA-induced ODC activity, HPx production, and DNA synthesis, although to a lesser degree than the heterogenous TA samples tested before (Table IV). In contrast to TA, EA, and PG, castalagin and vescalagin are weak antioxidants and pentagalloylglucose fails to inhibit the HPx response to TPA (Table IV), suggesting that the presence of polygalloyl chains and/or some GA, digallic acid, and trigallic acid contaminants in heterogenous TA samples may play a significant role in the mechanism by which these HTs inhibit TPA-induced HPx production. Moreover, the ability of castalagin and vescalagin to interact with other organic molecules may be very low as compared to most other tannins.

Castalagin (R $_1$=H, R $_2$=OH)
Vescalagin (R $_1$=OH, R $_2$=H)

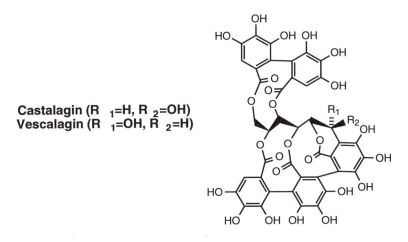

Figure 10. Examples of purified ellagitannins.

Mechanisms of Anti-Tumor Promotion by HTs

The mechanism by which TA inhibits the effects of TPA-type tumor promoters is unknown (*26–29*).

Inhibition of ODC Induction. The inhibition of TPA-induced ODC activity by TA is reversible and cannot be explained on the basis of cytotoxicity, pH fluctuation, or traces of polyphenols interacting directly with components of the enzymic assay, suggesting that TA interferes with the action of TPA and/or the molecular pathways regulating enzyme activities (*26*). Since TA is more effective against ODC induction when applied before rather than after TPA treatment (*26*), it might inhibit the binding of TPA to the phorbol ester receptor, Ca^{2+} mobilization, protein kinase C (PKC) activation or down-regulation, ODC mRNA expression, or

Table IV. Inhibitory Effects of Pure Ellagi- and Gallotannins on TPA-induced Ornithine Decarboxylase Activity (ODC), Hydroperoxide (HPx) Production and DNA Synthesis in Mouse Epidermis *In Vivo*

Treatment[a] (dose/application)	ODC activity[b]		HPx production[c]		DNA synthesis[d]	
	nmol CO_2/hr/ mg protein	% of TPA	nmol H_2O_2/4 h/ mg protein	% of TPA	cpm per µg DNA	% of TPA
Control	0.44 ± 0.06	—	11.7 ± 0.5	—	64.2 ± 2.0	—
TPA (8.5 nmol)	12.56 ± 1.29	100	32.6 ± 1.3	100	382.6 ± 15.5	100
+ pentagalloyl glucose (5 µmol)	4.17 ± 1.04	31	35.3 ± 2.2	113	254.1 ± 38.1	60
+ castatagin (5 µmol)	8.95 ± 1.34	70	27.2 ± 1.0*	74		
+ vescalagin (5 µmol)	9.57 ± 1.53*	75	26.2 ± 0.2	70		

[a]Ellagi- and gallotannins were applied 20 min before each TPA treatment; [b]Five h after a single TPA treatment, values are the mean ± S.D. from 6 samples; [c]Sixteen h after 2 TPA treatments 48 h apart, values are the mean ± S.D. from 4 samples; [d]Sixteen h after a single TPA treatment, values are the mean ± S.D. from 3 samples.
*Significantly different from TPA value, p<0.005, smaller than TPA.

protein synthesis (*37–40*). EA and TA afford total protection against TPA-induced HPx production for 16 h (*29*), much longer than the duration of their effectiveness against ODC induction (*26*). Since doses of EA unable to affect the ODC response to TPA totally inhibit TPA-induced HPx production, HTs do not inhibit the induction of ODC activity by TPA because of their potent antioxidant activities (*26,29*). TA, EA, and GA have decreasing M.Ws. and decreasing inhibitory effects on tumor promotion, suggesting that oligomeric HTs have more anti-tumor promoting activities than monomeric HTs because their more highly polymerized phenolic structures allow for the formation of more H bonds, increasing their binding affinity to macromolecules. Interestingly, TA treatments have also been shown to stimulate or facilitate the activities of epidermal enzymes involved in xenobiotic detoxification (*14,20,41*), suggesting that the inhibitory effects of TA *in vivo* are not solely due to nonspecific protein complexation and enzyme inactivation.

Inhibition of HPx Production. TPA induces HPx formation because it inhibits the antioxidant protective system and stimulates the activities of various endogenous enzymic and non-enzymic sources of reactive O_2 species (*5,6,25,31, 34,42*). TA, which inhibits TPA-induced HPx production, may decrease oxidative stress and enhance antioxidant protection. Tannins, which have a high reducing power and form complexes with various metal ions and cofactors (*1–3*), may chelate Fe and inhibit the Fe-catalyzed reactions generating free radicals involved in multistage carcinogenesis (*6*). After TA treatment, the TPA-stimulated levels of HPx are reduced by almost 100% up to 16 h and by about 35% thereafter (*29*). This sequence suggests a dual effect of TA. At first, this compound may both inhibit the TPA-activated enzymes generating HPx by 35% and directly scavenge the HPx produced by 100%. Later, when TA is metabolized or eliminated from the epidermis, its scavenging activity disappears while the enzymic production of HPx remains inhibited by 35%. Because the XO-inducing and HPx-producing activities of the TPA-treated epidermis appear to be linked to DNA synthesis, hyperplasia, and stage 2 promotion (*25,31,34*), it is significant to show that HTs are natural polyphenolic antioxidants capable of inhibiting HPx production, complete tumor promotion by TPA, and stage 2 tumor promotion by mezerein.

Inhibition of DNA Synthesis and Other Effects of HTs. TA fails to alter TPA-induced DNA synthesis at 16 h when it is applied 15 h after the tumor promoter, i.e. 20 min before the 40-min period of pulse-labeling with ^{3}H-thymidine. Thus, HTs appear to inhibit the mechanism by which TPA stimulates DNA synthesis without interfering directly with the incorporation of precursor into this nucleic acid. *In vitro*, TA inhibits the activity of poly(ADP-ribose) glycohydrolase, an enzyme involved in gene expression and cell proliferation and differentiation (*43*). Moreover, the dimeric ellagitannin agrimoniin may inhibit the growth of transplantable tumors by enhancing the immune response of the hosts (*44*). TA and several other HTs and CTs are also immunopotentiators *in vitro* (*45,46*). The failure of Aleppo gall TA and commercial TA to inhibit the first stage of tumor promotion is a little bit odd since TPA-induced protease activity and DNA synthesis are essential for this process (*4,47*). Perhaps 20 μmol of sumac leaf TA would have been effective since this extract is significantly more potent against complete tumor promotion. Theoretically, a single TPA treatment might be sufficient for stage 1 promotion, suggesting that the remaining TA-inhibited effects of 4 TPA treatments might still be enough to fully trigger this stage of tumorigenesis.

Antitumor-Promoting Effects of CTs

Structural and functional differences between HTs and CTs are blurred by the fact that some CTs contain GA substituents [For example, catechin gallate and epigallocatechin gallate (EGCG)]. EGCG, the main polyphenolic constituent of tannin in green tea, inhibits the specific binding of ^{3}H-TPA to mouse skin, the number of phorbol ester receptors, the activation of PKC by teleocidin, TPA-induced ODC activity, H_2O_2 formation, and oxidative DNA damage, and skin tumor promotion by TPA, teleocidin, or UV-B radiation in DMBA-initiated mice (48–51). But (+)-catechin does not inhibit tumor promotion or the activation of PKC by teleocidin (52). And in contrast to PG, (+)-catechin and (-)-epicatechin fail to inhibit, or inhibit to a lesser degree, TPA-induced epidermal lipoxygenase and ODC activities, DNA synthesis, and skin tumor promotion (53,54), suggesting that the GA moiety of EGCG may be essential for its antitumor-promoting activity. The galloylation of flavonoids strongly influences the degree of astringency of the tannin molecule. Addition of galloyl groups as esters at C-3 of the pyran ring increases the cytotoxicity and antiviral activity of various CTs (3). Moreover, the activation of human leukocytes and HL-60 cells by EGCG may be attributable to the special location of the galloyl group relative to that of the B-ring in the flavone skeleton (45). Interestingly, ellagitannins inhibit poly(ADP-ribose) glycohydrolase activity more than gallotannins whereas CTs have no significant inhibitory effects (55). Therefore, it is crucial to determine if several samples of pure or heterogenous CTs originating from various sources can also mimic the inhibitory effects of HTs on the biochemical markers of TPA promotion in mouse epidermis *in vivo*.

Heterogenous CT Samples from Various Sources. Many CT extracts are composed of oligomers or polymers with an average M.W. corresponding to 5–9 flavonoid units (3). CTs from guamuchil, loblolly pine, and southern red oak barks are mixtures of related oligomers with various numbers of C-15 flavan units. The M.W. distributions of these heterogenous CTs are unknown, making it difficult to compare their effectiveness as inhibitors of TPA-induced ODC activity and HPx production. When applied topically 20 min before TPA, these oligomeric CTs all inhibit to various degrees the ODC and HPx responses linked to skin tumor promotion (Table V). Loblolly pine bark CT, especially, is an effective antioxidant. In contrast, catechin even at a dose of 20 μmol (6 mg) fails to inhibit the HPx response to TPA, suggesting that the antioxidant activity of CTs may increase with the number of monomer units attached by interflavonoid linkage in their molecules. The metal-chelating, antioxidant, and free radical-scavenging activities of CTs may be associated with the reducibility of the B-ring, which is proportional to the increasing number of OH substituents in phenol, catechol, or pyrogallol (3). Indeed, the antioxidant and antiviral activities of CTs increase with the number of pyrogallol moieties in the molecule and the degree of polymerization (56).

Flavonoid Derivatives. Phytoalexins are natural biocides produced by plants in response to attack by pathogens, particularly microorganisms. Flavonoids are the most important chemical class of phytoalexins (57,58). Two sets of phytoalexin analogues of CT have been synthesized and their antifungal and antibacterial activities characterized (Figure 11). Catechin dialkyl ketals are prepared by the reaction of catechin with the appropriate ketone (*57*). Epicatechin-4-alkylsulphides are synthesized by acidic thiolysis of CT with alkylthiols (*58*). The antibacterial activities of these semisynthetic CT biocides may be linked to their number of OH groups, length of their alkyl chain, and their degree of lipophilicity. Catechin undecanone and epicatechin-4-dodecyl-sulfide applied topically 20 min before TPA both decrease the HPx-producing activity of the epidermis and inhibit TPA-induced ODC activity to a much greater degree than catechin or epicatechin (Table VI). Catechin acetone, catechin nonanone, catechin dihexylketone, catechin-2-tridecanone, epicatechin-4-hexysulfide, epicatechin-4-nonylsulfide, epicatechin-4-decylsulfide, and epicatechin-4-hexadecylsulfide (Figure 11) are also much more effective than catechin or epicatechin against TPA-induced ODC activity but have weak antioxidant activities (data not shown). These results suggest that certain monomeric flavonoid derivatives may have a higher degree of reactivity and/or lipophilicity than their naturally occurring parent molecules.

Catechin diakyl ketals

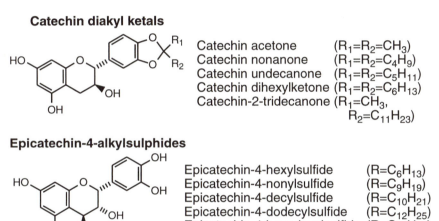

Catechin acetone $(R_1=R_2=CH_3)$
Catechin nonanone $(R_1=R_2=C_4H_9)$
Catechin undecanone $(R_1=R_2=C_5H_{11})$
Catechin dihexylketone $(R_1=R_2=C_6H_{13})$
Catechin-2-tridecanone $(R_1=CH_3,$
                        $R_2=C_{11}H_{23})$

Epicatechin-4-alkylsulphides

Epicatechin-4-hexylsulfide $(R=C_6H_{13})$
Epicatechin-4-nonylsulfide $(R=C_9H_{19})$
Epicatechin-4-decylsulfide $(R=C_{10}H_{21})$
Epicatechin-4-dodecylsulfide $(R=C_{12}H_{25})$
Epicatechin-4-hexadecylsulfide $(R=C_{16}H_{33})$

Figure 11. Examples of monomeric flavonoid derivatives.

Conclusions

HTs, especially TA, may be universal inhibitors of the multistage process of tumorigenesis. They were already known to inhibit tumor initiation and now we have demonstrated that they are also very effective against complete and stage 2 tumor promotion. Moreover, there is preliminary evidence to suggest that various CTs and their analogues might also inhibit tumor promotion. Both HTs and CTs, therefore, might inhibit tumorigenesis when applied directly to the target organ. It is premature, however, to suggest that tannins could be used to develop new means of cancer prevention and/or therapy. Studies of the systemic effects of tannins are required to demonstrate that, in spite of the molecular sizes of polymerized tannins and tannin-protein complexes limiting solubility and incorporation into cells, the

Table V. Inhibitory Effects of Various Condensed Tannins (CTs) on TPA-induced Ornithine Decarboxylase Activity (ODC) and Hydroperoxide (HPx) Production in Mouse Epidermis *In Vivo*

Treatment[a] (dose/application)	ODC activity[b]			HPx production[c]		
	nmol CO_2/h/ mg protein	% of control	% of TPA	nmol H_2O_2/4 h/ mg protein	% of control	% of TPA
Control	0.33 ± 0.03	100	—	10.2 ± 0.1	100	—
TPA (8.5 nmol)	10.08 ± 0.96	3055	100	29.1 ± 0.5	285	100
+ catechin (6 mg)	5.22 ± 0.53	1582	50	29.3 ± 1.6	287	101
+ guamuchil bark CT (6 mg)	6.12 ± 0.54	1855	59	25.2 ± 1.1***	247	80
+ loblolly pine bark CT (6 mg)	5.85 ± 0.42*	1773	57	17.0 ± 1.5****	166	36
+ southern red oak bark CT(6 mg)	6.97 ± 0.83**	2112	68	23.9 ± 1.5	235	73

[a]CTs applied 20 min before each TPA treatment; [b]Five h after a single TPA treatment, mean $\pm$ S.D. (n=6); [c]Sixteen h after 2 TPA treatments 48 h apart, mean $\pm$ S.D. (n=4). *Significantly different from TPA + southern red oak bark CT, $p<0.025$; **Significantly different from TPA, $p<0.0005$; ***Significantly different from TPA, $p<0.005$; ****Significantly different from TPA + southern red oak bark CT, $p<0.005$.

Table VI. Inhibitory Effects of Monomeric Flavonoid Derivatives on TPA-induced Ornithine Decarboxylase Activity (ODC) and Hydroperoxide (HPx) Production in Mouse Epidermis *In Vivo*

Treatment[a] (dose/application)	ODC activity[b]			HPx production[c]		
	nmol CO_2/h/ mg protein	% of control	% of TPA	nmol H_2O_2/4 h/ mg protein	% of control	% of TPA
Control	0.31 ± 0.04	100	—	13.2 ± 0.7	100	—
TPA (8.5 nmol)	14.34 ± 1.98	4626	100	30.7 ± 0.5	233	100
+ catechin (20 µmol)	8.25 ± 1.10	2661	57	31.3 ± 1.9	237	103
+ catechin undecanone (20 µmol)	1.81 ± 0.19	584	11	21.9 ± 1.6	166	50
+ epicatechin (20 µmol)	6.88 ± 0.96	2219	47	30.9 ± 1.2	234	101
+ epicatechin-4-dodecyl sulfide (20 µmol)	2.54 ± 0.40	819	16	22.6 ± 1.4	171	54

[a]Monomeric flavonoids were applied 20 min before each TPA treatment; [b]Five h after a single TPA treatment, mean $\pm$ S.D. (n=6); [c]Sixteen h after 2 TPA treatments 48 h apart, mean $\pm$ S.D. (n=4).

uptake of HTs and CTs administered i.p., i.v., or orally is sufficient to protect distant tissues from various tumor-promoting events without causing undue acute or delayed toxicity.

Acknowledgments

This work was supported by the American Cancer Society (Grant CN-24), the Department of Health and Human Services, National Cancer Institute (Grant CA56662), the Wesley Foundation of Wichita (Wesley Scholar Program: Molecular Biology and Cell Growth Regulation), BioServe Space Technologies, and the Center for Basic Cancer Research, Kansas State University.

Literature Cited

1. Haslam, E. In *Biochemistry of Plant Phenolics*; Swain, T.; Harborne, J. B.; Van Sumere, C. F., Eds.; Recent Advances in Phytochemistry; Plenum Press: New York, 1979; Vol. 12, pp 475–523.
2. Zucker, W. V. *Am. Nat.* **1983**, *121*, 335–365.
3. *Chemistry and Significance of Condensed Tannins*; Hemingway, R. W.; Karchesy, J. J., Eds.; Plenum Press: New York, 1989.
4. Slaga, T. J. In *Mechanisms of Tumor Promotion*; Slaga, T. J., Ed.; CRC Press: Boca Raton, FL, 1985; Vol. 2, pp 189–196.
5. Perchellet, J. P.; Perchellet, E. M. *ISI Atlas Sci. Pharmacol.* **1988**, *2*, 325–333.
6. Perchellet, J. P.; Perchellet, E. M. *Free Radical Biol. Med.* **1989**, *7*, 377–408.
7. Wood, A. W.; Huang, M. T.; Chang, R. L.; Newmark, H. L.; Lehr, R. E.; Yagi, H.; Sayer, J. M.; Jerina, D. M.; Conney, A. H. *Proc. Natl. Acad. Sci. USA* **1982**, *79*, 5513–5517.
8. Lesca, P. *Carcinogenesis* **1983**, *4*, 1651–1653.
9. Del Tito, B. J.; Mukhtar, H.; Bickers, D. R. *Biochem. Biophys. Res. Commun.* **1984**, *114*, 388–394.
10. Mukhtar, H.; Das, M.; Del Tito, B. J.; Bickers, D. R. *Biochem. Biophys. Res. Commun.* **1984**, *119*, 751–757.
11. Mukhtar, H.; Del Tito, B. J.; Marcelo, C. L.; Das, M.; Bickers, D. R. *Carcinogenesis* **1984**, *5*, 1565–1571.
12. Chang, R. L.; Huang, M. T.; Wood, A. W.; Wong, C. Q.; Newmark, H. L.; Yagi, H.; Sayer, J. M.; Jerina, D. M.; Conney, A. H. *Carcinogenesis* **1985**,*6*, 1127–1133.
13. Zee-Cheng, R. K. Y.; Cheng, C. C. *Drugs of the Future* **1986**, *11*, 1029–1033.
14. Mukhtar, H.; Das, M.; Bickers, D.R. *Cancer Res.* **1986**, *46*, 2262–2265.
15. Teel, R. W.; Martin, R. M.; Allahyari, R. *Cancer Lett.* **1987**, *36*, 203–211.
16. Das, M.; Mukhtar, H.; Bik, D. P.; Bickers, D. R. *Cancer Res.* **1987**, *47*, 760–766.
17. Das, M.; Khan, W. A.; Asokan, P.; Bickers, D. R.; Mukhtar, H. *Cancer Res.* **1987**, *47*, 767–773.
18. Mukhtar, H.; Das, M.; Khan, W. A.; Wang, Z. Y.; Bik, D. P.; Bickers, D. R.*Cancer Res.* **1988**, *48*, 2361–2365.
19. Hayatsu, H.; Arimoto, S.; Negishi, T. *Mutat. Res.* **1988**, *202*, 429–446.
20. Athar, M.; Khan, W. A.; Mukhtar, H. *Cancer Res.* **1989**, *49*, 5784–5788.
21. Vance, R. E.; Teel, R. W. *Cancer Lett.* **1989**, *47*, 37–44.
22. Das, M.; Bickers, D. R.; Mukhtar, H. *Intl. J. Cancer* **1989**, *43*, 468–470.

23. Stoner, G. D.; Mandal-Chaudhuri, S.; Daniel, E. M.; Yeur, Y. H. *Proc. Am. Assoc. Cancer Res.* **1991**, *32*, 472.
24. Wilson, T.; Lewis, M. J.; Cha, K. L.; Gold, B. *Cancer Lett.* **1992**, *61*, 129–134.
25. Perchellet, E. M.; Jones, D.; Perchellet, J. P. *Cancer Res.* **1990**, *50*, 5806–5812.
26. Gali. H. U.; Perchellet, E. M.; Perchellet, J. P. *Cancer Res.* **1991**, *51*, 2820–2825.
27. Perchellet, J. P.; Gali, H. U.; Perchellet, E. M.; Klish, D. S.; Armbrust, A. D. In *Plant Polyphenols: Biogenesis, Chemical Properties, and Significance*; Hemingway, R.W., Ed.; Plenum Press: New York, 1992; pp 783–801.
28. Gali, H. U.; Perchellet, E. M.; Klish, D. S.; Johnson, J. M.; Perchellet, J. P. *Carcinogenesis* **1992**, *13*, 715–718.
29. Gali. H. U.; Perchellet, E. M.; Klish, D. S.; Johnson, J. M.; Perchellet, J. P. *Intl. J. Cancer* **1992**, *51*, 425–432.
30. Perchellet, J. P.; Boutwell, R. K. *Cancer Res.* **1981**, *41*, 3918–3926.
31. Perchellet, E. M.; Perchellet, J. P. *Cancer Res.* **1989**, *49*, 6193–6201.
32. Perchellet, J. P.; Boutwell, R. K. *Cancer Res.* **1981**, *41*, 3927–3935.
33. Perchellet, J. P.; Perchellet, E. M.; Gali, H. U. *Proc. Am. Assoc. Cancer Res.* **1992**, *33*, 186.
34. Pence, B. C.; Reiners, J. J. *Cancer Res.* **1987**, *47*, 6388–6392.
35. Mole, S.; Waterman, P. G. *Phytochemistry* **1987**, *26*, 99–102.
36. Scalbert, A.; Duval, L.; Peng, S.; Monties, B.; DuPenhoat, C. *J. Chromatogr.* **1990**, *502*, 107–119.
37. Blumberg, P. M. *Cancer Res.* **1988**, *48*, 1–8.
38. Verma, A. K.; Pong, R. C.; Erickson, D. *Cancer Res.* **1986**, *46*, 6149–6155.
39. Gilmour, S. K.; Avdalovic, N.; Madara, T.; O'Brien, T. G. *J. Biol. Chem.* **1985**, *260*, 16439–16444.
40. Droms, K. A.; Malkinson, A. M. *Mol. Carcinogenesis* **1991**, *4*, 1–2.
41. Das, M.; Bickers, D. R.; Mukhtar, H. *Carcinogenesis* **1985**, *6*, 1409–1413.
42. Perchellet, E. M.; Abney, N. L.; Perchellet, J. P. *Cancer Lett.* **1988**, *42*, 169–177.
43. Tanuma, S.; Sakagami, H.; Endo, H. *Biochem. Int.* **1989**, *18*, 701–708.
44. Miyamoto, K.; Kishi, N.; Koshiura, R. *Japan. J. Pharmacol.* **1987**, *43*,187–195.
45. Sakagami, H.; Hatano, T.; Yoshida, T.; Tanuma, S. I.; Hata, N.; Misawa,Y., Ishii, N.; Tsutsumi, T.; Okuda, T. *Anticancer Res.* **1990**, *10*, 1523–1532.
46. Sakagami, H.; Asano, K.; Tanuma, S. I.; Hatano, T.; Yoshida, T.; Okuda,T. *Anticancer Res.* **1992**, *12*, 377–388.
47. Kinzel, V.; Loehrke, H.; Goerttler, K.; Fürstenberger, G.; Marks, F. *Proc. Natl. Acad. Sci. USA* **1984**, *81*, 5858–5862.
48. Yoshizawa, S.; Horiuchi, T.; Fujiki, H.; Yoshida, T.; Okuda, T.; Sugimura, T. *Phytother. Res.* **1987**, *1*, 44–47.
49. Wang, Z. Y.; Agarwal, R.; Bickers, D.; Mukhtar, H. *Carcinogenesis* **1991**,*12*, 1527–1530.
50. Wang, Z. Y.; Huang, M. T.; Ferraro, T.; Wong, C. Q.; Lou, Y. R.; Reuhl, K.; Iatropoulos, M.; Yang, C. S.; Conney, A. H. *Cancer Res.* **1992**, *52*, 1162–1170.
51. Agarwal, R.; Katiyar, S. K.; Zaidi, S. I. A.; Mukhtar, H. *Cancer Res.* **1992**, *52*, 3582–3588.
52. Horiuchi, T.; Fujiki, H.; Hakii, H.; Suganuma, M.; Yamashita, K.; Sugimura, T. *Japan. J. Cancer Res. (Gann)* **1986**, *77*, 526–531.
53. Kozumbo, W. J.; Seed, S. L.; Kensler, T. W. *Cancer Res.* **1983**, *43*, 2555–2559.
54. Nakadate, T.; Yamamoto, S.; Aizu, S.; Kato, R. *Gann* **1984**, *75*, 214–222.

55. Tsai, Y. J.; Abe, H.; Maruta, H.; Hatano, T.; Nishina, H.; Sakagani, H.; Okuda, T.; Tanuma, S. I. *Biochem. Int.* **1991**, *24*, 889–897.
56. Laks, P. E. In *Chemistry and Significance of Condensed Tannins*; Hemingway, R. W.; Karchesy, J. J., Eds.; Plenum Press: New York, 1989; pp 249–263.
57. Laks, P. E.; Pruner, M. S. *Phytochemistry* **1989**, *28*, 87–91.
58. Laks, P. E. *Phytochemistry* **1987**, *26*, 1617–1621.

RECEIVED June 22, 1993

PHYTOCHEMICALS IN SOYBEANS

Chapter 26

Soybean Saponin and Isoflavonoids

Structure and Antiviral Activity against Human Immunodeficiency Virus In Vitro

Kazuyoshi Okubo[1], Shigemitsu Kudou[2], Teiji Uchida[2], Yumiko Yoshiki[1], Masaki Yoshikoshi[3], and Masahide Tonomura[1]

[1]Department of Applied Biological Chemistry, Faculty of Agriculture, Tohoku University, 1−1 Tsutsumidori, Amamiyamachi, Aoba-ku, Sendai 981, Japan
[2]Kanesa Company, Ltd., 202 Hamada, Tamagawa, Aomori 030, Japan
[3]Nestle Japan, Ltd., 1−16, 7−Chome, Goko-Dori, Chuo-ku, Kobe 651, Japan

Nine kinds of isoflavone glycosides were isolated from the hypo-cotyls of soybean seeds. Three were proved to be malonylated soybean isoflavones named 6″-O-malonyldaizin, 6″-O-malonylglycitin and 6″-O-malonylgenistin. Soyasaponins are divided into three groups according to their respective type of aglycone, soyasapo-genol A, B and E. Bb, major constituent of group B saponins, completely inhibited HIV-induced cytopathic effects and virus-specific antigen expression 6 days after infection at concentration greater than 0.25 mg/ml, but had no direct effect on HIV reverse transcriptase activity. Bb also inhibited HIV-induced cell fusion in the MOLT-4 cell system.

Soybean seeds have been consumed by humans for thousands of years because they are rich in protein and their oil is of good nutritional quality. Most East Asians consume soybean seeds regularly from childhood via a variety of soybean products. The incidence of breast and colon cancer in Oriental people is considerably lower than in those living in Western countries (*1*), who seldom eat soybean products. Additionally, vegetarians, who are also at decreased risk of breast and colon cancer, frequently consume soybean-based meat substitutes (*2*). These associations suggest that soybeans may play a role in reducing breast and colon cancer risk.

Relative to other food stuffs, soybean seeds contain high levels of four classes of compound with demonstrated anticancer activity: glycosides, phytosterols, protease inhibitors, and phytic acid. Soybean seeds contain about 2% glycosides, which are composed of several kinds of soyasaponins and isoflavonoids (*3*). These two groups of glycosides will be the focus of this presentation, especially their chemical structures and their antiviral activity — such as against human immunodeficiency virus (HIV) and Epstein-Barr virus (EBV).

Chemical Structure and Distribution

Saponins. Among the various edible legume beans, soybeans contain the highest amount of saponins (*3*). Depending upon the chemical nature of their aglycones, saponins can be divided into triterpene saponins and steroidal saponins. Furthermore, depending upon the skeleton of the aglycone, triterpene saponins are subdivided into oleanane, urasane, dammarane and cycloartane types (*4*). Soyasaponins are oleanane-type triterpene saponins. Two nomenclature systems (from Kitagawa and Okubo, Table I) have been used to identify various soyasaponins. For the sake of consistency, only the Okubo system will be used here.

Table I. Soyasaponin Nomenclature Assigned by the Research Groups of Kitagawa and Okubo

	Kitagawa	Okubo
A group — acetylated[a]		
glc-gal-glcUA-A-ara-xyl (2,3,4-triAc)	acetylA4	Aa
glc-gal-glcUA-A-ara-glc (2,3,4,6-tetraAc)	acetylA1	Ab
rha-gal-glcUA-A-ara-glc (2,3,4,6-tetraAc)	—	Ac
glc-ara-glcUA-A-ara-glc (2,3,4,6-tetraAc)	—	Ad
gal-glcUA-A-ara-xyl (2,3,4-triAc)	acetylA5	Ae
gal-glcUA-A-ara-glc (2,3,4,6-tetraAc)	acetylA2	Af
ara-glcUA-A-ara-xyl (2,3,4-triAc)	acetylA6	Ag
ara-glcUA-A-ara-glc (2,3,4,6-tetraAc)	acetylA3	Ah
A group — deacetylated[a]		
glc-gal-glcUA-A-ara-xyl (2,3,4-triAc)	A4	deacetyl Aa
glc-gal-glcUA-A-ara-glc (2,3,4,6-tetraAc)	A1	deacetyl Ab
gal-glcUA-A-ara-xyl (2,3,4-triAc)	A5	deacetyl Ae
gal-glcUA-A-ara-glc (2,3,4,6-tetraAc)	A2	deacetyl Af
rha-gal-glcUA-A[b]	—	—
ara-glcUA-A-ara-xyl (2,3,4-triAc)	A6	deacetyl Ag
ara-glcUA-A-ara-glc (2,3,4,6-tetraAc)	A3	deacetyl Ah
B group[c]		
glc-gal-glcUA-B	V	Ba
rha-gal-glcUA-B	I	Bb
rha-ara-glcUA-B	II	Bc
gal-glcUA-B	III	Bb'
ara-glcUA-B	IV	Bc'
E group[d]		
glc-gal-glcUA-E		Bd
rha-gal-glcUA-E		Be

[a] Sugar chains on the left of A are linked to 3-*O* and those on the right to 22-*O*.
[b] Isolated by Curl *et al.* and given the name soyasaponin A3.
[c] Soyasapogenol B contains a hydroxyl moiety at C-22.
[d] Soyasapogenol E contains a ketone moiety at C-22.

Recently, we have determined the structures of genuine saponins in soybean seeds, which are named soyasaponins α_g, β_g, β_a, γ_g and γ_a and are characterized as the aglycones of soyasaponins Ba, Bb, Bc, Bb′ and Bc′, respectively, with 2,3-dihydro-2,5-dihydroxy-6-methyl-4*H*-pyran-4-one (DDMP) attached by acetal linkage to C-22 (see Kudou *et al.*, Chapter 27). The structures of typical soyasaponins are shown in Figure 1; these are not only confined to soybeans but may also be found in other beans and peas, and in alfalfa. Soyasaponins are subdivided into three groups — A, B and E — depending upon whether the aglycone is olean-12-en-3,21,22,24-tetraol (soyasapogenol A), olean-12-en-3,22, 24-triol (soyasapogenol B), or olean-12-en-3,24-diol-22-one (soyasapogenol E). Group A saponins which contain two ether-linked sugar chains attached to C-3 and 22 are termed bis-desmosides, and groups B and E which contain sugars attached to C-3 alone are thus mono-desmosides.

Five kinds of soyasaponins (Bb, Bc, Bb′, deacetylated Ab and deacetylated Af) were isolated and completely characterized by Kitagawa *et al.* (*5–7*) in 1985. More recently, using a different method (without saponification), Okubo *et al.* (*8–9*) isolated and identified 11 kinds of intact soyasaponins (six group A and five group B) from the hypocotyl of the seeds.

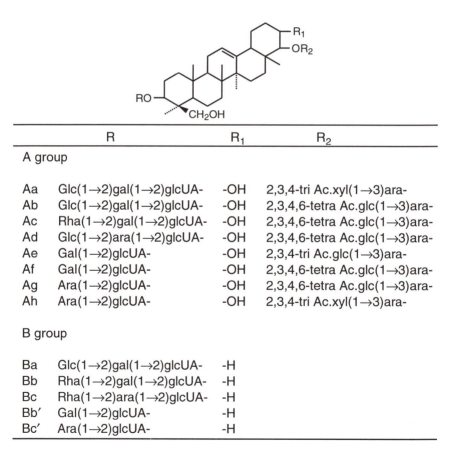

	R	R$_1$	R$_2$
A group			
Aa	Glc(1→2)gal(1→2)glcUA-	-OH	2,3,4-tri Ac.xyl(1→3)ara-
Ab	Glc(1→2)gal(1→2)glcUA-	-OH	2,3,4,6-tetra Ac.glc(1→3)ara-
Ac	Rha(1→2)gal(1→2)glcUA-	-OH	2,3,4,6-tetra Ac.glc(1→3)ara-
Ad	Glc(1→2)ara(1→2)glcUA-	-OH	2,3,4,6-tetra Ac.glc(1→3)ara-
Ae	Gal(1→2)glcUA-	-OH	2,3,4-tri Ac.glc(1→3)ara-
Af	Gal(1→2)glcUA-	-OH	2,3,4,6-tetra Ac.glc(1→3)ara-
Ag	Ara(1→2)glcUA-	-OH	2,3,4,6-tetra Ac.glc(1→3)ara-
Ah	Ara(1→2)glcUA-	-OH	2,3,4-tri Ac.xyl(1→3)ara-
B group			
Ba	Glc(1→2)gal(1→2)glcUA-	-H	
Bb	Rha(1→2)gal(1→2)glcUA-	-H	
Bc	Rha(1→2)ara(1→2)glcUA-	-H	
Bb′	Gal(1→2)glcUA-	-H	
Bc′	Ara(1→2)glcUA-	-H	

Figure 1. Structures of group A and B saponins.

The sugars found in the oligosaccharides of soyasaponins are D-galactose, L-arabinose, L-rhamnose, D-glucose, D-xylose and D-glucuronic acid. Chain lengths of 2 or 3 saccharide units are the most common. The terminal monosaccharide of the C-22 sugar chain of intact soyasaponin A is usually acetylated (*10*).

The majority of the hypocotyl saponins were found to be group A soyasaponins. Aa and Ab are the major soyasaponins in group A, whereas Ba and Bb are two major soyasaponins in group B. In contrast to group A and Ba, which are present only in the hypocotyl, groups B (except Ba) and E are present in both the hypocotyl and the cotyledon. On an equal weight basis, the hypocotyl contains a much greater amount of soyasaponins than the cotyledon (*11*). The hull does not contain soyasaponins (*12*).

Isoflavonoids. Kudou *et al.* (*13*) isolated and characterized nine kinds of iso-flavonoids from soybean hypocotyl. They are daizin, glycitin, genistin, acetylated daizin, acetylated glycitin, acetylated genistin, malonylated daizin, malonylated glycitin and malonylated genistin (Figure 2). The malonylated glucosides were found to be heat labile; at 80°C, the majority of the glucosides were converted to daizin, glycitin and genistin. The non-malonylated compounds and their aglycones were stable at that temperature. Eldridge and Kwolek (*14*) reported six kinds of isoflavonoids in soybean seed. They did not observe the malonylated glucosides, presumably because heat was applied in the extraction procedure.

Isoflavonoids, like soyasaponins, are concentrated in the hypocotyl; the isoflavonoid content of the soybean hull is very low. On an equal weight basis, the hypocotyl contains 5 to 6 times more total isoflavonoids than the cotyledon.

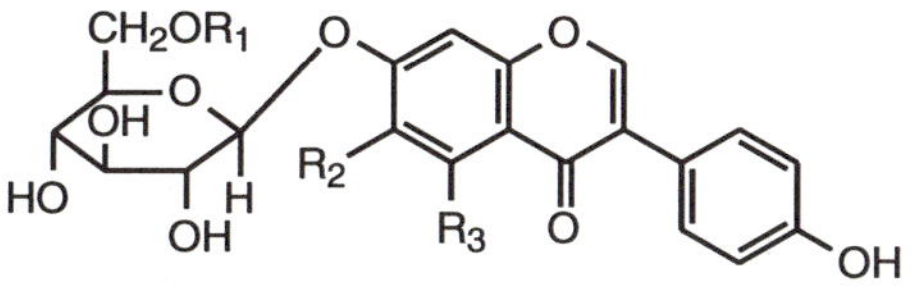

Compound	R_1	R_2	R_3
Daizin	-H	-H	-H
Glycitin	-H	$-OCH_3$	-H
Genistin	-H	-H	-OH
6″-*O*-Malonyldaizin	$-COCH_2COOH$	-H	-H
6″-*O*-Malonylglycitin	$-COCH_2COOH$	$-OCH_3$	-H
6″-*O*-Malonylgenistin	$-COCH_2COOH$	-H	-OH
6″-*O*-Acetyldaizin	$-COCH_3$	-H	-H
6″-*O*-Acetylglycitin	$-COCH_3$	$-OCH_3$	-H
6″-*O*-Acetylgenistin	$-COCH_3$	-H	-OH

Figure 2. Structures of the isoflavones in soybean seeds.

Factors Affecting Saponin and Isoflavonoid Content

The total and relative amounts of soyasaponins vary, depending upon the variety of soybean. Kitagawa *et al.* (*15–16*) found one Chinese variety to be richer in total soyasaponins (0.3% Bb plus deacetylated Ab') than those from Japan, Canada and the United States. The amounts of group A and B saponins in the hypocotyl were 0.36–3.41% and 0.26–2.75% respectively. In a detailed study by Shiraiwa *et al.* (*17*), 457 varieties of soybeans were analyzed, and classified into seven types according to their hypocotyl group A saponin composition. Of the types, three were predominant: the Aa type, with soyasaponin Aa (16.6%); the Ab type, with soyasaponin Ab (76.2%); and the Aa-Ab type, with soyasaponins Aa and Ab (5.5%). Two kinds of group A saponin, Aa and Ab, are present as the main constituent in the hypocotyl of soybean seed.

The saponin composition and content in F1 and F2 seeds derived from the crossing of Aa and Ab types were analyzed. The group A saponin was of Aa-Ab type in all the F1 seeds, and the ratio of Aa type:Aa-Ab type:Ab type was 1:2:1 in the F2 seeds. From these results, it appears that Aa and Ab were controlled by codominant allelic alternatives at a single locus (*18*).

The saponin level and composition in the seed is different from the iso-flavonoid level and composition varies tremendously not only by variety but also by year harvested and geographic location, even among the same variety. One study reported a three-fold variation in total isoflavonoid content among four varieties of soybeans.

There are few studies concerning the effects of cooking or processing on soyasaponins and isoflavonoids. Fermentation was found to reduce the saponin content (*19*), due to enzymatic degradation by microorganisms (*20*). Germination of soybean seeds changed the composition of soyasaponins, especially in the sprout (*21*). Germination under light caused a slight increase in the amount of soyasaponin Bb in the cotyledon, while in the sprout, the amounts of soyasaponins Ba, Bb and Bc (especially Bb) increased greatly. Germination under a 12 hr light/dark cycle caused more dramatic changes. For example, the content of soyasaponin Bb in the sprout was increased 12-fold (on dry weight basis) in 8 days. Kitagawa *et al.* analyzed soy products for soyasaponin content. On a dry weight basis, tofu, dried tofu and soymilk were found to contain more soyasaponins, 0.301–0.407%, than soybeans (0.3%), while the fermentation products miso and natto contained less, 0.148% and 0.264%, respectively. Liener (*22*) reported the same results.

Anticancer Activity

Isoflavonoids have activity against tyrosine protein kinase (*23*), DNA topo-isomerase (*24*) and ribosomal S6 kinase (*25*) and induce specific cytochrome P450s (*26*). In one animal study, carcinogen-induced mammary tumorigenesis was inhibited by feeding the animals isoflavonoid-rich soy products. In postmenopausal women with low levels of estrogen, however, isoflavonoids were found to have estrogenic effect (*27*).

The saponins most extensively investigated for anticancer activity have been glycyrrhizin and its aglycone (*28*). More recently, oleanolic acid (*29–31*), ursolic acid and saponins isolated from other plants (*32*) have also been shown to have anticancer activity. Dietary saponins were found to enhance natural killer cell activity (*33*), presumably with an initial effect on the mucosal immune system due to its poor absorption (*34*). Saponins were also found to be cytotoxic to sarcoma 37 cells (*35*), to inhibit DNA synthesis in tumor cells (*36*), and to decrease the growth of human epidermal carcinoma cells (*37*) and human cervical carcinoma cells (*38*). Glycyrrhetinic acid has also been found to inhibit the specific binding of 12-*O*-tetra-decanoylphorbol-13-acetate to mouse epidermal membrane receptors in a dose- and time-dependent manner associated with increased K_d, but without affecting the number of binding sites (*39*).

Recently, Konoshima and Kozuka (*40*) isolated wistariasaponins D and G, and dehydrosoyasaponin I (Be). They tested the inhibitory effects of these saponins on the activation of Epstein-Barr virus early antigen induced by TPA (32 pmol). Soyasapogenol E exhibited remarkably inhibitory effects at 5×10^2 mol ratio (100% inhibition) and preserved Raji cell viability even at higher doses (1×10^3 mol ratio). These results suggest inhibitory effects by group B saponins.

Inhibitory Effect of Saponins from Soybean on the Infectivity of HIV *In Vitro*

The chemical structure of soyasaponins is very similar to that of glycyrrhizin, a major licorice saponin (*41–42*), which is a known antitumor promoter (*43–46*). Soybean saponins isolated from soybean seeds were investigated for their antiviral activity on HIV, using an HTLV-1-carrying cell line, MT-4 (*47*). Saponin B1 completely inhibited HIV-induced cytopathic effects and virus-specific antigen expression 6 days after infection at concentrations greater than 0.5 mg/ml. Saponin B2 also inhibited HIV infection, although less potently. Both B1 and B2 had no direct effect on the reverse transcriptase activity of HIV. Saponin B1 also inhibited HIV-induced cell fusion in the MOLT-4 cell system. The results of this study suggest that soybean saponins, especially saponin B1, have inhibitory activity against HIV infection.

Furthermore, we examined the inhibitory effect of isolated saponins on the infectivity and cytopathic activity of HIV and considered the relationship between the structure and anti-HIV action of soybean saponins (*48*). Bb, a major constituent of group B saponins, completely inhibited HIV-induced cytopathic effects and virus-specific antigen expression 6 days after infection at concentration greater than 0.25 mg/ml (Figures 3 and 4). Bb had no direct effect on the reverse transcriptase activity of HIV. Also, Bb inhibited HIV-induced cell fusion in the MOLT-4 cell system. Group A saponins, another group of soybean saponins, also showed inhibitory effects against HIV infection *in vitro*, but these effects were minor as compared with that of Bb.

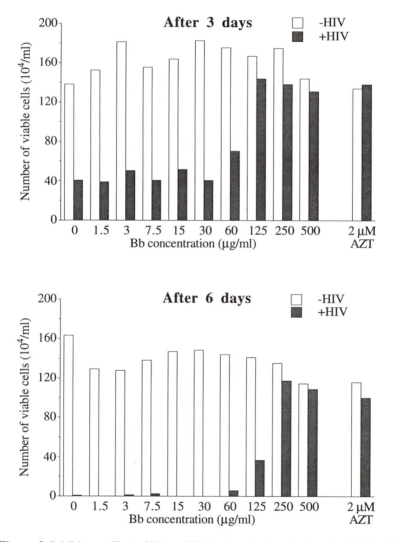

Figure 3. Inhibitory effect of Bb on HIV-induced cytopathology in MT-4 cells.

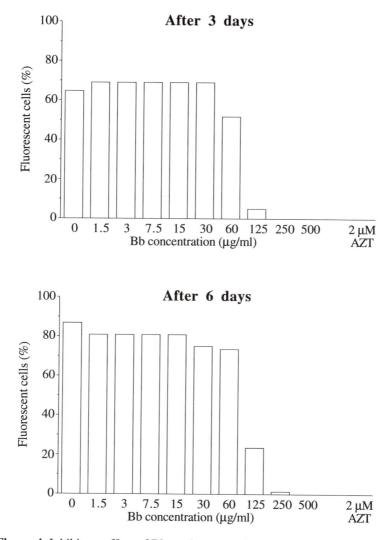

Figure 4. Inhibitory effect of Bb on the expression of HIV-specific antigens in MT-4 cells.

Literature Cited

1. Kurihara, M.; Aoki, K.; Hisamichi, F. UICC Publication, Nagoya University Press: Nagoya, Japan, 1989.
2. Nair, P. N.; Turjuman, N.; Kessie, G.; Calkins, B.; Goodman, G. T.; Davitvitz, H.; Nimmagadda, G. *Am. J. Clin. Nutr.* **1984**, *40*, 927–930.
3. Fenwick, G. R.; Price, R.; Tsukamoto, C.; Okubo, K. *Toxic Substances in Crop Plants* D'Mello, J. P. F.; Duffus, C. M.; Duffus, J. H.; Royal Society of Chemistry: Cambridge, 1991; pp 285–327.
4. Scheline, R. R. *CRC Handbook of Mammaliam Metabolism of Plant Compounds* CRC Press: Boca Raton, FL, 1991; pp.208–213.
5. Kitagawa, I.; Yoshikawa, M.; Wang, H. K.; Saito, M.; Tosirisuk, V.; Fujiwara, T.; Tomita, K. *Chem. Pharm. Bull.* **1982**, *30*, 2294.
6. Kitagawa, I.; Saito, M.; Taniyama, T.; Yoshikawa, M. *Chem. Pharm. Bull.* **1985**, *33*, 1069.
7. Kitagawa, I.; Saito, M.; Taniyama, T.; Yoshikawa, M. *Chem. Pham. Bull.* **1985**, *30*, 598.
8. Shiraiwa, M.; Kudo, S.; Shimoyamada, M.; Harada, K.; Okubo, K. *Agric. Biol. Chem.* **1991**, *55*, 315–322.
9. Shiraiwa, M.; Harada, K.; Okubo, K. *Agric. Biol. Chem.* **1991**, *55*, 911–917.
10. Okubo, K. *Kagaku to Seibutu* **1987**, *25*, 421–423.
11. Shimoyamada, M.; Kudo, S.; Okubo, K.; Yamauchi, F.; Harada, K. *Agric. Biol. Chem.* **1990**, *54*, 77–81.
12. Taniyama, T.; Yoshikawa, M.; Kitagawa, I. *Yakugaku Zasshi* **1988**, *108*, 562–57.
13. Kudou, S.; Fleury, Y.; Welti, D.; Magnolato, D.; Uchida, T.; Kitamura, K.; Okubo, K. *Agric. Biol. Chem.* **1991**, *55*, 2227–2233.
14. Eldridge, A. C.; Kwolek, W. F. J. *Agric. Food Chem.* **1983**, *31*, 394–396.
15. Kitagawa, I.; Yoshikawa, M.; Hayashi, I.; Taniyama, T. *Yakugaku Zasshi* **1984**, *104*, 162–168.
16. Kitagawa, I.; Yoshikawa, M.; Hayashi, I.; Taniyama, T. *Yakugaku Zasshi* **1984**, *104*, 275–279.
17. Shiraiwa, M.; Harada, K.; Okubo, K. *Agric. Biol. Chem.* **1991**, *55*, 323–331.
18. Shiraiwa, M.; Harada, K.; Okubo, K. *Agric. Biol. Chem.* **1990**, *54*, 1347–1352.
19. Price, K. R.; Johnson, I. T.; Fenwick, G. R. *CRC Critical Reviews in Food Science and Nutrition* **1987**, *26*, 27–135.
20. Kudou, S.; Ojima, S.; Okubo, K.; Yamauchi, F.; Fujinami, H.; Ebinw, H. *Proc. International Symposium on New Technology of Vegetable Proteins, Oil and Starch Processing, Part I* **1987**, 148–160.
21. Shimoyamada, M.; Okubo, K. *Agric. Biol. Chem.* **1991**, *55*, 577–579.
22. Nishino, H. *NCI Workshop on Cancer Prevention, February 2–5,1991, La Jolla, California* p. 40.
23. Akiyama, T.; Ishida J.; Nakagawa, S.; Ogawara, H.; Watanabe, S.; Itoh, N.; Sibuya, M.; Fukami, V. J. *Biol. Chem.* **1987**, *262*, 5592–5595.
24. Naim, M.; Gestetner, B.; Bondi, A.; Birk, Y. J. *Agric. Food Chem.* **1976**, *24*, 1174.
25. Linassier, C.; Pierre, M.; Le Pecq, J.-B.; Pierre, J. *Biochem. Pharmacol.* **1990**, *39*, 187–193.
26. Sariaslani, F. S.; Kunz, D. A. *Biochem. Biophys. Res. Commun.* **1986**, *141*, 405–410.
27. Messina, M; Barness, S. J. *Natl. Cancer Inst.* **1991**, *83*, 541–546.

28. Nishino, H.; Yoshioka, K.; Iwashima, A.; Takizawa, H.; Konishi, S.; Okamoto, H.; Okabe, H.; Shibata, H.; Fujiki, H.; Sugimura, T. *Jpn.J. Cancer Res. (Gann)* **1986**, *77*, 33–38.

29. Muto, Y.; Ninomiya, M.; Fujiki, H. *Jpn. J. Clin. Oncol.* **1990**, *20*, 219–224.

30. Ohigashi, H.; Takamura, H.; Koshimizu, K.; Tokuda, H.; Ito, Y. *Cancer Letters* **1986**, *30*, 143–151.

31. Tokuda, H.; Ohigashi, H.; Koshimizu, K.; Ito, Y. *Cancer Letters* **1986**, *33*, 279–285.

32. Tokuda, H.; Konoshima, T.; Kozuka, M.; Kimura, T. *Cancer Letters* **1988**, *40*, 309–317.

33. Chavali, S. R.; Barton, L. D.; Campbell, J. B. *Clin. Exp. Immunol.* **1988**, *74*, 339–343.

34. Chavali, S. R., Barton, L. D.; Campbell, J. B. *Int. J. Immunopharmac.* **1987**, *9*, 675–683.

35. Huang, H. P.; Cheng, C. F.; Lin, W. Q.; Yang, G. X.; Song, J. Y.; Ren, G. Y. *Acta Pharmacologica Sinica* **1982**, *3*, 286–288.

36. Zhang, Y. D.; Shen, J. P.; Song, J.; Wang, Y. L.; Shao, Y. N.; Li, C. F.; Zhou, S. H.; Li, Y. F.; Li, D. X. *Yao Hsue Hsueh Pao* **1985**, *19*, 619–621.

37. Aswal, B. S.; Bhakuni, D. S.; Goel, A. K.; Kar, K.; Mehrotra, B. N.; Mukherjee, K. C. *Indian J. Experimental Biology* **1984**, *22*, 312–332.

38. Sati, O. P.; Pant, G.; Nohara, T.; Sato, A. *Pharmazie* **1985**, *40*, 586.

39. Kitagawa, K.; Nishino, H.; Iwashima, A. *Oncology* **1986**, *43*, 127–130.

40. Konoshima, T.; Kozuka, M. *J. Natural Products* **1991**, *54*, 830–836.

41. Barré-Sinoussi, F.; Chermann, J. C.; Rey, F.; Nugeyre, M. T.; Chamaret, S.; Dauguet, C.; Axler-Blin, C.; Vézinet-Brun, F.; Rouzioux, C.; Rozenbaum, W.; Montagnier, L. *Science* **1983**, *220*, 868.

42. Levy, J. A; Hoftman, A. D.; Kramer, S. M.; Landis, J. A., Shimabukuro, J. M; Oshiro, L. S. *Science* **1984**, *225*, 840.

43. Propovic, M.; Sarngadharan, M. G.; Read, E.; Gallo, R. C. *Science* **1984**, *224*, 497.

44. Mitsuya, H.; Weinhold, K. J.; Furman, P. A.; St. Claire, M. H.; Nusinoff-Lehrmann, S.; Gallo, R. C.; Bolognesi, D.; Barry, D. W.; Broder, S. *Proc. Natl. Acad. Sci. USA* **1985**, *82*, 7096–7100.

45. Nakashima, H.; Matsui, T.; Harada, S. *Antimicrob. Agents Chemothr.* **1986**, *30*, 933.

46. Mitsuya, H.; Broder, S. *Proc. Natl. Acad. Sci. USA* **1986**, *83*, 1911.

47. Nakashima, H.; Okubo, K.; Honda, Y.; Tamura, T.; Matsuda, S.; Yamamoto, N. *AIDS* **1989**, *3*, 655–658.

48. *Conference of Soybean Processing and Utilization* **1991**, 95–102.

RECEIVED August 20, 1993

Chapter 27

Structural Elucidation and Physiological Properties of Genuine Soybean Saponins

Shigemitsu Kudou[1], Masahide Tonomura[1], Chigen Tsukamoto[1], Teiji Uchida[2], Masaki Yoshikoshi[3], and Kazuyoshi Okubo[1]

[1]Department of Applied Biological Chemistry, Faculty of Agriculture, Tohoku University, 1–1 Tsutsumidori, Amamiyamachi, Aoba-ku, Sendai 981, Japan
[2]Kanesa Company, Ltd., 202 Hamada, Tamagawa, Aomori 030, Japan
[3]Nestle Company, Ltd., 2–4–5 Azabudai, Minato-ku, Tokyo 106, Japan

The composition and the structures of group B saponins in native soybean seeds were investigated. Five kinds of saponins named soyasaponins α_g, β_g, β_a, γ_g and γ_a according to elution order from HPLC were isolated, and the structures were characterized as having a 2,3-dihydro-2,5-dihydroxy-6-methyl-4H-pyran-4-one (DDMP) group attached by acetal linkage to C-22 of the aglycones soyasaponin V, I, II, III and IV respectively. These DDMP conjugated saponin are considered to be the genuine saponins in native soybean seeds, and soyasaponins I, II, III, IV and V are artifacts produced by heating during the extraction or concentration process. Maltol produced by amino-carbonyl reaction was identified after heating an aqueous solution of DDMP conjugated saponins. In addition, soyasaponin β_g is same as the phytochrome killer chromosaponin I. DDMP conjugated saponins possess several biological activities.

Many kinds of saponins have been isolated from soybean seeds. They are divided into three groups, group A, B and E saponins, on the basis of their aglycone structures (soyasapogenol A, B and E) (*1-2*). It has been reported that soybean saponins possess many physiological activities—hypolipidemic (*3*), antioxidative (*3*), anti-tumor promoting (*4*) and anti-HIV infection (*5*) properties. Soyasaponin I, II, III (*6-7*) and IV (*8*) from whole soybean seeds and soyasaponins (*6-8*) from hypocotyl are group B saponins that have been isolated and their structures fully characterized (Figure 1). In a recent paper (*9*), we isolated a genuine saponin, BeA, and showed that the structure had an unusual sugar 2,3-dihydro-2,5-dihydroxy-6-methyl-4H-pyran-4-one (DDMP) attached by acetal linkage to C-22 of the aglycone of soyasaponin I. We further suggested that soyasaponin I, II, III, IV and V also exist as DDMP conjugated forms in soybean seeds.

Here we discuss our investigations of the composition of group B saponins, and the structures of the soyasaponins α_g, β_g, β_a, γ_g and γ_a, which contain DDMP, are described. Furthermore, the physiological properties of the unstable and reactive DDMP conjugated saponins are discussed.

0097–6156/94/0546–0340$06.00/0

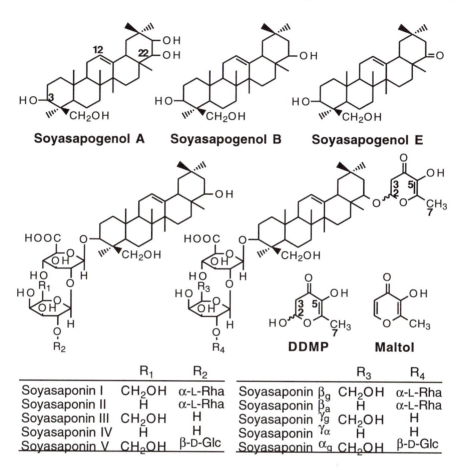

Figure 1. Structures of the group B saponins in soybean seeds.

	R_1	R_2		R_3	R_4
Soyasaponin I	CH_2OH	α-L-Rha	Soyasaponin $β_g$	CH_2OH	α-L-Rha
Soyasaponin II	H	α-L-Rha	Soyasaponin $β_a$	H	α-L-Rha
Soyasaponin III	CH_2OH	H	Soyasaponin $γ_g$	CH_2OH	H
Soyasaponin IV	H	H	Soyasaponin $γ_α$	H	H
Soyasaponin V	CH_2OH	β-D-Glc	Soyasaponin $α_g$	CH_2OH	β-D-Glc

The Composition of Group B Saponins in Whole Soybean Seeds

The composition of group B saponins in whole soybean seeds was analyzed by HPLC (Figure 2). Five peaks were detected that were presumed to correspond to DDMP conjugated saponins, because all of these compounds showed the same UV spectra with a maximum absorption at 292 nm due to the DDMP moiety. These compounds were named soyasaponins $α_g$ and $β_g$ (tentatively named BdA and BeA, respectively, in reference 9), $β_a$, $γ_g$ and $γ_a$ according to their HPLC elution order. In our previous paper, we showed that the structure of soyasaponin $β_g$ has the unusual sugar DDMP attached by acetal linkage to C-22 of the aglycone of soyasaponin I.

Structural Elucidation of Genuine Soybean Saponin

Heating the soyasaponins $α_g$, $β_g$, $β_a$, $γ_g$ and $γ_a$ at 100°C for 1 hr to examine the effect of heating on them revealed that they are completely converted into soyasaponins V, I, II, III and IV, respectively. This result shows that soyasaponins I, II, III, IV and V are artifacts produced by heating during the extraction or concentration process. It also strongly suggested that soyasaponins α_g, β_g, β_a, γ_g and γ_a are the DDMP conjugated soyasaponins V, I, II, III and IV, respectively.

The five genuine saponins — α_g, β_g, β_a, γ_g and γ_a — were isolated to determine their chemical structures. Acid hydrolysis gave a glucose, a galactose and a glucuronic acid from α_g; a rhamnose, an arabinose and a glucuronic acid from β_a; a galactose and a glucuronic acid from γ_g; an arabinose and a glucuronic acid from γ_a; and soyasapogenol B was detected from all of these saponins. Mild alkaline hydrolysis of soyasaponins α_g, β_a, γ_g and γ_a gave corresponding soyasaponins V, II, III and IV and another product. The product was identified by HPLC and a UV spectrum, which showed a maximum absorption at 274 nm, and was completely consistent with that of an authentic sample (Figure 3).

The molecular formulas of soyasaponins α_g, β_a, γ_g and γ_a were determined to be $C_{54}H_{84}O_{22}$ (M_r 1084), $C_{53}H_{82}O_{20}$ (M_r 1038), $C_{48}H_{74}O_{17}$ (M_r 922) and $C_{47}H_{72}O16$ (M_r 892), respectively, by high resolution FAB-MS, indicating the conjugation of a $C_6H_6O_3$ fragment and the corresponding soyasaponins V, II, III and IV. FAB-MS (negative-ion mode) gave a molecular ion at m/z 1083 [M-H]- and fragment ions m/z at 921 [M-H-Glc]- and 759 [M-H-Glc-Gal]- for soyasaponin α_g; a molecular ion at m/z 1037 [M-H]- and fragment ions at m/z 891 [M-H-Rha]- and 759 [M-H-Rha-Ara]- for soyasaponin β_a; a molecular ion at m/z 921 [M-H]- and a fragment ion at m/z 759 [M-H-Gal]- for soyasaponin γ_g; a molecular ion at m/z 891 [M-H]- and a fragment ion at m/z 759 [M-H-Ara]- for soyasaponin γ_a (Table I). None of the DDMP saponins gave a fragment ion due to the loss of DDMP, but a intense fragment ion at m/z 127 [DDMP-H2O+H]- was observed.

The assignments of soyasaponins α_g, β_a, γ_g and γ_a were established by ^{13}C-^{1}H COSY spectra, ^{1}H-^{1}H COSY spectra, NOE and HMBC (10), and are compared with the spectral data of soyasaponins I, II, III and soyasaponin β_g (which has been already assigned in previous paper) in Tables II and III. The ^{13}C-NMR spectra of soyasaponins α_g, β_a, γ_g and γ_a showed six signals at 185.2, 152.5, 132.9, 96.6 or 96.7, 41.5 and 15.2. The ^{1}H-NMR also indicated the presence of one hydroxy group (7.45 or 7.44, 1H, br), one methane (5.35 or 5.38, 1H, dd), one unequivalent methylene group (2.93, 1H, dd; 2.35 or 2.36, 1H, dd) and one methyl group (1.90, 3H, s). These ^{13}C- and ^{1}H-NMR data indicate the presence of DDMP moiety on soyasaponins α_g, β_a, γ_g and γ_a as well as soyasaponin β_g.

The structure of the DDMP moiety was further confirmed by the HMBC experiment and the NOE difference spectra of soyasaponin β_g. The HMBC experiment of soyasaponin β_g exhibited cross peaks C-2' and H-3', C-4' and H-2', C-4' and H-3', C-5' and H-7', C-6 and H-2', C-6' and H-7', C-22 and H-2', respectively. When the signal at δ 2.36 (H-3'b) was irradiated, NOE was observed at δ 5.38 (H-2')and δ 2.93 (H-3'a). Irradiation at δ 2.93 (H-3'a) yielded NOE at δ 5.38 (H-2') and δ 2.36 (H-3'b), and irradiation at δ 5.38 (H-2') resulted in NOE at δ 3.35 (H-22), δ 2.93 (H-3'a), δ 2.36 (H-3'b) and δ 1,90 (H-7'). These results support that the compound linked to C-22 hydroxy group of soyasapogenol B is a 2,3-dihydro-2,5-dihydroxy-6-methyl-4H-pyran-4-one and is attached by acetal linkage to the C-22 position of soyasapogenol B. The chemical shifts (δ) of the aglycone moieties of soyasaponins α_g, β_a, γ_g and γ_a were the almost same as that of soyasaponin β_g. The chemical shifts (δ) of the sugar chain moiety linked to the oxygen at C-3 of the aglycone of soyasaponins α_g and β_a were almost identical with those of soyasaponin V and II, respectively.

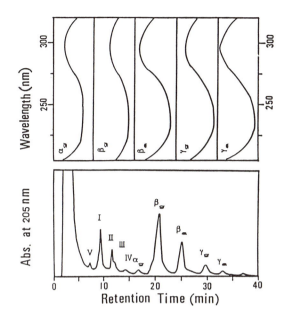

Figure 2. HPLC pattern of group B saponins in soybean seeds and their UV spectra.

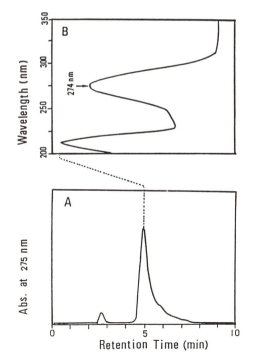

Figure 3. HPLC analysis of a compound produced by alkaline hydrolysis of DDMP conjugated saponins.

Table I. Main Peaks in FAB-MS of DDMP Saponins

	α_g	β_g	β_a	γ_g	γ_a
Positive-ion mode					
$[M+H]^+$	1085	1069	1039	923	893
$[M+Na]^+$	1107	1091	1061	945	915
$[M+K]^+$	1123	1107	1077	961	931
Negative-ion mode					
$[M-H]^-$	1083	1067	1037	921	891
$[(M-H)-146]^-$		921	891		
$[(M-H)-162]^-$	921			759	
$[(M-H)-132]^-$					759
$[(M-H)-146-162]^-$		759			
$[(M-H)-162-162]^-$	759				
$[(M-H)-146-132]^-$			759		

Table II. ^{13}C-NMR Spectral Assignments for DDMP Saponins

	V	I	II	α_g	β_g	β_a	γ_g	γ_a
Aglycone moiety								
C-3	89.5	90.1	89.8	89.3	90.0	89.7	89.4	89.2
C-12	121.6	121.6	121.5	121.8	121.8	121.8	121.8	121.8
C-13	144.1	144.1	144.1	143.7	143.7	143.7	143.7	143.7
C-22	74.1	74.1	74.0	81.1	81.0	81.0	81.0	81.1
C-24	62.4	62.4	62.1	62.4	62.3	62.3	62.3	62.1
3-O-β-D-glucuronopyranosyl								
C-1″	103.3	104.0	104.0	103.2	103.9	103.8	103.3	103.8
C-2″	77.3	76.6	76.4	77.2	77.1	76.6	79.8	79.5
C-3″	74.9	74.7	75.0	74.8	74.6	75.3	74.4	74.5
C-4″	71.3	74.6	73.5	71.7	73.6	73.5	71.7*	71.8*
C-5″	76.1	75.3	75.2	76.5*	75.3	75.3	76.6	76.5
C-6″	170.7	170.6	170.5	171.6	171.6	171.1	171.4	171.3
2″-O-β-D-galactopyranosyl								
C-1‴	101.1	100.0		100.6	99.9		103.6	
C-2‴	82.0	75.7		82.1	75.7		71.9*	
C-3‴	72.6	70.6		72.7	70.6		73.4	
C-4‴	68.2	69.3		68.2	69.3		68.7	
C-5‴	76.1	75.0		76.1*	74.6		75.4	
C-6‴	60.2	59.8		60.2	59.8		60.4	

Continued on next page

Table II (Continued). ^{13}C-NMR Spectral Assignments for DDMP Saponins

	V	I	II	α_g	β_g	β_a	γ_g	γ_a
2″-*O*-α-L-arabinopyranosyl								
C-1‴			100.2			100.2		103.4
C-2‴			75.7			75.7		71.6*
C-3‴			70.5*			70.6		72.8*
C-4‴			68.7			68.8		68.2
C-5‴			65.4			65.5		66.0
2‴-*O*-α-L-rhamnopyranosyl								
C-1⁗		100.0	100.3		100.3	102.0		
C-2⁗		70.6	70.7*		70.6	70.6		
C-3⁗		72.0	72.0		72.4	72.3		
C-4⁗		72.4	72.3		72.5	72.3		
C-5⁗		68.0	68.2		68.0	68.2		
C-6⁗		17.9	17.9		17.9	17.9		
2‴-*O*-β-D-glucopyranosyl								
C-1⁗	104.5			104.5				
C-2⁗	74.1			74.1				
C-3⁗	70.0			69.8				
C-4⁗	70.0			69.8				
C-5⁗	74.8			74.9				
C-6⁗	61.0			61.0				
22-*O*-DDMP**								
C-2′				96.6	96.7	96.6	96.6	96.6
C-3′				41.5	41.5	41.5	41.5	41.5
C-4′				185.2	185.2	185.2	185.2	185.2
C-5′				132.9	132.9	132.9	132.9	132.9
C-6′				152.5	152.5	152.5	152.5	152.5
C-7′				15.2	15.2	15.2	15.2	15.2

* Assignments for these signals within the same column may be interchanged.
** 2,3-dihydro-2,5-dihydroxy-6-methyl-4*H*-pyran-4-one

Table III. ^{1}H-NMR Chemical Shift Values (δ) of Anomeric Protons in the Sugar and DDMP Moieties of DDMP Saponins

	V	I	II	α_g	β_g	β_a	γ_g	γ_a
3-O-β-D-glucuronopyranosyl								
C-1''	4.37d (7.7[a])	4.26d (7.7)	4.22d (7.3)	4.38d (8.0)	4.16d (7.6)	4.25d (6.6)	4.34d (7.3)	4.34d (8.0)
2''-O-β-D-galactopyranosyl								
C-1'''	4.82d (6.6)	4.76d (6.6)		4.82d (5.9)	4.76d (7.0)		4.50d (7.0)	
2''-O-α-L-arabinopyranosyl								
C-1'''	4.68d (6.3)			4.69d (6.2)		4.44d (6.2)		
2''-O-α-L-rhamnopyranosyl								
C-1''''	4.95s		4.93s		4.93s	4.93s		
2'''-O-β-D-glucopyranosyl								
C-1'''''	4.42d (7.7)			4.34d (7.3)				
22-O-DDMP[b]								
C-2'				5.38dd (3,3)	5.35dd (3,3)	5.38dd (3,3)	5.38dd (3,3)	5.38dd (3,3)
C-3'α				2.93dd (14,3)	2.93dd (14,3)	2.93dd (14,3)	2.93dd (14,3)	2.93dd (14,3)
C-3'β				2.36dd (14,3)	2.35dd (14,3)	2.36dd (14,3)	2.36dd (14,3)	2.36dd (14,3)
C-7' (CH3)				1.90s	1.90s	1.90s	1.90s	1.90s
OH				7.45br	7.45br	7.45br	7.44br	7.45br

[a] Number in parentheses is J value. [b] 2,3-dihydro-2,5-dihydroxy-6-methyl-4*H*-pyran-4-one

Assignment of sugar chain moieties of soyasaponins γ_g and γ_a were done by comparison with those of soyasaponins β_g and β_a, considering the upfield shifts (ca. 4 ppm) of C-2''' signals of galactose and arabinose moieties by deglycosidation of the rhamnose moiety from soyasaponins β_g and β_a. Therefore, the structures of soyasaponins α_g, β_a, γ_g and γ_a were deduced to be a 2,3-dihydro-2,5-dihydroxy-6-methyl-4H-pyran-4-one attached by acetal linkage to the oxygen at C-22 of the aglycones soyasaponin V, II, III and IV, respectively (Figure 1).

Physiological Properties of DDMP Conjugated Saponins

Heating DDMP conjugated saponins changed the solution to a brown color and produced a compound that was identified as a maltol by HPLC and a UV spectrum, which had a maximum absorption at 274 nm, compared to an authentic sample. It has been reported that the maltol is formed during cooking of soybeans and contributes to the sweet aroma of soybean products. Although it has been reported that maltol is produced by amino-carbonyl reaction, the formation of maltol from saponins during heating may be an important phenomenon in soybean processing.

In our preliminary experiments, it was suggested that many kinds of legumes, such as Indian potato (*Apios tuberosa* Moench), chickpea (*Cicer arietinum* L.), scarlet runner bean (*Phaseolus coccineus* L.), kidney bean (*Phaseolus vulgaris* L.), pea (*Pisum sativum* L.), mung bean [*Vigna mungo* (L.),Hepper] and cowpea [*Vigna sinensis* (L.) Hassk] contain DDMP saponins. This result suggests that group B saponins might be widely distributed in the legumes as DDMP conjugated forms.

Yokota *et al.* (*11*) isolated phytochrome killer, a substance which instantaneously causes *in vitro* spectral denaturation of phytochrome, from the methanol extract of etiolated pea shoots and identified it as soyasaponin I. Recently, Tsurumi *et al.* (*12*) reported that a chromosaponin I containing DDMP with the same structure as soyasaponin was isolated from *Pisum sativam* and soyasaponin I might be an artifact derived from chromosaponin I. In seven-day-old etiolated pea seedlings, chromosaponin I was found in all parts of seedlings, but at higher concentrations in the hook and root tip than in other non growth tissues.

We have reported that germination of soybean seeds changed the composition of soyasaponins, especially in sprouts (*13*). Germination under light caused a slight increase in the amounts of soyasaponins V, I and II (especially V). Germination under a 12 hr light/dark cycle caused more dramatic changes. Yamaura *et al.* have reported participation of photochrome in the photoregulation of terpenoid synthesis in thyme seedlings (*14*).

The knots of *Wistaria brachybotrys* Sieb.et Zucc. (Leguminosae) have been used in Japanese folk medicine for the treatment of gastric cancer. It has been reported that group B saponins isolated from this plant had an inhibitory effect on Epstein-Barr virus early antigen induced by the tumor promoter 12-*O*-tetradecanoyl-phorbol-13-acetate (TPA). Furthermore, DDMP conjugated saponins, which are suggested to possess anticancer activities, showed radical oxygen removal activity similarly to superoxide dismutase (SOD) (Okubo, K., unpublished data).

Literature Cited

1. Fenwick, G. R.; Price, K. R.; Tsukamoto, C.; Okubo, K. In *Toxic Substances in Crop Plants*; D'Mello; Deffus, Eds.; Royal Society of Chemistry, 1991; pp 285–327.

2. Curl, C. L.; Price, K. R.; Fenwick, G. R. *J. Sci. Food Agric.* **1988**, *43*, 101–107.

3. Ohminami, H.; Kimura, Y.; Okuda, H.; Arich, S.; Yoshikawa, M.; Kitagawa, I. *Planta Medica* **1984**, *46*, 440–441.

4. Konoshima, T.; Kozuka, M.; Haruna, M.; Ito, K. *J. Natural Prod.* **1991**, *54*, 830–836.

5. Nakashima, H.; Okubo, K.; Honda, Y.; Tamaru, T.; Matsuda, S.; Yamamoto, N. *AIDS* **1989**, *3*, 655–658.

6. Kitagawa, I.; Yosikawa, M.; Wang, H. K.; Tosirusuk, V.; Fujiwara, T.; Tomita, K. *Chem. Pharm. Bull.* **1982**, *30*, 2294–2217.

7. Taniyama, T.; Yosikawa, M.; Kitagawa, I. *Yakugaku Zasshi* **1988**, *108*, 562–571.

8. Burrow, J. C; Price, K. R.; Fenwick,G.R. *Phytochem.* **1987**, *26*, 1214–1215.

9. Kudou, S.; Tonomura, M.; Tsukamoto, C.; Shimoyamada, M.; Uchida, T.; Okubo, K. *Biosci. Biotech. Biochem.* **1992**, *56*, 142–143.

10 Bax, A.; Summers, M. F. *J. Am. Chem. Soc.* **1988**, *108*, 2093–2094.

11. Yokota, T.; Baba, J.; Konomi, K.; Shimazaki, Y.; Takahashi, N.; Furuya,M. *Plant Cell Physiol.* **1982**, *23*, 265–271.

12. Tsurumi, S.; Takagi, T.; Hashomoto, T. *Phytochem.* **1992**, *31*, 2435–2438.

13. Shimoyamada, M.; Okubo,K. *Agric. Biol. Chem.* **1991**, *55*, 577–579.

14. Yamaura, T.; Tanaka, S.; Tabata, M. *Plant Cell Physiol.* **1991**, *32*, 603–607.

RECEIVED October 12, 1993

Chapter 28

Fermentation-Derived Anticarcinogenic Flavor Compound

Michael W. Pariza

Food Research Institute, Department of Food Microbiology and Toxicology, University of Wisconsin—Madison, Madison, WI 53706

Diets supplemented with Japanese-style fermented soy sauce inhibit carcinogen-induced forestomach neoplasia in mice. An active anti-carcinogenic principal was identified as 4-hydroxy-2(or 5)-ethyl-5(or 2)-methyl-3(2H)-furanone (HEMF), a principal flavor/aroma compound. HEMF was effective in inhibiting benzo[a]pyrene-induced mouse forestomach neoplasia when fed at a level of 25 ppm (4 mg/kg body weight/day).

Our studies began soon after publication of a report by Wakabayashi *et al.* (*1*), indicating that Japanese-style fermented soy sauce (shoyu) contains substances which under acidic conditions react with nitrite to form direct-acting bacterial mutagens. The investigators postulated that the formation of such mutagens within the human digestive tract might be a contributing factor in the relatively high rate of stomach cancer that occurs among Japanese.

We explored this possibility by investigating the effect of dietary soy sauce, given with or without nitrite (administered in drinking water), on the initiation and promotion of benzo[a]pyrene-induced forestomach neoplasia in mice (2). Contrary to expectations, we found that soy sauce inhibited forestomach carcinogenesis. Moreover, nitrite appeared to enhance the inhibitory effect. The cross-over design of the study made it possible to draw a tentative conclusion that the inhibitory effect occurred during the tumor promotion stage. To our knowledge, this was the first report of an anticarcinogenic effect that was enhanced by nitrite.

In those initial studies (2) we did not prepare the shoyu-supplemented diets on a daily basis. Hence, oxidation of the product occurred during the feeding trials. We hypothesized that anticarcinogenic factor(s) also might have oxidized with loss of activity, and that nitrite, a reducing agent without antioxidant activity, might have then regenerated the anticancer activity. In subsequent studies (3) we took precautions to minimize shoyu oxidation. Under these conditions, dietary shoyu by itself produced a significant reduction in benzo[a]pyrene-induced forestomach neoplasia (Table I). Co-administration of drinking water containing nitrite neither enhanced nor diminished the anticancer effect of minimally-oxidized soy sauce.

0097–6156/94/0546–0349$06.00/0

HEMF as an Anticarcinogen

Shoyu is produced through a complex microbial fermentation of soy beans and wheat (*4*). During the fermentation process innumerable chemical reactions occur, resulting in the generation of many new substances. Characteristic flavor and aroma compounds are thus produced, as well as amino-carbonyl condensation products that are responsible for the deep-brown color of soy sauce.

We fractionated shoyu with ethyl acetate (*5*). The ethyl acetate-soluble fraction contained flavor and aroma compounds, whereas the ethyl acetate-insoluble fraction (representing over 99% of the mass) contained the amino-carbonyl condensation products. Both fractions were fed to mice and found to possess anticarcinogenic activity. Because of the complexity and bulk of the ethyl acetate-insoluble fraction, however, we focused our efforts on the ethyl acetate-soluble fraction.

HEMF (Figure 1) is known to be a principal flavor component of shoyu (*6,7*). We verified that the ethyl acetate-soluble fraction contained HEMF, and then tested the substance for anticarcinogenic activity. As shown in Table II, HEMF displays potent anticarcinogenic activity against benzo[*a*]pyrene-induced mouse forestomach neoplasia. It was effective when fed at just 25 ppm in the diet, the equivalent of 4 mg/kg body weight/day. By contrast, the phenolic antioxidant BHA (2-*tert*-butyl-4-hydroxyanisole) was reported to reduce the number of tumors in this system by 75% when fed at 5400 ppm (*8*).

Table I. Inhibition of Benzo[*a*]pyrene-induced Mouse Forestomach Neoplasia by Dietary Soy Sauce

	Control diet	Soy sauce diet
No. of mice	137	138
Neoplasms/mouse[a]	9.1 ± 0.5	2.5 ± 0.2
Tumor incidence	98%	72%

SOURCE: Adapted from ref. 3.
[a]Mean ± S.E.

Table II. Inhibition of Benzo[*a*]pyrene-induced Mouse Forestomach Neoplasia by HEMF Administered in the Diet

HEMF in diet	Number of mice	Tumor incidence[a]	Body weight[a]
0	27	5.90 ± 0.52	41.3 ± 1.32
25 ppm	26	1.85 ± 0.35	43.5 ± 1.49
50 ppm	25	0.92 ± 0.23	44.1 ± 1.56
75 ppm	28	1.92 ± 0.37	39.4 ± 1.34

SOURCE: Adapted from ref. 5.
[a]Mean ± S.E.

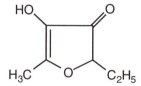

Figure 1. Chemical structure of HEMF {4-hydroxy-2(or 5)-ethyl-5(or 2)-methyl-3[2*H*]-furanone, a major flavor/aroma component of Japanese-style fermented soy sauce. The tautomer ratio is about 3:2, the predominant form is shown.

Mechanism of Anticancer Effect

As indicated, the experimental protocol employed in our first study of the effects of shoyu on induced forestomach neoplasia was designed to separate anti-initiation from anti-promotion effects (*2*). The results indicated that shoyu exerted its inhibitory effect during the tumor promotion stage. It was then established (*5*) that HEMF also inhibits forestomach neoplasia when fed after administration of the carcinogen, i.e., during the tumor promotion stage. We consider this finding to be important in that it suggests that HEMF might be effective against a range of initiating agents.

In many tissues, the induction of ornithine decarboxylase (ODC) (EC 4.1.1.17) activity is an early biochemical event that occurs a few hours after administration of a tumor promoter (*9*). We developed a system for studying changes in ODC activity in mouse forestomach produced by the intubation of test substances (*3*). Not surprisingly, ODC activity was induced in this system by TPA (12-*O*-tetradecanoylphorbol-13-acetate), a potent tumor promoter for mouse epidermis. Contrary to expectations, however, intubating shoyu also induced ODC activity. The effect was replicated in part by administering a dose of NaCl equivalent to that present in the shoyu, in keeping with the observation that NaCl is a tumor promoter in the rat glandular stomach (*10*). Clearly, the anti-promotion activity of shoyu does not involve the inhibition of ODC induction. Rather, it would appear that the chemopreventive effects of shoyu are exerted at one or more biochemical steps that follow.

An aspect of HEMF chemistry possibly important to its chemopreventive activity is the observation that, like BHA and many other anticarcinogens, HEMF is an antioxidant (*5,8*). For example, when dissolved in water at a concentration of 250 ppm (the level found in shoyu), HEMF exhibits a specific activity of 15.2 (nanoequivalents of antioxidant activity per mg). By contrast, a 10,000 ppm solution of ascorbic acid exhibits a specific activity of 10.1 (*5*). The precise mechanistic relationship between antioxidant and anticarcinogenic activity is not yet understood. HEMF may prove to be a valuable tool in this regard, given the extent of its anticarcinogenic and antioxidant potency.

Calorie restriction is one of the most powerful known anti-promoters (*11*). It was therefore of interest to find that mice fed shoyu-containing diets consumed fewer calories and gained less weight than control mice (*2,3*). One possibility for the anticarcinogenic effect of shoyu was calorie restriction induced by the dietary regimen itself.

To test this possibility, we maintained detailed records on feed intake and body weight gain in mice fed ethyl acetate-soluble and -insoluble shoyu fractions, and mice fed HEMF (*5*). Calorie restriction and depressed weight gain were associated only with the diet containing the ethyl acetate-insoluble shoyu fraction. Both the ethyl acetate-soluble fraction and synthetic HEMF inhibited neoplasia, but neither interfered with feed consumption nor weight gain. Hence, the anti-promotion effect of HEMF is directly related to its chemistry (Table II).

In experiments to be published elsewhere, we found that doses of HEMF sufficient to inhibit neoplasia in mouse forestomach do not induce glutathione transferase activity (Kataoka and Pariza, unpublished). This finding, taken in conjunction with the lack of observed adverse effects on feed intake or body weight gain, indicates that HEMF may inhibit carcinogenesis at dose levels low enough to be essentially non-toxic. It is thus a reasonable candidate for further study with an eye toward possible eventual human application.

Acknowledgements

This work was supported in part by the College of Agricultural and Life Sciences, University of Wisconsin-Madison; grants from the Kikkoman Corporation and the World Health Organization; and gift funds administered through the Food Research Institute, University of Wisconsin-Madison.

Literature Cited

1. Wakabayashi, K.; Ochiai, M.; Saito, H.; Tsuda, M.; Suwa, Y.; Nagao, M.; Sugimura, T. *Proc. Natl. Acad. Sci. USA* **1983**, *80*, 2912–2916.
2. Benjamin, H.; Storkson, J.; Tallas, P. G.; Pariza, M. W. *Fd. Chem. Toxic.* **1988**, *26*, 671–679.
3. Benjamin, H.; Strokson, J.; Nagahara, A.; Pariza, M.W. *Cancer Res.* **1991**, *51*, 2940–2942.
4. Nelson, J. H.; Richardson, G. H. In *Microbial Technology* Peppler, H. J. (Ed.); Reinhold Publishing Corp.; New York, 1967; pp 82–106.
5. Nagahara, N.; Benjamin, H.; Storkson, J.; Krewson, J.; Sheng, K.; Liu, W.; Pariza, M. W. *Cancer Res.* **1992**, *52*, 1754–1756.
6. Nunomura, N.; Sasaki, M.; Yokotsuka, T. *Agric. Biol. Chem.* **1980**, *44*, 339–351.
7. Nunomura, N.; Sasaki, M.; Asao, Y.; Yokotsuka, T. *Agric. Biol. Chem.* **1976**, *40*, 491–495.
8. Wattenberg, L. W.; Coccia, J. B.; Lam, L. K. T. *Cancer Res.* **1980**, *40*, 2820–2823.
9. O'Brien, T. G.; Simsiman, R. C.; Boutwell, R. K. *Cancer Res.* **1975**, *35*, 1662–1670.
10. Takahashi, M.; Kokubo, T.; Furukawa, F.; Kurokawa, Y.; Hayashi, Y. *Gann* **1984**, *75*, 494–501.
11. Pariza, M. W. *Ann. Rev. Nutrit.* **1988**, *8*, 167–183.

RECEIVED September 16, 1993

Chapter 29

Antioxidative Activity of Fermented Soybean Products

H. Esaki[1], H. Onozaki[1], and Toshihiko Osawa[2]

[1]Department of Food and Nutrition, Sugiyama Jogakuen University, Chikusa, Nagoya 464, Japan
[2]Department of Food Science and Technology, Nagoya University, Chikusa, Nagoya 464-01, Japan

The antioxidative activity and chemical components of the traditional fermented soybean products incubated with *Aspergillus oryzae, Bacillus natto* and *Rhizopus oligosporous* were investigated. These products are popular as "miso," "natto" and "tempeh," respectively, and have proved to be more stable against lipid peroxidation than steamed soybeans. This result indicates that antioxidative compounds can be produced by fermentation. Using HPLC analysis, no tocopherols were found to increase, but the amount of free isoflavones such as daidzein and genistein increased in miso and tempeh. These isoflavones are presumed to be the principal antioxidants in miso and tempeh. In the case of natto, there was little increase in the amount of free isoflavones, but the antioxidative activity of a water-soluble antioxidative fraction increased dramatically during incubation with *Bacillus natto*. From the evaluation of the antioxidative activities of aqueous ethanol (80%) extracts from soybeans fermented using 18 different kinds of *Aspergillus* strains, it was found that *Aspergillus* No. 13 and 14 showed the strongest antioxidative activity.

There are many traditional fermented soybean products, especially in Asian countries. Tempeh, a fermented product of *Rhizopus oligosporous*, is famous in Indonesia. On the other hand, natto — manufactured by fermentation of steamed soybeans using *Bacillus natto* — is a typical soybean food in Japanese diets. Traditional natto is eaten with hot rice, soy sauce and mustard. These fermented soybean products have been reexamined because of the increase in digestible protein content and other nutritive value. These soybean products can be used as good substitutes for meat in the diet. On the other hand, miso is very popular seasoning in traditional Japanese soup, and for many other Japanese dishes. Recently, some antioxidative components in the diet have attracted special interest because they can protect the human body from free radicals (*1*) which may cause many diseases, including

0097–6156/94/0546–0353$06.00/0

cancer, and aging (2,3). With this background, we started our project to isolate and characterize the antioxidative components in fermented soybean products.

Lipid Stability of Fermented Soybean Products

The lipid peroxidation rate of the lyophilized powder of natto, tempeh and miso was measured by a modified method of Ikehata *et al.* (4). Each powdered sample (400 mg) of natto, tempeh, miso or S.S.B. was divided into a 100 ml Erlenmeyer flask, and the flasks, open to the air, were stored at 40°C in the dark. The extent of lipid peroxidation in natto, tempeh, miso and S.S.B. was monitored at regular intervals by the thiocyanate method of Mitsuda *et al.* (5). The lipids in natto, tempeh and miso are much more stable against autoxidation than those in unfermented steamed soybeans (S.S.B.) (Figure 1).

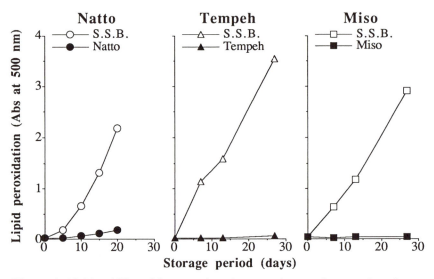

Figure 1. Lipid stability of fermented soybean products and steamed soybeans (S.S.B.). The extent of lipid peroxidation in natto, tempeh, miso compared to S.S.B. was determined at 500 nm by the thiocyanate method.

Chemical Analysis of Antioxidative Tocopherols

The first attempt to evaluate the antioxidative activities of these three fermented soybean products was made by comparative quantitative analyses of the main antioxidative tocopherols. Tocopherols were extracted from each lyophilized powder with *n*-hexane containing BHT and 6-hydroxy-2,2,5,7,8-pentamethylcroman (internal standard), and were determined by high performance liquid chromatography (HPLC) according to the method of Kanematsu *et al.* (6). HPLC analysis was carried out using a Develosil SI 60-5 column (φ4.6 x 250 mm, Nomula Chem. Co., Ltd.) with a mobile phase of *n*-hexane, dioxane and isopropylalcohol (98.7:1.0:0.3) at a flow rate of 1.0 ml/min. A UV detector at 298 nm was used for the determination of tocopherols. The amounts of tocopherols were determined by comparison to the peak areas of external standards which were injected after each analysis.

From HPLC analysis of these three fermented soybean products, none of the tocopherols was found to increase (Figure 2). In natto, α-, β-, γ- and δ-tocopherol concentrations were 2.0, 0.4, 28.9 and 17.3 mg/100 g on dry basis, respectively, while in S.S.B. they were 2.2, 0.5, 29.0 and 17.6 mg/100 g on dry basis, respectively. This result agrees with the data reported by Kanno *et al.* (*7*). In addition, α-, β-, γ- and δ-tocopherol concentrations in tempeh were 1.1, 0.3, 11.9 and 9.2 mg/100 g on dry basis, respectively, while in S.S.B. they were 1.2, 0.3, 14.9 and 9.2 mg/100 g on dry basis, respectively. In the case of natto and tempeh, therefore, the tocopherols in S.S.B. were unmodified by the fermentation process. On the other hand, in miso, tocopherols in S.S.B. were inclined to decrease during the fermentation process. Concentrations of α-, β-, γ-, and δ-tocopherol in miso were 1.7, 0.2, 21.8 and 8.4 mg/ 100 g on dry basis, respectively, while in S.S.B. they were 2.4, 0.3, 28.3 and 17.9 mg/100 g on dry basis, respectively.

Chemical Analysis of Antioxidative Isoflavonoids

Antioxidative free isoflavones in the three fermented soybean products were also quantified using HPLC according to the method of Murakami *et al.* (*8*). Isoflavones and their glucosides were extracted from each lyophilized powder with 80% methanol containing *n*-butyrophenone (internal standard). HPLC analysis was carried out using a Develosil ODS-7 column (ϕ4.6 x 250 mm, Nomura Chem. Co., Ltd.) with a linear gradient of methanol in water from 20 to 60% in 60 min. The solvent flow rate was 0.7 ml/min and the absorption was measured at 262 nm.

The amounts of isoflavones and their glucosides are shown in Figure 3. In natto, there was little increase in the amount of free isoflavones such as daidzein and genistein. On the other hand, the content of daidzein and genistein in tempeh increased by approximately 5 times compared with that in S.S.B. In miso, daidzein and genistein increased dramatically — by 25 times — from fermentation, and no isoflavone glucosides, daidzin and genistin, remained.

Antioxidative Activity of Fermented Soybean Products

György *et al.* (*9*) reported 6,7,4'-trihydroxyisoflavone as an antioxidant in tempeh. While this isoflavone has proved to be a potent antioxidant in aqueous solution at pH 7.4, it was not effective at preventing autoxidation of soybean oil and soybean powder (*4*). Murakami *et al.* (*8*) reported that the main isoflavones responsible for the antioxidative activity in tempeh were daidzein and genistein, which are liberated from daidzin and genistin in soybeans by β-glucosidase from *Rhizopus oligosporous* (Figure 4). Chemical analysis of the antioxidative components in tempeh showed tocopherol levels were not modified by fermentation, but lipophilic aglycones of isoflavone glucosides were liberated by β-glucosidase during fermentation. In the case of miso, some melanoidins (*10*) and peptides (*11*) are presumed to have antioxidative activity. But the principal antioxidants in miso have not been isolated. From HPLC analysis, tocopherols in S.S.B. tended to be metabolized during fermentation, but a large amount of lipophilic antioxidative aglycones can be produced in fermented product. These results suggest that the

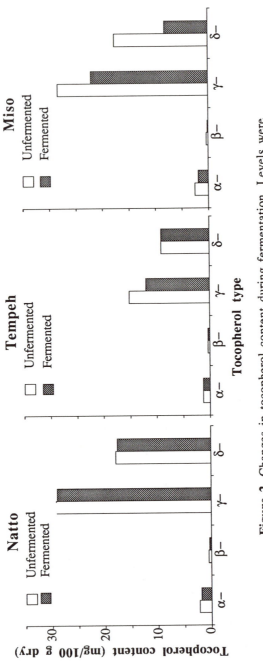

Figure 2. Changes in tocopherol content during fermentation. Levels were measured by HPLC analysis.

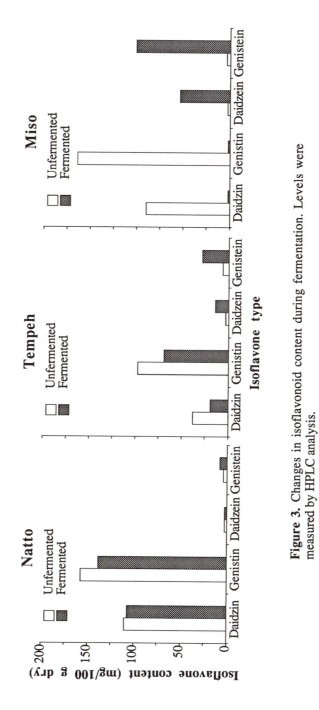

Figure 3. Changes in isoflavonoid content during fermentation. Levels were measured by HPLC analysis.

lipophilic aglycones such as daidzein and genistein are mainly responsible for the stabilities of tempeh and miso.

On the other hand, no significant change was observed in the amount of antioxidative tocopherols and free isoflavones in either natto or S.S.B, and the antioxidant potential in natto has not been ascertained. Each 80% methanol extract from natto and S.S.B. was fractionated by Toyopearl HW-40 chromatography, and the antioxidative activity of each fraction was determined by the buffer-ethanol system described by Osawa *et al.* (*12*). The water-soluble fraction exhibiting antioxidative activity increased dramatically in natto (*13*). Isolation and identification of the antioxidative substances present in natto are now in progress.

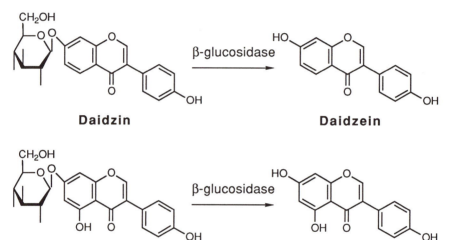

Figure 4. Scheme for the liberation of the antioxidants daidzein and genistein from daidzin and genistin.

Aspergillus strains have been used in manufacturing miso, "shoyu" (soy sauce), "sake," etc. S.S.B. was inoculated with 18 different kinds of *Aspergillus* strains and incubated at 30°C for 96 hrs. The fermented soybeans were dried and powdered. The antioxidative activities of aqueous ethanolic (80%) extracts from these fermented soybeans, in addition to natto, tempeh and miso groups were evaluated by the determination of lipid peroxidation in liposomes (*14*). Lipid peroxidation was induced by 2,2'-azobis(2-amidinopropane) hydrochloride and determined spectrometrically by an increase in thiobarbituric acid reacting substances (*15*). As may be seen in Figure 5, soybeans fermented with *Aspergillus* No. 13 and 14 have the strongest antioxidative activity.

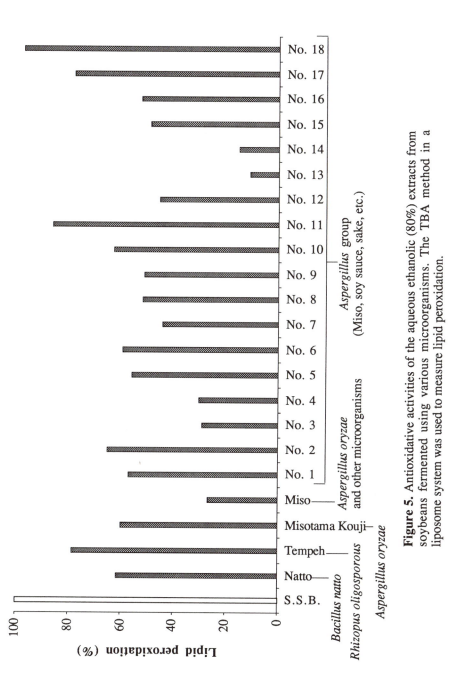

Figure 5. Antioxidative activities of the aqueous ethanolic (80%) extracts from soybeans fermented using various microorganisms. The TBA method in a liposome system was used to measure lipid peroxidation.

Acknowledgements

The authors wish to thank C. Matsuda, S. Suzuki, A. Ohori and Y. Yokoe for their cooperation. They also thank Fujisawa Pharmaceutical Co. Ltd., Marutake Co. Ltd., and Ichibiki Co. Ltd. for generously providing the fermented soybeans incubated with *Aspergillus* strains, natto and miso, respectively, used in this study.

Literature Cited

1. Osawa,T.; Kawakishi, S.; Namiki, M. In *Antimutagenesis and Anticarcino–genesis Mechanisms II*; Kuroda, Y.; Shankel, D.M.; Waters, M.D. Eds.; Plenum: New York, 1990; pp 139-153.
2. Player, T. In *Free Radicals, Lipid Peroxidation and Cancer;* McBrien, D. C. H.; Slater, T. F., Eds.; Academic Press: London, 1982, pp 173–195.
3. Osawa, T.; Ide A.; Su, J. D. and Namiki, M. *J. Agric. Food. Chem.* **1987**, *35*, 808–812.
4. Ikehata, H.; Wakaizumi, M.; Murata, K. *Agric. Biol. Chem.* **1968**, *32*, 740– 746.
5. Mitsuda H.; Yasumoto K.; Iwami K. *Eiyo To Shokuryo* **1966**, *19*, 210–214.
6. Kanematsu H.; Ushigusa T.; Murayama T.; Niiya I.; Matsumoto T. *Yakugaku* **1983**, 32, 51–53.
7. Kanno, A.; Takamatsu, H.; Tuchihashi, N.; Watanabe, T.; Takai, Y. *Nippon Shokuhin Kogyo Gakkaishi* **1985**, *32*, 754–758.
8. Murakami, H.; Asakawa, T.; Terao, J.; Matsushita, S. *Agic. Biol. Chem.* **1984**, *48*, 2971–2975.
9. György, P.; Murata, K.; Ikehata, H. *Nature* **1964**, *203*, 870–872.
10. Yamaguchi N.; Fujimaki M. *Nippon Shokuhin Kogyo Gakkaishi* **1973**, *20*, 507–512.
11. Okamoto M.; Honma S.; Fujimaki M. *Kaseigaku Zashi* **1982**, *33*, 585–590.
12. Osawa T.; Namiki M. *Agric. Biol. Chem.* **1981**, *45*, 735–739.
13. Esaki, H.; Nohara, Y.; Onozaki, H.; Osawa, T. *Nippon Shokuhin Kogyo Gakkaishi* **1990**, *37*, 474–477.
14. Pelle E.; Maes D.; Padulo G. A.; Kim, EK.; Smith, W.P. *Arch. Biochem. Biophys.* **1990**, *283*, 234–240.
15. Buege J. A.; Aust S. D. *Methods in Enzymology* **1978**, *52*, 302–310.

RECEIVED April 4, 1993

Chapter 30

Chemopreventive Phytochemicals in Soy and Licorice Diets Affecting Key Rat Enzyme Systems

T. E. Webb[1], P. C. Stromberg[2], H. Abou-Issa[3], M. Moeschberger[4], H. F. Pierson[5,7], and R. W. Curley, Jr.[6]

Departments of [1]Medical Biochemistry, [2]Veterinary Pathobiology, College of Veterinary Medicine, [3]Surgery, and [4]Preventive Medicine, College of Medicine, The Ohio State University, Columbus, OH 43210
[5]National Cancer Institute, Bethesda, MD 20892
[6]Division of Medicinal Chemistry and Pharmacognosy, College of Pharmacy, The Ohio State University, Columbus, OH 43210

As a component of a feeding study of the possible chemopreventive diet additives soybean meal and licorice root extract, simplified extraction and HPLC methods were developed for the analysis of the soy isoflavones genistein and daidzein and the licorice triterpenoids glycyrrhizic acid and glycyrrhetinic acid. In the diet containing 25% soybean meal, genistein and daidzein were present at about 2–5 μg/g of diet although some variability suggests these isoflavones, especially genistein, may not be stable in frozen diet extracts. Markers glycyrrhizic acid and glycyrrhetinic acid showed the 3% licorice extract containing diet to be uniformly mixed and stable with final concentrations of 300 and 20 μg/g of diet each respectively. Of these markers, only glycyrrhetinic acid was reliably detected in the plasma of rats fed the appropriate diet with an observed concentration of 5.83 μg/ml.

Recent studies suggest that diet has a marked impact on the incidence of cancer (*1*) and that this may be due to protective agents in foods, many of which are phytochemicals (*2*). Soybeans are known to contain potential chemopreventive isoflavones and protease inhibitors (*3*) while licorice root contains various flavonoids (*4*) and triterpenoids (*5*) of interest. Of particular interest in this study has been the impact of these phytochemical-rich diet additives on the induction of protective enzymes or the suppression of enzymes which may increase cancer risk (*6*). Such enzymes could serve as intermediate end point markers in chemoprevention studies or provide clues to possible mechanisms of action of food additive phytochemicals. We have evaluated the effects of soybean meal and licorice root extract on twenty

[7]Current address: Preventive Nutrition Consultants, 19508 189th Place Northeast, Woodinville, WA 98072

0097–6156/94/0546–0361$06.00/0

potentially important enzyme systems in the male rat as well as the impact of the diets on histopathological and clinical chemistry parameters. The results, reported in detail elsewhere (7), will be summarized below. Of importance for interpreting these results was the need to establish the uniformity of food additive mixing, the stability of the phytochemicals in these modified diets, and the oral absorption of important agents from these diets. Our efforts in this regard are described in more detail herein.

Chemicals Chosen for Study

Phytochemicals to be used as analytical markers for diet mixing, stability, and absorption were chosen based on the likelihood of a significant concentration in the blended diet as well as a possibility that the chemicals may be contributors to potential chemopreventive effects of the food additives. In the case of soybean, the major isoflavones genistein (**1**, Figure 1) and daidzein (**2**) were selected for monitoring. These isoflavones have been reported to show estrogenic (8), antifungal (9), and antioxidant (10) activities as well as the ability to induce cytochrome P450 in *Streptomyces griseus* (11). Thus, these compounds have a broad range of biological activities, many of which may be due to regulation of key enzyme systems of potential impact in chemoprevention.

The principal triterpenoid in licorice root is the acidic diglucuronide glycyrrhizic acid (**3**). This triterpenoid is believed to have estrogenic, antiulcer, and glucocorticoid activities (12) and has been suggested to cause "pseudoaldosteronism" when ingested in large doses (13). The acidic diglucuronide **3** is readily hydrolyzed, especially *in vivo* (13), and thus **3** and its aglycone glycyrrhetinic acid (**4**) — which has shown chemopreventive potential (5) — are the two major triterpenoids from licorice root extract which were estimated.

A number of novel flavonoids have been isolated from licorice root (14–16). From among these, we chose to focus on the isoflavone formononetin (**5**) and the chalcone licochalcone A (**6**) because of their presence in reasonable concentration in the mixed Chinese and Russian licorice root extract used in these studies (Jeffcoat, A.R., Research Triangle Institute, personal communication). The licochalcones have shown radical scavenging/antioxidant activity which may be due to regulation of (per)oxidizing enzymes and could be relevant to chemoprevention (16).

The principal method we employed for approximate quantitation of the dietary and blood levels of these six agents was high performance liquid chromatography (HPLC) separation and comparison of chromatographic peak areas with standard curves prepared using an appropriate range of known analyte concentrations.

Materials and Methods

Standard glycyrrhizic acid and glycyrrhetinic acid were from Aldrich Chemical (Milwaukee, WI), genistein from ICN Biochemicals (Cleveland, OH), daidzein from Spectrum Chemical (Gardena, CA), formononetin from Indofine Chemical (Somerville, NJ), and licochalcone A from Arthur D. Little (Cambridge, MA). All solvents and buffers were HPLC grade from Fisher Scientific (Pittsburgh, PA). Nylon, 0.45 μm syringe filters were from Fisher Scientific. Ultraviolet spectra were performed on a Beckman Instruments (San Ramon, CA) DU-40 spectrophotometer. HPLC analyses were performed on a Beckman Instruments Model 332 gradient liquid chromatograph system equipped with a Beckman Model 164 variable-wavelength UV detector and Kipp and Zonen (Delft, Holland) BD 41 dual channel

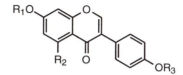

1 Genistein R$_1$ = H, R$_2$ = OH, R$_3$ = H
2 Daidzein R$_1$ = R$_2$ = R$_3$ = H
5 Formononetin R$_1$ = H, R$_2$ = H, R$_3$ = CH$_3$

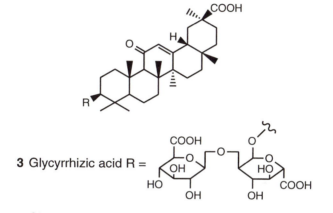

3 Glycyrrhizic acid R =

4 Glycyrrhetinic acid R = OH

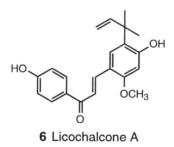

6 Licochalcone A

Figure 1. Structure of assayed phytochemicals.

recorder. Separations were performed on Zorbax-ODS 250 x 4.6 mm columns (Mac-Mod Analytical, Chadds Ford, PA) equipped with a matching pre-column. Mobile phase flow rates were 1 ml/min and all sample handling and chromatography was performed under yellow light at ambient temperature.

Animals and Diets. Male Fischer 344 rats (Harlan Labs, Indianapolis, IN) were entered into the studies at approximately 50 days old and a weight of 175–180 g. Powdered diets were replenished weekly and water provided *ad libitum*. Powdered AIN-76A diet was from Dyets (Bethlehem, PA). Toasted, defatted soybean meal was from Central Soya (Fort Wayne, IN). Licorice root extract was supplied through A. D. Little by McAndrew's and Forbes (Camden, NJ). Essentially isocaloric diets were prepared from control diet to contain (w/w) 25%, 12.5%, or 3.13% soybean meal (SBM), 3%, 1.5%, or 0.38% licorice root extract (LIC), and all nine combinations thereof. Diets were prepared by mixing in a Reynolds commercial food mixer for 15 min, removed and mixed manually, then blended an additional 10 min in the mixer. Diets were stored for a maximum of two weeks. Diets were fed for 1 and 3 months to individual groups of rats.

Diet Extraction Procedures. Briefly, 1 g of chosen diet was shaken for 1 hr at 25°C with 5 ml hexane, centrifuged, and the hexane (containing no analyte by UV/HPLC analysis) discarded. Diet residue was resuspended in methanol and extracted as above, centrifuged, and the methanol layer removed. The methanol extract was syringe filtered, rotary evaporated, and stored at -20°C for reconstitution in 1 ml methanol prior to HPLC analysis of a suitable aliquot.

Blood Extraction Procedures. Plasma aliquots (500 μl) from two rats in each experimental group were combined, shaken 30 sec with 1 ml hexane, centrifuged, the analyte free hexane layer discarded, and the process repeated. Methanol (10 ml) was added to the aqueous phase, mixed for 30 min, centrifuged, and the liquid phase removed. The extract was rotary evaporated and stored at -20°C for reconstitution in 1 ml methanol prior to HPLC analysis.

HPLC Analysis Conditions. The phytochemicals were separated and quantitated by reverse phase HPLC using 10 mM ammonium acetate-containing mobile phases of (a) methanol/water 60:40 for **1**, **2**, and **3**, (b) methanol/water 85:15 for **4**, (c) methanol/water 65:35 for **5**, and (d) methanol/water 80:20 for **6**. Analyses were performed under the direction of a Good Laboratory Practices consultant (Ms. Kathleen M. Zajd) who required that samples be run in duplicate with two injections (runs) per sample. If intra- and inter-sample variability was less than 5%, then the average of averages was also recorded. If intersample variability was greater than 5%, a third duplicate injection was done. If analysis of this sample agreed with one of the previous two, it replaced the other and the average of averages was recorded, otherwise all three results were reported. Analyte concentrations were estimated by the standard curve method. That is, an appropriate range of known analyte amounts were chromatographed and plotted versus corresponding peak area and the curve determined by linear regression. Chromatographic peak areas were approximated manually according to the relationship that peak area equals peak height at maximum times peak width at half height: $A = h_{max} \times w_h$. This method has been estimated to be 94% accurate and show 2.6% precision (*17*). Standard curves were verified after every twelve HPLC samples.

Summary of Biological Results

While described in detail elsewhere (*7*), a summary of the impact of the diets on the rats and on key enzymes in the liver and intestinal mucosa is appropriate here.

Food Consumption and Body Weights. Food consumption for the rats fed the individual diets was similar to controls except that it was 8% lower for those fed the 3% LIC diet and 23% higher for animals consuming the 25% SBM. In contrast, all animals in these groups showed weight gain of 5–15% above those of controls.

For the combined diets, consumption was higher than the control group for all groups with the highest being 20% above control for the 25% SBM + 0.38% LIC fed rats. Weight gain ranged from 6% less than control for the 25% SBM + 0.38% LIC to a 13% increase with the 12.5% SBM + 0.38% LIC. Most groups were essentially identical to the controls. None of the differences were statistically significant.

Histopathology and Blood Chemistry. The consumption of the diets for either one or three months caused no anatomical lesions, observable by histopathological analysis, which were attributable to the SBM or LIC diet additives. The same was true for the hematological profile of these animals. The SBM-containing diets, however, were found to cause a concentration-dependent decrease in serum cholesterol and increases in serum alkaline phosphatase, blood urea nitrogen, and phosphorous concentrations.

Enzymology. A limited number of enzyme systems assessed showed changes in activity due to addition of SBM or LIC to the diet. Thus, both SBM and LIC addition caused concentration-related inductions (up to 50%) in the activities of hepatic glutathione transferase, catalase, and protein kinase C. Likewise, most SBM and LIC containing diet caused reductions (up to 50%) in liver ornithine decarboxylase activity. None of the diet additives caused other than marginal effects on enzyme activity in the intestinal mucosa nor were any additive, synergistic, or antagonistic effects observed upon combining the diet additives.

Analysis of Compounds of Interest

Soy Diet Extract Analysis. Because of their similar chromatographic behavior, extracts of SBM-containing diet could be prepared and analyzed simultaneously for the contained quantity of **1** and **2**.

Results of the HPLC analysis for the quantity of **2** in the extracts are tabulated in Table I. Samples were subjected to a variety of storage conditions and were drawn from different layers in the food mixer to assess the uniformity of diet blending. While there is substantial intersample variability for this analyte, there does not appear to be deterioration of **2** or nonuniform diet mixing.

Analysis of these same extracts for the presence of **1** is presented in Table II. As opposed to the previous compounds analyzed, **1** showed a surprising variation in apparent concentration in different replicate samples. For example, this is the only compound which seemed to show any variation with regard to region of the diet mixer from which feed sample was drawn. That is, there appeared to be a significantly higher concentration of **1** at the mixer bottom. While this might appear unlikely in view of the relatively uniform distribution of the soy flavonoid **2** (Table

I), the third samples for this species showed consistently much higher concentrations of **1** than the first two samples. These third samples were prepared and analyzed sooner after extraction than the earlier extracts (which were prepared ca. 2 months before analysis). Since these earlier samples were stored frozen before analysis, it is tempting to speculate that the resorcinol-like structure of **1** might make it significantly more susceptible to oxidative degradation than the simple phenolic **2**.

Table I. Amount of Daidzein (2) Detected in Diet Extracts

Diet	Amount (μg/g diet)			
	Sample 1	Sample 2	Sample 3	Average
25% SBM diet (-20°C storage)	1.66	1.66		1.66
25% SBM diet (top of mixer, -20°C)	5.79	3.42	8.99	
25% SBM diet (middle of mixer, -20°C)	2.77	3.42	8.05	
25% SBM diet (bottom of mixer, -20°C)	3.81	31.00	3.48	
25% SBM diet (4°C 2 wks, then -20°C)	1.66	1.63		1.65
25% SBM diet (RT 1 wk, then -20°C)	2.51	2.61		2.56
SBM (4°C 4 wks, then -20°C)	10.12	12.19	21.17	
SBM (-20°C)	10.69	33.50	—	

Table II. Amount of Genistein (1) Detected in Diet Extracts

Diet	Amount (μg/g diet)			
	Sample 1	Sample 2	Sample 3	Average
25% SBM diet (-20°C storage)	1.66	1.70		1.68
25% SBM diet (top of mixer, -20°C)	0.96	0.61	5.29	
25% SBM diet (middle of mixer, -20°C)	0.54	0.90	5.98	
25% SBM diet (bottom of mixer, -20°C)	5.79	5.57		5.68
25% SBM diet (4°C 2 wks, then -20°C)	1.53	1.62	4.34	
25% SBM diet (RT 1 wk, then -20°C)	5.54	5.33		5.44
SBM (4°C 4 wks, then -20°C)	7.22	7.12		7.17
SBM (-20°C)	7.75	10.42	—	

As a test of the above hypothesis, some of the third samples, now frozen for ca. 6 weeks, were reanalyzed for content of **1**. The samples from the top, middle, and bottom of mixer showed 4.27, 4.65, and 4.21 μg/g of **1** respectively. In addition, a fresh set of fourth samples analogous to those above were prepared and showed 4.87, 4.95, and 4.29 μg/g of **1** respectively. Lastly, the sample above containing 4.21 μg/g of **1** was bubbled with compressed air for one hour and upon reanalysis showed 3.54 μg/g of genistein.

The above offered rationale for variability of the concentration of **1** is plausible based on observations about sample age and the apparent impact of compressed air on the sample. These experiments are by no means conclusive, however, and one might suggest possible alternative explanations. For example, the earlier samples analyzed for content of **1** spent a considerably longer time as dried extracts at -20°C in glass vials. Perhaps this flavonoid binds avidly to glass in this type of situation, effectively reducing its concentration. Alternatively, the protein-binding ability of these soy flavonoids could conceivably result in sample concentration variability, although we have no direct experimental confirmation of either supposed binding phenomena.

In summary, we do not believe there is any significant variation in the concentration of **1** in the diet mix nor any decomposition of this flavonoid in the diet mix, but there may be some decomposition of this component over time when stored as a frozen diet extract.

An overlay of chromatograms representing extracts of control diet, diet spiked with **1** and **2**, and diet containing 25% SBM is shown in Figure 2.

Licorice Extract Diet Analysis. Extracts of 3% LIC-containing diet were prepared and analyzed for the presence of **3**. Again, samples were subjected to several storage conditions and drawn from different layers of the food mixer to assess the uniformity of diet blending. The results in Table III suggest that while there was some intersample variability, there was no significant degradation of **3** under any of the conditions and **3** appeared to be uniformly mixed in the diet. Note that the LIC samples were diluted 30-fold for analysis because of the high concentration of **3** and to approximate the 3% LIC diet concentrations.

Similarly, the diet extracts were analyzed for the presence of the triterpenoid aglycone **4**. As for **3**, the results shown in Table IV suggested some intersample variability but no evidence of non-uniform mixing of diet nor significant degradation of **4** under any of the conditions.

Table III. Amount of Glycyrrhizic Acid (3) Detected in Diet Extracts

Diet	Amount (µg/g diet)			
	Sample 1	Sample 2	Sample 3	Average
3% LIC diet (-20°C storage)	296.59	291.88		294.23
3% LIC diet (top of mixer, -20°C)	378.06	307.02	210.77	
3% LIC diet (middle of mixer, -20°C)	241.30	383.12	206.51	
3% LIC diet (bottom of mixer, -20°C)	393.71	280.41	242.41	
3% LIC diet (4°C 2 wks, then -20°C)	303.94	304.09		304.01
3% LIC diet (RT 1 wk, then -20°C)	404.53	368.65	251.60	
LIC (4°C 4 wks, then -20°C)	400.27	314.48	259.39	
LIC (-20°C)	549.53	—	339.19	

It was not decided until late in this study to employ **5** and **6** as flavonoid markers of the LIC diet and its consumption. Nonetheless, we did find that as with all the other phytochemical markers investigated, when these two compounds were added to control diet our extraction procedures readily removed the analyte from the spiked diet.

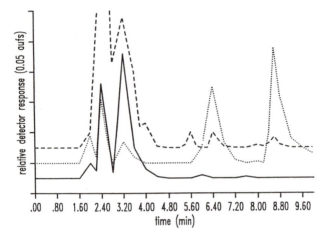

Figure 2. HPLC overlay of control diet extract (———), extract of diet spiked with **1** and **2** (·····), and extract of 25% SBM diet (– – – –).

While we did not have sufficient time to quantitate either of these two materials in the experimental diets as was done for **3** and **4** (Table III and IV), using the appropriate HPLC method described above, both **5** and **6** were found to be present in the LIC-containing diet.

Table IV. Amount of Glycyrrhetinic Acid (4) Detected in Diet Extracts

Diet	Amount (μg/g diet)			
	Sample 1	Sample 2	Sample 3	Average
3% LIC diet (-20°C storage)	15.46	25.77	18.03	
3% LIC diet (top of mixer, -20°C)	20.12	25.42	23.09	
3% LIC diet (middle of mixer, -20°C)	20.03	17.81	21.78	
3% LIC diet (bottom of mixer, -20°C)	21.06	17.86	24.44	
3% LIC diet (4°C 2 wks, then -20°C)	18.90	25.69	27.52	
3% LIC diet (RT 1 wk, then -20°C)	18.19	20.25	30.11	
LIC (4°C 4 wks, then -20°C)	109.36	107.66		108.51
LIC (-20°C)	136.66		142.52	139.59

Combined Diet Extract Analysis. Due to time constraints and because **4** was the only substance we could reliably detect in rat blood extracts (see below), a brief survey of the combined LIC-SBM diet was conducted to determine whether there was any unusual interactions that occurred with this diet combination leading to any enhanced/reduced recovery of **4**. This approach was chosen to maximize the opportunity to acquire quality data of importance to the study. With this in mind, two samples of the diet prepared as a 25% SBM + 3% LIC combination were extracted and found to contain 26.46 and 27.20 μg/g of **4** respectively. Comparing these results with the data in Table IV, it would appear there is little influence on the observed concentration of this substance in the diet after combining the SBM and LIC in a single diet.

Blood Extract Analyses. As with the diet extracts, preliminary experiments were conducted in which it was found that all six of our chosen analytes when added to control rat plasma were extractable from control rat plasma using our methods. Further experiments were conducted on the plasma samples from rats receiving the low concentrations of feed additive in the preliminary study. As opposed to the diet extracts, the presence of interfering substances extracted from the plasma posed a more significant problem using our simple techniques in the face of apparently very low concentrations of the compounds under investigation. In these early surveys, it appeared that perhaps only **4** could be reliably identified under our established protocols and, with perhaps some difficulty, **2** might also be quantified.

Thus, we turned to study of plasma extracts from the rats fed the diets containing the high dietary additive conditions for one month. In these samples, while it was somewhat easier to observe **4** and even **2** in the appropriate samples, the remaining compounds remained difficult to unequivocally identify and quantitate in the absence of feeding radiolabeled test compound. In the case of **2**, we have observed the compound in 2 of 6 of the plasma samples investigated with estimated concentrations of 2.67 and 2.68 µg/ml respectively (detection limit estimated to be 1.25 µg/ml). The aglycone **4** has clearly been observed in all 4 samples evaluated with an apparent concentration of 5.84 ± 0.43 µg/ml.

Since **4** was the only marker we could reliably detect in the plasma of LIC-fed rats, as with the diet combination, we explored the possibility that feeding the combined 25% SBM + 3% LIC diet to rats might have some impact on the observed concentration of **4**. Thus, two of these rat blood samples were extracted and found to contain 7.63 and 8.04 µg/ml of **4** respectively. While insufficient samples were analyzed to make any firm statistical comparison, these values do not appear to differ dramatically from those found above for the LIC diet fed rats alone.

Summary and Conclusions

Long-term feeding to Fisher 344 rats of SBM- or LIC-containing diets had little remarkable deleterious effect on the animals. At least four hepatic enzyme systems potentially important with regard to chemoprevention, however, experienced changes in their level of activity because of the diet additives (7). Increasing doses of SBM- and LIC-containing diets caused concentration-dependent inductions (up to 50%) of liver glutathione transferase, catalase and protein kinase C activities and reductions (up to 50%) of hepatic ornithine decarboxylase activity. These enzymes are generally thought to be protective or indicative of lowered risk for cancer.

Using the triterpenoids **3** and **4** as markers and as representative agents likely to influence the chemopreventive activity of licorice root extract, HPLC analysis suggested the experimental LIC-containing diets were evenly mixed and the phytochemical components were stable in the diet. The final concentrations of **3** and **4** in these diets were about 300 and 20 µg/g of 3% LIC-containing diet respectively. The potentially important licorice root flavonoids **5** and **6** were also found to be present in the LIC diet but were not quantitated. The important isoflavones **1** and **2** were used as markers for phytochemical stability in the soybean meal containing diets. For both of these components, there was greater intersample variability in the apparent concentration of the compounds. Thus, while both isoflavones showed concentrations of about 2–5 µg/g in 25% SBM-diet, some evidence gathered suggested that long-term storage of frozen SBM-diet extracts resulted in apparent decreases in isoflavone content. While not clear, this observation may be due to oxidative degradation, binding to the glass vial walls or avid protein-binding of these isoflavones. The mixing of these two dietary additives in a single feed appeared to have little additional effect on the animals nor the concentration or stability of the assayed phytochemicals.

Finally, after long-term feeding of the diets containing the high SBM or LIC concentrations, using our methods few of the target phytochemicals could be reliably detected in the plasma of the treated rats. Only triterpenoid **4** was readily detected as a marker for LIC phytochemical absorption with a concentration of 5.84 ± 0.43 µg/ml. The soy isoflavone **2** was detected less consistently and showed a concentration of about 2.68 µg/ml in about one-third of the plasma samples investigated. It would appear that given our protocols and methods, detection of absorbed dietary compounds **1–6** is difficult, with the exception of **4**, in the absence of more sensitive techniques such as the feeding of radiolabelled phytochemicals.

Acknowledgment

This study was supported by the Division of Cancer Prevention and Control, National Cancer Institute under Contract NO1-CN-05261-01.

Literature Cited

1. Doll, R.; Peto, R. *J. Natl. Cancer Inst.* 1981, *66*, 1191–1308.
2. Ames, B. *Science* **1983**, *221*, 1256–1258.
3. Messina, M.; Barnes, S. *J. Natl. Cancer Inst.* **1991**, *83*, 541–546.
4. Mitscher, L. A.; Drake, S.; Gollapudi, S. R.; Harris, J. A.; Shankel, D. M. In *Antimutagenesis and Anticarcinogenesis Mechanisms*; Shankel, D. M., Ed.; Plenum: New York, 1986; pp 153–165.
5. Wang, Z. Y.; Agarwal, R.; Zhou, Z. C.; Bickers, D. R.; Mukhtar, H. *Carcinogenesis* **1991**, *12*, 187–192.
6. Sparins, V. C.; Chuan, J.; Wattenberg, L. W. *Cancer Res.* **1982**, *42*, 205–209.
7. Webb, T. E.; Stromberg, P. C.; Abou-Issa, H.; Curley, R. W., Jr.; Moeschberger, M. *Nutr. Cancer*, **1992**, *18*, 215–230.
8. Drane, H. M.; Patterson, D. S. P.; Roberts, B. A.; Saba, N. *Food Cosmet. Toxicol.* **1980**, *18*, 425–427.
9. Kramer, R. P.; Hindorf, H.; Jha, H. C.; Kallage, J.; Zilliken, F. *Phytochemistry* **1984**, *23*, 2203–2205.
10. Pratt, D. E.; Birac, P. M. *J. Food Sci.* **1979**, *44*, 1720–1722.
11. Saviaslani, F. S.; Kunz, D. A. *Biochem. Biophys. Res. Commun.* **1986**, *141*, 405–410.
12. Duke, J. A. *Handbook of Medicinal Herbs*; CRC Press: Boca Raton, FL, 1985; pp 215–216.
13. Sakiya, Y.; Akada, Y.; Kawano, S.; Miyauchi, Y. *Chem. Pharm. Bull.* **1979**, *27*, 1125–1129.
14. Litivenko, V. I. *Dokl. Akad. Nauk SSR* **1964**, *155*, 600–606.
15. Mitscher, L. A.; Park, Y. H.; Clark, D.; Beal, J. L. *J. Nat. Prods.* **1980**, *43*, 259–269.
16. Hatano, T.; Kagawa, H.; Yasuhara, T.; Okuda, T. *Chem. Pharm. Bull.* **1988**, *36*, 2090–2097.
17. Yost, R. W.; Ettre, L. S.; Conlon, R. D. *Practical Liquid Chromatography: An Introduction*; Perkin-Elmer: Norwalk, CT, 1980; pp 216–241.

RECEIVED July 27, 1993

Chapter 31

Genetic Improvement of Saponin Components in Soybean

Chigen Tsukamoto[1], Akio Kikuchi[2], Shigemitsu Kudou[1], Kyuya Harada[3], Tsutomu Iwasaki[4], and Kazuyoshi Okubo[1]

[1]Laboratory of Food Biotechnology, Department of Applied Biological Chemistry, Tohoku University, 1–1 Tsutsumidori, Amamiyamachi, Aoba-ku, Sendai 981, Japan
[2]Tohoku National Agricultural Experiment Station, MAFF, Kariwano, Nishisenboku-cho, Senboku-gun, Akita 019–21, Japan
[3]National Institute of Agrobiological Resources, MAFF 2–1–2, Kannondai, Tsukuba, Ibaraki 305, Japan
[4]Taishi Foods Company Ltd., 68, Okinaka, Kawamorita, Sannohe-machi, Sannohe-gun, Aomori 039–01, Japan

Group B and group E soybean saponins are known to have inhibitory effects against the infectivity of the AIDS virus (HIV) and the activation of Epstein-Barr virus early antigen. Group A acetyl saponins have a strong undesirable taste. The purpose of our research is to manipulate the content of saponin components by genetic methods. A soybean (*Glycine max*) *c v.* Nattoshoryu contained high amounts (about 6%) of group B and E saponins in the seed hypocotyl. About 70% of 154 wild soybean (*Glycine soja*) accessions contained arabinosides that are not found in cultivated soybeans, and one accession lacked the group A acetyl saponins. Increased health benefits and decreased undesirable taste in soybeans therefore seems possible.

Soybean seeds have been consumed as food for thousands of years by Oriental people including Japanese, Koreans, Chinese and Indonesians. The incidence of breast and colon cancer in Oriental peoples is considerably lower than in those living in Western countries (*1*). Vegetarians, who frequently consume soybean based meat substitutes, are also at decreased risk of breast and colon cancers (*2*). Additionally, soybean saponins have inhibitory effects against the infectivity of the AIDS virus (HIV) (*3*) and the activation of Epstein-Barr virus early antigen (*4*). These results suggest that soybean saponins may play an important role in human health as functional components of soybean foods.

Soybeans have not generally been recognized as a staple food in spite of their known nutritional benefits. The reasons for this are twofold. First, soybean seeds contain enzymes called lipoxygenases that generate undesirable flavors during processing; second, soybeans contain an undesirable bitter and astringent taste caused by saponins and isoflavonoids. The problems created by lipoxygenase have been largely solved by genetic improvement and production of a soybean lacking lipoxygenase (*5–7*). The improved soybean varieties lack all three lipoxygenase isozymes (L-1, L-2 and L-3) and do not produce "grassy" flavors. The second problem has been decreased by removal of hypocotyl parts from the seed and by the extraction of soymilk at low temperature. This technique has been applied industrially (*8,9*), but this method may also sacrifice beneficial saponin components. Reducing the saponins possessing undesirable characteristics and increasing the others having health benefits is an important goal. Therefore, genetic improvement of soybeans used as starting material is clearly required if the levels of these saponins are to be optimized. In this report, we address the possibility of genetically improving the balance of key saponins in the soybean.

Characteristics of Soybean Saponins

Many soybean saponins have been isolated and characterized (*10,11*). They are divided into three groups — A, B and E (Figure 1) — according to their respective aglycone, soyasapogenol A, B or E (*12–14*). They have a common structure that contains a glucuronic acid residue attached at the C-3 position of soyasapogenol.

Group A saponins are bis-desmoside saponins that contain two sugar chains attached at the C-3 and C-22 positions of soyasapogenol A.

Group B and E saponins had been thought to be mono-desmoside saponins that contain a sugar chain attached to the C-3 position alone, but recently it has been found that they contain a 2,3-dihydro-2,5-dihydroxy-6-methyl-4*H*-pyran-4-one (DDMP) moiety at the C-22 position of soyasapogenol B (Figure 2) (*11*). Saponin Bb belongs to the group B saponins, and saponin Be is included in the group E saponins. Soyasaponin βg (saponin BeA) contains DDMP at the C-22 position of saponin Bb, and must be the genuine saponin of "soybean saponin Bb and Be," because the purified soyasaponin βg fraction produces saponin Bb and Be after heating (*15*). Furthermore, other DDMP saponins, αg, βa, γg and γa, which correspond to each of the four group B saponins, Ba, Bc, Bb' and Bc', respectively, have now been identified (Kudou, S., *et al.*, Kanesa Co. Ltd., Aomori, Japan, *Biosci. Biotech. Biochem.* 1993, in press). This means that group B and group E saponins must have been derived from the saponins which contain DDMP at the C-22 position of soyasapogenol B.

Group B saponins (saponins Ba and Bb) inhibit the infectivity of AIDS virus (HIV) (*3*), and the group E saponin, saponin Be, inhibits the activation of Epstein-Barr virus early antigen (*4*). As mentioned above, group B and E saponins exist as DDMP conjugated forms in soybean seeds. Much attention is now being given to the physiological and pharmacological activities of DDMP saponins. These studies are addressed by coworker elsewhere in this volume.

Distribution of DDMP Saponins in the Legume

DDMP-conjugated saponins easily release the DDMP moiety during extraction under high temperature, and therefore the extraction process must be conducted at room temperature or below. Using 70% aqueous ethanol extracts, the existence of DDMP saponins in the edible part of legumes (21 varieties of 12 species in 8 genera), which appear to contain group B saponins, was examined by HPLC, TLC, FAB-MS and chemical analyses. These saponins may differ from soybean DDMP saponins according in their sugar sequence, composition, or position of attachment to soyasapogenol B. DDMP saponins were detected in the tubers of the Indian

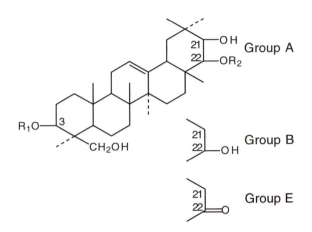

Figure 1. Differences in the chemical structures of the three saponin groups.

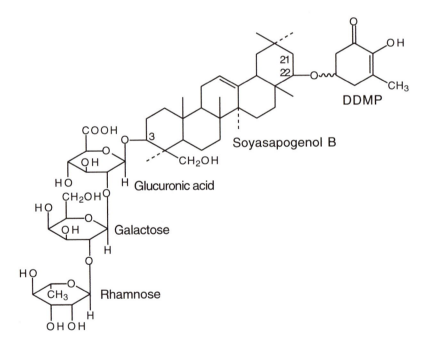

Figure 2. Chemical structure of the DDMP-conjugated saponin soyasaponin βg.

potato (*Apios tuberosa* Moench), and the whole seeds of the chick pea (*Cicer arietinum* L.), scarlet runner bean (*Phaseolus coccineus* L.), kidney bean (*Phaseolus vulgaris* L.), pea (*Pisum sativum* L.), mung bean [*Vigna mungo* (L.) Hepper], and cowpea [*Vigna sinensis* (L.) Hassk] as well as in soybean [*Glycine max* (L.) Merr.] and wild soybean (*Glycine soja* Sieb. & Zucc.) (*16*). These data suggest that group B saponins might be widely distributed in legumes as DDMP conjugated forms.

DDMP saponins could not be detected in the whole seeds of the cluster bean [*Cyamopsis tetragonoloba* (L.) Trub.], broad bean (*Vicia faba* L.) or azuki bean [*Vigna angularis* (Willd.) Ohwi & Ohashi] (*16*). Although it is possible that they contain little or no DDMP saponins, it is also possible that the analytical conditions used were inappropriate due to a difference in their sugar chain structures. Further studies are clearly required to detect and to elucidate the chemical structure of individual DDMP saponins. It has recently been reported that the DDMP saponin soyasaponin βg (called chromosaponin I in the report) has been isolated from peas (*Pisum sativum*) (*17*). This result provides one piece of evidence that DDMP saponins exist not only in the soybean but also in other plants and that DDMP saponins may play a significant physiological role in plants.

DDMP Saponin Content in Soybean

The wild soybean (*Glycine soja* Sieb. & Zucc.) is considered to be the ancestor of the cultivated soybean [*G. max* (L.) Merr.], and the hybrid plant obtained by a cross between them usually gives fertile seeds (*18*). As the wild soybean seems to carry more genetic variation than the cultivated one, it may provide potential material for genetic improvement. The cultivated soybean seed contains about 0.2–0.5% soybean saponins (*10*), but most saponins are localized in the hypocotyl-radicle axis (hypocotyl part) of the seed (*19*).

We have examined the saponin content of the seed hypocotyl part of 414 cultivars and 149 wild soybean accessions of NIAR germplasm collections. As shown in Table I, on average the wild soybean contained a larger quantity of saponins than the cultivated ones. One cultivar, however, Nattoshoryu, contained about 6% DDMP saponins. Although a few varieties of the wild soybean contained DDMP saponins at about 7% of the seed hypocotyl part, the maximum value was 7.2% in Col/Nara/1983/Nobuoka-2.

The Nattoshoryu strain has maintained a high saponin content for three seasons of cultivation. It has been reported that the saponin content in soybean seed depends more on the variety than on the cultivation conditions (*13*). The seed of the Nattoshoryu variety is very small and round in shape, and is a special variety used for "natto," a traditional soybean food in Japan produced by nonsalted fermentation with the microorganism *Bacillus natto*. It is not clear whether there is a relationship between the high DDMP content and the special use.

Genetic Analysis of Sugar Chain Sequence of Soybean Saponins

It has been observed that the saponin composition in the soybean seed is not affected by a difference in cultivation conditions but is peculiar to the variety (*13*). Soybean saponins Aa and Ab are the major constituents of group A saponins. Most soybean varieties contain either soybean saponin Aa or Ab (*13*). Saponin Aa contains a xylose residue at the C-22 position of soyasapogenol A, whereas saponin

Ab contains a glucose residue at this position (*12*). The existence of soybean saponins Aa and Ab is controlled by codominant alleles at a single locus (*20*), and so most soybean varieties can be divided into the xylose (X) and glucose (G) types.

Mikuriyaao, a traditional Japanese cultivar, belongs to the X type. It does not contain saponins Ab, Ba and Bd, and accumulates saponin Af (*13*). This variety is thought to be deficient in the enzyme glucosyltransferase, which catalyzes the glucosylation of a galactosyl moiety of the sugar chain attached at the C-3 position of soyasapogenols (*21*). This characteristic is a recessive trait (*22*). The existence of an arabinose residue next to the glucuronic acid at the C-3 position also appears to be controlled by an enzyme, although the dominant gene coding for this enzyme does not affect at seed hypocotyl part of cultivated soybeans (*23*). Thus, the sugar chain sequence of soybean saponins is thought to be controlled by certain enzymes capable of utilizing soyasapogenol glycosides as substrates. Therefore, total saponin composition and the chemical structure of 'unknown' soybean saponins are able to be assessed from the combination of the sugar sequence of 'known' soybean saponins as shown in Table II. All soybeans, however, do not contain every saponin component. It depends on the variety as well as the part of the soybean.

Variation of Saponin Components in Wild Soybeans

The wild soybean accessions could be classified into 6 types, according to the saponin composition as shown in Table III. The wild soybeans contained either saponin Aa or Ab in the seed hypocotyl part, and therefore they were classified into the X, G, and XG types as are cultivated soybeans.

Of 154 accessions of wild soybeans, 72.7% contained saponin Bc, even though saponin Bc has not been detected in the seed hypocotyl part of the cultivated soybean (*13*). As saponin Bc contains an arabinose residue in the C-3 position sugar chain, we added the letter 'A' to each type when the strain contains saponin Bc in the seed hypocotyl part. The XG type is thought to be a mixture of the X and G types because of their codominant alleles. Although a mixture of the X and GA types gave a different saponin composition from that of the XA and G types, we did not subdivide them further and present them in this table as the XGA type.

Both saponin Ad and Bc contain an arabinose residue in the C-3 position sugar chain, and their detection in the seed hypocotyl part completely correlated. Although there are five cultivars that contain saponin Ad in the seed hypocotyl part (*13*), we confirmed that all contained saponin Bc as well. The existence of saponin Bc in the seed hypocotyl part is controlled by a dominant gene (*23*). As the X and G types are codominant, the XA type soybean varieties should contain an unknown group A saponin, saponin Ax, instead of saponin Ad as in the GA type (see Table II). We have already identified saponin Ax and some another saponins from some wild soybeans, and these findings will be reported in detail in another communication.

Discovery of a Group A Acetyl Saponin-Deficient Mutant and the Possibility of Genetic Improvement of Soybean

The group A saponins, in which the aglycone is soyasapogenol A, have a stronger undesirable taste than group B and E saponins (*8*). This is caused by the acetylation of the hydroxyl groups of the terminal sugar of a sugar chain attached at the C-22

Table I. Soybean Saponin Content in the Seed Hypocotyl Part of Cultivated and Wild Soybeans

	Number of samples	Saponin content (%, mean ± S.D.)	
		Group A saponin	DDMP saponin[a]
cultivated soybeans[b]	414	1.62 ± 0.60	1.83 ± 0.60
Wild soybeans[b]	149	4.30 ± 1.00	4.35 ± 0.85
Nattoshoryu[c]	1	3.72 ± 0.07	6.05 ± 0.11
Col/Nara/1983/Nobuoka-2[c]	1	5.84 ± 0.23	7.20 ± 0.22

[a] DDMP saponin content was calculated as the total amount of group B and group E saponins.
[b] Average ± S.D. of non-replicates.
[c] Average ± S.D. of three replicates

Table II. The Nomenclature of Known and Unknown Soybean Saponins According to the Sugar Sequence Attached at the C-3 and C-22 Position Hydroxyl Groups of Soyasapogenols

C-3 position OH group	C-22 position OH group				
	Soyasapogenol A		B		E
	-Ara-AcXyl	-Ara-AcGlc	-DDMP[a]	-OH	=O
-GlcUA-Gal-Glc	Aa	Ab	(αg)	Ba	Bd
-GlcUA-Gal-Rha	(Au)*	Ac	βg	Bb	Be
-GlcUA-Gal	Ae	Af	(γg)	Bb′	(Be′)
-GlcUA-Ara-Glc	(Ax)	Ad	(αa)*	(Bx)	(Bf)
-GlcUA-Ara-Rha	(Ay)*	(Az)*	(βa)	Bc	(Bg)
-GlcUA-Ara	Ag	Ah	(γa)	Bc′	(Bg′)

[a] 2,3-dihydro-2,5-dihydroxy-6-methyl-4*H*-pyran-4-one
Unpublished saponins are shown in parenthesis, and * shows unidentified ones.

Table III. Classification of Saponin Types of 154 Wild Soybeans According to the Saponin Composition

Saponin Type	Frequency [Number (%)]	Group A						Group B				Group E	
		Aa	Ab	Ac	Ad	Ae	Af	Ba	Bb	Bb′	Bc	Bd	Be
X	24 (15.6)	+	-	-	-	+	-	+	+	+	-	+	+
G	13 (8.4)	-	+	+	-	-	+	+	+	+	-	+	+
XG	5 (3.3)	+	+	+	-	+	+	+	+	+	-	+	+
XA	96 (63.0)	+	-	-	-	+	-	+	+	+	+	+	+
GA	4 (2.6)	-	+	+	+	-	+	+	+	+	+	+	+
XGA	11 (7.1)	+	+	+	±	+	+	+	+	+	+	+	+

+: detected; -: not detected; ±: depended on the independent accession

position of soyasapogenol A (24). Therefore, a group A acetyl saponin-deficient mutant would be expected to improve the taste of soybean foods. Although much effort had been expended to identify such mutants from cultivated soybeans, this has not yet been successful. A spontaneous mutant lacking group A acetyl saponin in a wild soybean strain, however, has been previously identified (25). This mutant lacks a dominant gene that codes for acetylation of the hydroxyl groups of the C-22 position terminal sugar (26).

The data presented in this report provide us with the materials to improve the saponin components and their content in soybean seed. Studies are now under way that will allow us to use these characteristics in breeding programs designed to increase health benefits and decrease undesirable tastes in soybeans.

Acknowledgments

We thank Dr. G. R. Fenwick of AFRC Institute of Food Research, Norwich, UK, for his critical reading of this manuscript. This work was supported in part by a grant for Bio Renaissance Program (BRP92-VI-B-5) from the Ministry of Agriculture, Forestry and Fisheries (MAFF) of Japan, and a grant-in-aid for Scientific Research (Project No. 03304013, 1991) from the Ministry of Education, Science and Culture of Japan.

Literature Cited

1. Kurihara, M.; Aoki, K.; Hisamichi, F. In *Cancer Mortality Statistics in the World 1950–1985*; Kurihara, M.; Aoki, K.; Hisamichi, F., Eds.; The University of Nogoya Press: Nagoya, Japan, 1989; pp 1–93.
2. Nair, P. N.; Turjman, N. Kessie, G.; Calkins, B.; Goodman, G. T.; Davidovitz, H.; Nimmagadda, G. *Am. J. Clin. Nutr.* **1984**, *40*, 927-930.
3. Nakashima, H.; Okubo, K.; Honda, Y.; Tamura, T.; Matsuda, S.; Yamamoto, N. *AIDS*, **1989**, *3*, 655-658.
4. Konoshima, T.; Kozuka, M. *J. Nat. Prod.* **1991**, *54*, 830-836.
5. Kitamura, K. *Agric. Biol. Chem.* **1984**, *48*, 2339-2346.
6. Kitamura, K.; Kumagai, T.; Kikuchi, A. *Japan. J. Breed.* **1985**, *35*, 413-420.
7. Hajika, M.; Igita, K.; Kitamura, K. *Japan. J. Breed.* **1991**, *41*, 507-509.
8. Okubo, K.; Iijima, M.; Kobayashi, Y.; Yoshikoshi, M.; Uchida, T.; Kudou, S. *Biosci. Biotech. Biochem.* **1992**, *56*, 99-103.
9. Tsukamoto, C.; Kawasaki, Y.; Iwasaki, T.; Okubo, K. In *Proc. Intl. Conf. Soybean Processing and Utilization, Gongzhuling, China, June 25-29, 1990* Sendai Kyodo Printing; Sendai, Japan, 1991, pp 47-51.
10. Fenwick, G. R.; Price, K. R.; Tsukamoto, C.; Okubo, K. In *Toxic Substances in Crop Plants*; D'Mello, J. P. F.; Duffs, C. M.; Duffus, J. H., Eds.; The Royal Society of Chemistry: Cambridge, U. K., 1991; pp 285-327.
11. Kudou, S.; Tonomura, M.; Tsukamoto, C.; Shimoyamada, M.; Uchida, T.; Okubo, K. *Biosci. Biotech. Biochem.* **1992**, *56*, 142-143.
12. Shiraiwa, M.; Kudou, S.; Shimoyamada, M.; Harada, K.; Okubo, K. *Agric. Biol. Chem.* **1991**, *55*, 315-322.
13. Shiraiwa, M.; Harada, K.; Okubo, K. *Agric. Biol. Chem.* **1991**, *55*, 323-331.
14. Shiraiwa, M.; Harada, K.; Okubo, K. *Agric. Biol. Chem.* **1991**, *55*, 911-917.
15. Tomomura, M.; Kudou, S.; Tsukamoto, C.; Uchida, T.; Okubo, K. *Abstr. Ann. Meeting Jap. Soc. Biosci. Biotecnol. Agrochem.* **1992**, *66*, 565 (Abstr. 4Ba6).

16. Sugihara, K.; Tsukamoto, C.; Kudou, S.; Hoshikawa, K.; Okubo, K. *Abstr. Ann. Meeting Nihon Syokuhin Kogyo Gakkai* **1992**, *39*, 64 (Abstr. 2F6).
17. Tsurumi, S.; Takagi, T.; Hashimoto, T. *Phytochemistry* 1992, *31*, 2435-2438.
18. Hymowitz, T.; Singh, R. J. In *Soybeans: Improvement, Production and Use, 2nd Ed.*; Wilcox, J. R. Ed.; Agronomy No. 16; ASA CSSA SSSA: Madison, Wisconsin, 1987; pp 23-48.
19. Shimoyamada, M.; Kudou, S.; Okubo, K.; Yamauchi, F.; Harada, K. *Agric. Biol. Chem.* **1990**, *54*, 77-81.
20. Shiraiwa, M.; Yamauchi, F.; Harada, K.; Okubo, K. *Agric. Biol. Chem.* **1990**, *54*, 1347-1352.
21. Shimoyamada, M.; Harada, K.; Okubo, K. *Agric. Biol. Chem.* **1991**, *55*, 1403-1405.
22. Kikuchi, A.; Tsukamoto, C.; Sakai, S.; Okubo, K.; Murata, K. *Japan. J. Breed.* **1992**, *42* (Suppl. 1), 540-541.
23. Tsukamoto, C.; Kikuchi, A.; Harada, K.; Okubo, K. *Abstr. Ann. Meeting Jap. Soc. Biosci. Biotecnol. Agrochem.* **1992**, *66*, 565 (Abstr. 4Ba5).
24. Kitagawa, T.; Taniyama, T.; Nagahama, Y.; Okubo, K.; Yamauchi, F.; Yoshikawa, M. *Chem. Pharm. Bull.* **1988**, *36*, 2819-2828.
25. Tsukamoto, C.; Shimoyamada, M.; Harada, K.; Okubo, K. *Abstr. Ann. Meeting Jap. Soc. Biosci. Biotecnol. Agrochem.* **1991**, *65*, 83 (Abstr. 2Qa4).
26. Tsukamoto, C.; Kikuchi, A.; Kudou, S.; Harada, K.; Kitamura, K.; Okubo, K. *Phytochemistry* **1992**, *31*, 4139–4142.

RECEIVED July 6, 1993

MICRONUTRIENTS

Chapter 32

The Second Golden Age of Nutrition
Phytochemicals and Disease Prevention

Mark Messina and Virginia Messina

Nutrition Consultants, Mt. Airy, MD 21771

Nutritionists in the developed world have slowly moved away from concern over avoiding deficiency disease to focusing on the affluent diet and chronic disease. Until recently however, with the exception of fibre, only the nutrients in foods were considered to be important and traditional dietary advise focused on the entire diet, not individual foods. New insights into the many biologically active non-nutritive dietary components (phytochemicals), however, makes this perspective outdated. Foods can no longer be evaluated solely on their nutrient content and certain individual foods, because of their unique chemical constituents, may warrant being singled out for their health benefits. Phytochemicals may come to be viewed as the vitamins and minerals of the 21st century and phytochemical research as the beginning of the second golden age of nutrition.

The likelihood is that the participants of this symposium received their training, not in foods and nutrition, but in an assortment of related areas: biochemistry, pharmacology, pharmacognosy, and toxicology, for example. This diversity illustrates the multi-disciplinary aspect of the phytochemical field. But it illustrates something else that is even more important. That is, the traditional understanding of the diet and the foods we eat may be outdated in the face of our "new" awareness of food phytochemicals. This chapter highlights the ways in which phytochemicals may impact the traditional dietary approach to optimal health.

Phytochemicals, the Vitamins and Minerals of the 21st Century

Until recently, nutritionists could get along quite nicely without knowing anything about the role of phytochemicals in disease prevention. Although in many ways nutrition was somewhat simpler, things were also somewhat more boring. After all, how much can one discuss fat and fiber? In our opinion, the 1990s represents the beginning of the second golden age of nutrition. It is an era that promises not only to be exciting and challenging, but that will require the efforts of scientists across a

0097–6156/94/0546–0382$06.00/0

wide spectrum of diverse disciplines, as diverse as the backgrounds represented by those attending this symposium.

Of course, if we entering the second golden age of nutrition, it means that there was a first golden age. This period covered the years 1910 through 1940 or so, during which most of the vitamins were discovered (*1*). These discoveries stand as one of the great scientific achievements and resulted in the relief of much human suffering.

During those early years in the study of nutrition, the entire focus was placed on consuming enough vitamins and minerals to avoid deficiency diseases. Once adequate vitamin and mineral intake was assured, the rest of the diet was considered pretty much irrelevant. This attitude however, came back to haunt the nutrition community, because the paradigms that resulted from the discovery of the miraculous ways in which the vitamin and minerals worked are ill-suited to handle the diet-related chronic diseases which now plague the developed countries of the world.

Around the middle of this century, nutritionists started to appreciate that there was more to diet than vitamins and minerals and avoiding deficiency diseases. Focus switched to the effects of over-nutrition or the so-called affluent diet and its effects on chronic disease. The concept of diet was still largely limited to nutrients however, the exception being fibre, which came into vogue during the early and middle 1970s. But with the discovery of the phytochemicals, which in a sense may be thought of as the vitamins and minerals of the 21st century, there has been a gradual recognition of a whole new class of dietary components. Not only are these components not vitamins and minerals, but they are not even nutrients.

Defining Phytochemicals.   Phytochemicals may not be essential in the classic sense; that is, they are probably not necessary to sustain life. But it is likely that many are essential for optimal health. The second golden age of nutrition will focus on the role of phytochemicals in promoting optimal health and in preventing chronic disease. It is clear that it is no longer appropriate for nutritionists to evaluate foods on the basis of their nutrient content alone. It is also clear that under-standing the impact of phytochemicals on health and disease — at a level that is equivalent to our understanding of vitamins and minerals — will require even greater effort. For there are relatively few vitamins and minerals, and each has a relatively easily defined role. In contrast, there is likely to be a very large number of potentially important phytochemicals with biological effects much more subtle than those resulting from the classic deficiency diseases.

The word phytochemical has been mentioned several times, but it has not been defined. One's understanding of the word is likely to be related to the discipline in which that person is trained. Plant biochemists for example, may be more apt to think of phytochemicals as secondary metabolites or phytoalexins (*2*). Nutritionists may prefer the term "non-nutritive dietary components" because this term distinguishes between nutrients and non-nutrients, and thus conveys an important piece of information. But even the appropriateness of this distinction is lost in the case of nutrients such as β-carotene. Because, although β-carotene is a provitamin, recent interest in this nutrient stems from its role as an antioxidant, which is independent of its conversion to vitamin A (*3*). In a way, β-carotene has more in common with the phytochemicals than with the classic nutrients.

However one chooses to define phytochemicals, when talking about food phytochemicals, it is important to bear in mind that we eat foods, not phytochemicals. It may be that the most important benefits to be derived from research on phytochemicals is the development of drug-like compounds. As precedent, many of our most potent drugs were, at least initially, derived from plants. But because phytochemicals are found in commonly consumed foods, their dietary relevance or lack thereof, must always be considered. This is particularly true in a climate where both the media and public clamor after findings interpreted as suggesting that food will reduce cancer risk, promote hair growth, aid in weight loss, or enhance sex appeal.

Defining Functional Foods. Let us turn our attention to how our knowledge of food phytochemicals is likely to affect the traditional approach to diet and the foods we eat. We can start by looking at the term "functional food." Functional foods encompass both the ideas of phytochemicals and of diet.

For most of history, the function of food was to provide a source of non-toxic calories. If you ate it and lived, it performed its function. Food is defined as any substance taken into and assimilated by a plant or animal to keep it alive and enable it to grow. If food, by definition, enables growth and maintenance, then a functional food must perform a function above and beyond growth and maintenance. So what is this function, and when does a food leave the merely "food" category and become a "functional food?"

Perhaps some insight can be gained by considering another term often heard in connection with functional foods — "medicinal." Medicinal and functional are often used interchangeably. The term medicinal food in this context does not refer to the Food and Drug Administration's use of the word, which is much more narrowly defined (4). Webster defines medicinal as having the properties of medicine; curing, healing or relieving. Perhaps then, functional foods can be thought of as performing the function of medicine. And perhaps one day, physicians will be telling their patients to "take two soybeans and to call me in the morning."

If we accept the medicinal term, then functional foods are foods that are curing, healing or relieving. Of course, it is still necessary to define curing, healing or relieving, but at least these properties are distinct from those of maintenance and growth.

Now that we have at least a very general sense of what a functional food is, let's begin to examine some of the obstacles in the path of functional foods. Because if history is any judge, the concept of functional foods is in big trouble.

Consider the term junk food, which essentially is antithetical to functional foods. The nutrition community has really struggled with the term junk food. Some claim it is inappropriate to single out one particular food as good or bad, and make the point that all foods must be viewed in the context of the total diet. Potato chips, for example, are okay — just don't build your diet around them. Besides, look at all that vitamin C.

Others in the nutrition field have less of a problem with the term junk food and have tried to define it using various criteria — all of which were based on nutrient density and all of which met with only limited success. We have no problem with the term junk food, although the difficulty in defining precisely what constitutes a junk food is obvious. It is like pornography — it is hard to define it, but you know it when you see it.

In contrast to the so-called bad or junk foods, more recently, the American Heart Association attempted to start a program that was intended to label foods that were good for the heart. This program ran into many problems and never got off the ground (5). One major problem was the inability to define heart healthy foods. Foods that were really not all that desirable, such as corn oil, were slated to receive a heart healthy label. As with the term junk food, critics emphasized the need to focus on overall diet, not individual foods.

Another term, which sprang up in the 1960s, is "health food." Health foods can be thought of as the forerunners to functional foods. Health foods were supposed to be healthier than the normal or sick foods. The first so called health food that attracted the attention of mainstream consumers was yogurt.

If you think back to the early days of this food, you may recall that it was marketed on the idea that if you ate it, you would live as long as the spry elderly Soviet Georgians shown in the television commercials. There was something somewhat magical about unusual tasting food. But like junk foods, the term health food has been resoundingly rejected by the mainstream nutrition community. Today, even though the scientific community is taking a more serious look at folklore remedies and many of the foods touted by the health food community, calling something a health food is still considered quite unprofessional.

The point of these examples is a simple one: nutritionists have preferred to focus on overall diet, and generally have rejected the notion of focusing on individual foods, good or bad. So functional foods, which in some sense are the high tech version of health foods, are not going to have an easy time gaining acceptance by the nutrition community. And it is not clear yet that they warrant acceptance.

Individual Foods versus Overall Diet. Garlic, soybeans, licorice, citrus, coffee sesame, ginger, onion, tomato, Brassica vegetables, shallots, leeks, chives, ginseng, sage, barley, flaxseed. These foods are specifically mentioned in just the titles of the chapters in this book. Unquestionably, the list of foods under investigation will increase as research continues. It may be easier by the end of the decade, to list foods not being studied, than those that are.

So let us pause for a moment and consider whether thus far phytochemical research is suggesting anything more than the importance of eating plant foods. If it is, how can it be that so many foods are functional foods?

If the message is to eat a plant based diet, then this research is simply affirming a recommendation which is already touted by leading health agencies. This recommendation represents a whole diet approach to eating and is consistent with the traditional nutrition perspective discussed previously. This is not to diminish the importance of food phytochemicals, for if phytochemical research does nothing more than provide us with a better understanding of why it is important to eat plant foods, it will be well worth it.

It appears, however, that many phytochemical researchers are saying something more than eat a plant based diet. Many contend that certain foods, because of their unique chemical constituents, warrant a special place in the diet or perhaps in the diets of people with certain diseases, or those at high risk for certain diseases, such as cancer or heart disease.

To suggest that a food warrants a special dietary role implies that that food, when incorporated into a healthy diet, not only results in significant health benefits, but does so in a way that is relatively unique, or at least not easily duplicated by other foods. This second criteria is very restrictive, but a necessary one if the term "functional food" is to connote something of significance. After all, if vitamin C prevents cancer and/or heart disease, that would not make an orange a functional food, because there are many foods that contain appreciable amounts of vitamin C. But if isoflavones or lignans are beneficial, one might then be able to say that flax seed and soybeans are functional foods.

What will it take for the nutrition community to accept singling out individual foods? First, of course, the benefits of consuming a particular food must be obtainable when consuming reasonable amounts of that food. The mere presence of a potentially beneficial phytochemical does not necessarily say anything about that food, unless that phytochemical is present at a level sufficient to exert biological effects when that food is consumed in reasonable amounts and as part of an overall healthy diet. Second, one needs to always consider toxicity.

Public health policy makers are going to be very skittish about ever recommending specific foods that carry with them any demonstrable risk, an understandable precaution. When making dietary recommendation to 250 million Americans, even a slight chance of increasing cancer risk will be taken very seriously. The risk to benefit ratio of a food is going to have to be very favorable for it to gain acceptance (6). Importantly, it is likely that any food that is potent enough to produce benefits will have a potential flip side to it.

This is, of course, the case with all drugs, but drugs are used in the treatment of sick people — people for whom a higher risk is deemed acceptable. In the case of foods, we are usually talking about prevention, and chronic use, in the healthy general population, which is much different. Of course, most importantly, it will be necessary to demonstrate quite convincingly that a given food is beneficial. That is always difficult in the diet field, but particularly so when, as is so often the case with phytochemicals, we are talking about chronic disease prevention. To help build the case for the importance of phytochemicals and functional foods, we need to consider the role of epidemiology.

Research Needs

Many, if not all, of the major dietary hypotheses concerning chronic disease prevention that exist today were either initially proposed on the basis of epidemiologic work and/or have considerable epidemiologic support. Examples are dietary fat and breast cancer, and the inverse association between fruits and vegetables and cancer risk. In fact, the consistently protective effects of fruits and vegetables noted in case control studies have played a big part in fueling interest in the phytochemicals (7,8). It may be that the most important leads about potential functional foods or even phytochemicals will come from epidemiology.

Regardless of from where these leads initially come, if foods like garlic, flax and soybeans are thought to lower cancer risk because of their unique chemical constituents, people consuming garlic, flax and soybeans should have lower rates of cancer. If people who consume garlic, soybeans and flax do not get less cancer, those suggesting these foods are protective need to explain why. The burden of proof is on those suggesting these foods are protective.

Epidemiologic support will in many cases require phytochemical databases to more easily identify foods, and perhaps why these foods are protective. Establishing these kind of databases will not be easy. Phytochemical content will vary among cultivars, and is likely to be affected by factors such as time of harvesting and method of preparation. Equally important, phytochemical analysis is extremely complex. But clearly, a better understanding of the chemical classes that exist in plants is needed. In addition to a phytochemical databases, more information on the absorption and metabolism of phytochemicals is needed. There is a long list of research needs.

Practical Implications

Despite the very exciting and encouraging findings, much of the nutrition community will remain considerably reluctant to emphasize individual foods too heavily. But as more is learned about phytochemicals, certain foods, because of their unique chemical constituents, will very likely come to be seen as deserving a special place in a healthy diet. Even so, it is critical to continue emphasizing the "overall healthy diet" part of the equation.

We must be careful not to make it easier for the public to think some foods have magical qualities and that they can have their cake and eat it too, so to speak. But as we gain a better understanding of phytochemicals and their impact on health, it is likely we can provide the public with the best of both worlds.

Literature Cited

1. *Human Nutrition: Historic and Scientific*; Galdston I., Ed; Institute of Social and Historical Medicine, International Universities Press, Inc.: New York, 1960.
2. Smith, D. A.; Banks, S. W. *Phytochem.* **1986**, *25*, 979–995.
3. Krinsky, N. J. *Free Rad. Biol. Med.* **1989**, *7*, 617–635.
4. *Fed. Regist.* 21 CFR Part 101, November 27, 1991, pp 60377–60378.
5. Angier, N. *The New York Times* No. 48194, April 3, **1990**, 1.
6. Wattenberg, L. W. *Proc Nutr. Soc.* **1990**, *49*, 173–183.
7. Steinmetz, K. A.: Potter, J. D. *Cancer Causes Control* **1991**, *2*, 325–357.
8. Steinmetz, K. A.; Potter, J. D. *Cancer Causes Control* **1991**, *2*, 427–442.

RECEIVED July 27, 1993

Chapter 33

Synthesis of L-Ascorbic Acid and Its 2-Phosphate and 2-Sulfate Esters and Its Role in the Browning of Orange Juice Concentrate

Eldon C. H. Lee

Westreco, Inc., New Milford, CT 06776

The nutritional and health importance, synthesis, stability, and applications of L-ascorbic acid and its derivatives are reviewed. Syntheses and stabilities of the 2-sulfate and 2-phosphate esters of L-ascorbate are described, but only the 2-phosphate ester seems to possess vitamin C activity. L-Ascorbic acid is shown to contribute to the nonenzymatic browning of orange juice concentrate. The oxidative and anaerobic degradation of L-ascorbic acid may be its critical role in this respect.

L-Ascorbic acid is biosynthesized from carbohydrate precursors including glucose and galactose by a variety of plant and animal species. Humans, other primates, and guinea pigs, as well as insects, invertebrates, fishes, and certain bats and birds are not able to synthesize L-ascorbic acid due to the absence of the enzyme L-gulono-lactone oxidase (*1*). The prevention of scurvy has been accepted as thecriteria for estimating the minimal vitamin C requirements (*2*). Besides the widely accepted roles of L-ascorbic acid in preventing scurvy, facilitating amino acid metabolism, increasing iron absorption, collagen synthesis, and as a biological blocking agent against nitrosamine formation, a recent review on the nutritonal and health aspects of L-ascorbic acid (*3*) and an October 1991 review in *Science* by the American Association for the Advancement of Science indicated its important health aspects in increasing immunocompetence, faciliating drug metabolism, and reducing the risk of a wide variety of human disorders including cancer, heart disease, and atherosclerosis. The appropriate intake levels of vitamin C for each of its physiological functions have not yet been fully established.

Synthesis of L-Ascorbic Acid

Because of its vitamin C potency, useful reducing and antioxidant properties, and low toxicity for pharmaceutical, food, agricultural, and industrial applications (*4,5*), L-ascorbic acid production has grown continuously. The current commercial synthesis of L-ascorbic acid (*6,7*) and speculation on possible enzymic methods of preparation (*7*) have been recently reviewed.

In the Reichstein-Grussner five-step synthesis (*8*, Figure 1), D-glucose is reduced to sorbitol and then oxidized to L-sorbose by aerobic fermentation. Reaction of L-sorbose with acetone and acid gives 2,3:4,6-di-*O*-isopropylidene-L-xylo-2-hexulofuranose. The free primary alcohol group of the diacetone derivative is oxidized to 2,3:4,6-di-*O*-isopropylidene-L-xylo-2-hexulosonic acid, which is the converted to L-ascorbic acid by heating with an acid in a non-aqueous medium. The overall yield of L-ascorbic acid from D-glucose is about 50%.

A second synthesis is a two-stage fermentative process (*9*, Figure 2). Here, D-glucose is converted to 2,5-diketo-D-gluconate via a *Erwinia* sp. and the media is inactivated by adding sodium dodecyl sulfate without isolation. In the second fermentation step the intermediate is reduced to 2-keto-L-gulonic acid via a *Corynebacterium* sp. The overall conversion of D-glucose to L-ascorbic acid is about 73%.

Oxidation of L-Ascorbic Acid

L-Ascorbic acid can undergo a two-step oxidation to give dehydroascorbic acid by way of the intermediate monodehydroascorbic acid free radical (*10*). In aerobic organisms L-ascorbic acid protects biological tissue against activated radicals of oxygen.

The autoxidation of L-ascorbate with oxygen in the presence of transition metal ions (*11*) is important in aerobic systems found in tissues, pharmaceuticals, and foods. In the absence of catalysts, L-ascorbic acid reacts slowly with oxygen (*12*). The rate of aerobic oxidation of L-ascorbic acid is pH-dependent: the rate of oxidation is more rapid and the degradation is more extensive in an alkaline medium than in an acidic solution (*13*).

A number of enzymes, in particular ascorbic acid oxidase, in foods and biological systems accelerate the oxidation of ascorbic acid (*14*). These enzymes should be inactivated to prevent oxidative loss of L-ascorbic acid.

Synthesis and Stability of the 2-Sulfate and 2-Phosphate Esters of L-Ascorbate

Ester and ether derivatives of L-ascorbic acid have been important to determine its chemical structure and biological role, and to modify its solubility and stability (*13, 15,16*). The reactivity of ascorbic acid toward electrophiles is a function of the ionization and steric environments of the four hydroxyl groups at the C2, C3, C5, and C6 positions. Chemical substitution on the ene-diol hydroxyls stabilizes the molecule against oxidative decomposition.

In aqueous solution, the 3-OH and 2-OH have an ionization constant of $pK_1=4.17$ and $pK_2=11.79$, respectively (*17*). While the 3-OH readily ionizes, the delocalization of the electron density among the O(1)=C(1)-C(2)=C(3)-O(3) ring generating resonance results in low reactivity. Under more basic conditions, the ionization of the 2-OH occurs with the formation of the di-anion, which allows selective substitution of this position with electrophiles in the presence of free hydroxyls at C3, C5, and C6.

L-Ascorbate 2-sulfate (*18,19*) was synthesized in nearly quantitative yield by reacting ascorbate with trimethylamine-sulfur trioxide in alkali (pH 9.5–10.5) at 70°C (Figure 3). The 2-sulfate ester (*18,19*) was also obtained in 75% yield by

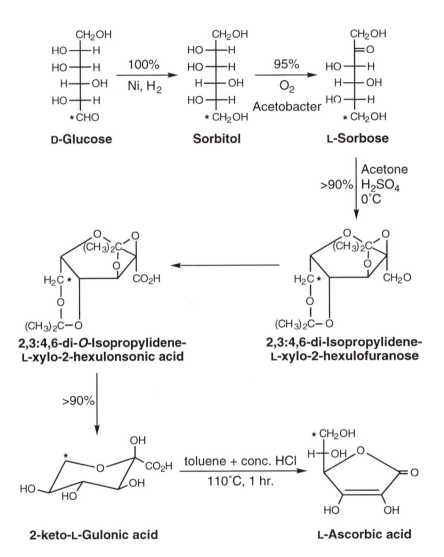

Figure 1. Synthesis of L-ascorbic acid by the Reichstein-Grussner synthesis.

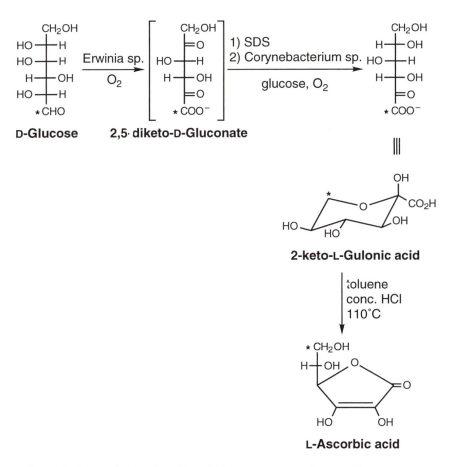

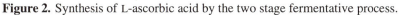

Figure 2. Synthesis of L-ascorbic acid by the two stage fermentative process.

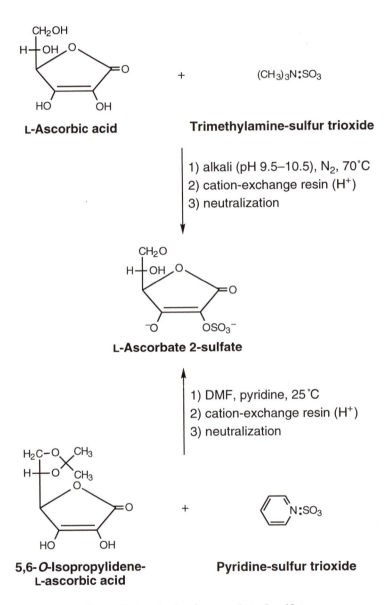

Figure 3. Synthesis of L-ascorbate 2-sulfate.

sulfation of 5,6-*O*-isopropylidene-L-ascorbic acid with pyridine-sulfur trioxide in DMF/pyridine followed by hydrolytic removal of the 5,6-*O*-isopropylidene blocking group.

L-Ascorbate 2-sulfate is approximately 20 times more stable than L-ascorbate towards oxygen in boiling water (*18,19*). L-Ascorbate 2-sulfate occurs in animals (*1*), but it has been reported to have vitamin C potency only in finfish (*20*), not in mammals (*21*). More recent data (*22*) contradicts the idea that L-ascorbate 2-sulfate is active in fish.

5,6-*O*-Isopropylidene-L-ascorbate 2-phosphate was prepared in almost quantitative yield by the action of phosphoryl chloride at 0–5°C on 5,6-*0*-isopropylidene-L-ascorbate in alkali (pH 12–13) containing a high concentration of pyridine (*23,24*). After hydrolytical cleavage of the 5,6-acetal group, L-ascorbate 2-phosphate was isolated in 70% yield as its crystalline tricyclohexylammonium salt (Figure 4). L-Ascorbate 2-phosphate has been shown to be an active source of vitamin C in the monkey and guinea pig (*25*). The 2-phosphate ester, like the 2-sulfate ester, is much more stable in air than L-ascorbic acid (*23*). The 2-phosphate ester might be used as a stable source of vitamin C for foods and feeds in which the ingredients are free of phosphatase activity.

Non-enzymatic Browning in Orange Concentrate

Presently, most fruit concentrates are frozen for long term storage in North America. With aseptic packaging, processors could eliminate enormous freezing process and shipping costs. Browning in fruit concentrates on storage at ambient temperatures, however, constitutes a major problem. Wagner (*26*) reported that the quality of fruit concentrates deteriorates in three or four months.

A great deal of research was carried out on the browning of various fruit products during and subsequent to World War II (*27*). Three hypotheses generally reported were: 1) Maillard or melanoidin condensation (*28*), 2) ascorbic acid oxidation, and 3) active-aldehyde reaction. Many conflicting results and conclusions, however, have been found.

Fruit concentrate browning is mainly due to non-enzymatic browning because the enzyme activity has been thermally inactivated. Orange concentrate, with relatively high levels of ascorbic acid, gives more serious browning than apple, pear, and grape concentrates, which have low levels of ascorbic acid. Therefore, orange concentrate was chosen to investigate ascorbic acid browning.

The browning of orange concentrate was studied by using cation and anion exchange fractionation of orange juice components and removal of residual oxygen with glucose oxidase. The study was directed to investigate the three types of reactions: 1) the oxidative browning reaction of ascorbic acid, 2) the anaerobic browning reaction of ascorbic acid at strongly acidic pH, and 3) the Maillard reaction of amino acids and reducing sugars.

Effect of Glucose Oxidase on the Oxidative Browning of Ascorbic Acid

The use of glucose oxidase for the removal of oxygen from beverages has been suggested (*29*). The glucose oxidase used (DeeO L-750, Miles Lab.) had an optimum pH range from 4.5 to 6.5 and optimum temperature range from 30 to 65°C. Acid-stable glucose oxidase (e.g. Fermo Biochemical Co., Illinois) might perform more effectively in low-pH orange concentrate.

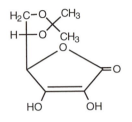

5,6-*O*-Isopropylidene-L-ascorbic acid

alkali (pH 12–13)
pyridine 2.5 M
0–5°C
POCl$_3$

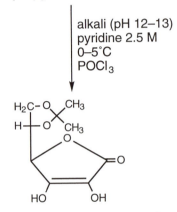

5,6-*O*-Isopropylidene-L-ascorbate 2-phosphate

1) cation-exchange resin (H$^+$)
2) neutralization

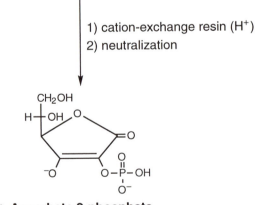

L-Ascorbate 2-phosphate

Figure 4. Synthesis of L-ascorbate 2-phosphate.

The effect of glucose oxidase on browning is given in Figures 5 and 6. The orange juice syrup with addition of glucose oxidase showed less browning than the untreated control. The aerobic oxidation of ascorbic acid occurs rapidly to dehydroascorbic acid when metal catalysts, particularly copper or iron, are present (*11*). Dehydroascorbic acid is converted to diketogulonic acid irreversibly, and then to furfural by decarboxylation and dehydration, with the formation of brown pigments by subsequent polymerization (*30*).

Effect of pH on the Anaerobic Browning of Ascorbic Acid

The effect of pH on the browning of orange juice syrup in Figure 7 showed that lower pH caused more browning. In strong acid the hydrogen ion catalyzed the decomposition of ascorbic acid by hydrolysis of the lactone ring, and further decarboxylation and dehydration to furfural and acids (*31*).

Cation Exchange Fractionation

The use of cation (H^+ form) exchange fractionation removed about 77% of total nitrogen and proteins (Table I), and 99.5% of free amino acids (Table II). Removal of the nitrogenous constituents would eliminate the non-enzymatic browning reactions from the condensation of the nitrogenous constituents, particularly amino acids with sugars (*28*), dehydroascorbic acid (*7*), and melanoidin intermediates (*27, 28*). The cation (H^+ form) exchange fractionation also removed the metallic cations (86% of total ash in Table I). Removal of the metallic ions, particularly iron and copper, reduces the rate of aerobic oxidation of ascorbic acid (*11,12*). The effect of cation (H^+ form) exchange fractionation on the browning of orange juice syrup is shown in Figures 5 and 6, and its effect on the browning of orange juice concentrate is shown in Figure 8.

Cation and Anion Exchange Fractionation

The effect of cation and anion exchange fractionation on orange concentrate is given in Table III. The cation (H^+ form) exchange resin removed the positive-charged proteins (77% of total proteins) and metallic ions whereas the anion (OH- form) exchange resin removed the negative-charged proteins (8% of total proteins), ascorbic acid, acidulants, and inorganic anions. The loss of ascorbic acid during cation and anion exchange fractionation in Table IV shows about 72%.

The use of cation and anion exchange fractionation gave the least browning of orange juice concentrate in Figure 8. Apparently, removal of the browning precursors, ascorbic acid, nitrogenous compounds particularly amino acids, and metallic ions, minimized the non-enzymatic browning.

Conclusion

This study has indicated the critical role of ascorbic acid in the non-enzymatic browning of orange juice concentrate and syrup. The contribution of each reaction pathway would depend on the relative amounts of the browning precursors, oxygen, pH, and metallic catalysts. Further study may be suggested to elucidate the reaction pathways and mechanisms. It would be challenge to develop practical solutions for preventing or minimizing the browning problem.

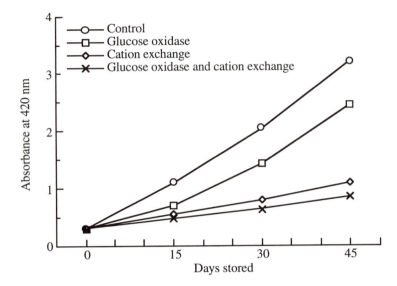

Figure 5. Effect of various treatments on the browning of orange juice syrup at pH 3.7 and 37°C.

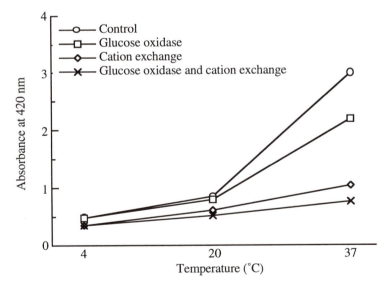

Figure 6. Effect of storage temperature on the browning of orange juice syrup at pH 3.7 for 45 days.

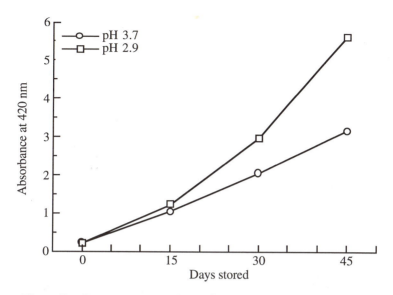

Figure 7. Effect of pH on the browning of orange juice syrup at 38°C.

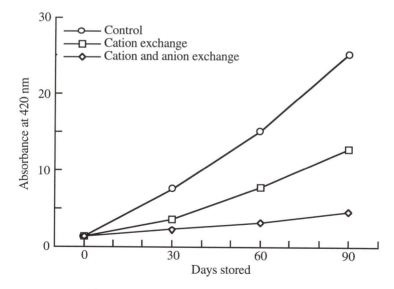

Figure 8. Effect of ion exchange fractionations on the browning of orange concentrate at pH 3.7 and 37°C.

Table I. Cation Exchange Fractionation of Orange Concentration

Analysis	Control	Treatment	Loss (%)
Total solids (%)	63.50	60.60	4.6
Total nitrogen (%)	0.61	0.14	77.4
Total proteins (%)	3.50	0.80	77.1
Total ash (%)	2.35	0.33 (3.10[a])	86.0
Brix (20°C)	64.00	60.30	5.8
Water activity	0.83	0.86	—
pH	3.70	2.00 (3.7[a])	—
Titratable acidity as citric acid, %	4.90	4.90	0

[a]After pH adjusted to 3.7 with KOH.

Table II. Effect of Cation Exchange Fractionation on the Free Amino Acids of Orange Concentrate

Amino acid	Control (mg/g)	Cation (H$^+$) (mg/g)
Lysine	0.18	<0.006
Histidine	0.07	,,
Arginine	1.62	0.012
Aspartic	PI	PI
Threonine	PI	PI
Serine	PI	PI
Glutamic	0.10	0.006
Proline	2.90	0.012
Glycine	0.09	<0.006
Alanine	0.48	,,
Valine	0.10	,,
Methionine	0.02	,,
Isoleucine	0.03	,,
Leucine	0.02	,,
Tyrosine	0.06	,,
Phenylalanine	0.17	,,
Tryptophan	<0.01	,,
Total amino acids	5.84	0.03
Free amino acids removed (%)		99.5

**Table III. Cation and Anion Exchange Fractionation
of Orange Concentrate**

Analysis	Control	Treatment	Loss (%)
Total solids (%)	63.50	47.50	25.2
Total nitrogen (%)	0.61	0.09	85.6
Total proteins (%)	3.50	0.52	85.2
Total ash (%)	2.35	0.04 (0.174[a])	98.2
Brix (20°C)	64.00	47.60	25.6
Water activity	0.83	0.87	—
pH	3.70	2.60 (3.7[a])	—
Titratable acidity as citric acid, %	4.90	0.51	89.6

[a]After pH adjusted to 3.7 with KOH.

**Table IV. Loss of Ascorbic Acid in
Ion Exchange Fractionated Orange Concentrate**

Treatment	Ascorbic acid (mg/100 g)	Loss (%)
Untreated control	300	—
Cation (H^+)	291	3.0
Cation (H^+) and anion (OH^-)	85	71.7

Literature Cited

1. Omaye, S. T.; Tillotson, J. A. In *Ascorbic Acid: Chemistry, Metabolism, and Uses*; Seib, P. A.; Tolbert, B. M., Eds.; American Chemical Society: Washington, D.C., 1982; pp 317–334.
2. *Recommended Dietary Allowances*; National Academy Sciences: Washington, D.C., 1974.
3. Brin, M. In *Ascorbic Acid: Chemistry, Metabolism, and Uses*; Seib, P. A.; Tolbert, B. M., Eds.; American Chemical Society: Washington, D.C., 1982; pp 369–379.
4. Bauernfeind, J. C.; Pinkert, D. M. *Adv. Food Res.* **1970**, *18*, 219–292.
5. Bauernfeind, J. C. In *Ascorbic Acid: Chemistry, Metabolism, and Uses*; Seib, P. A.; Tolbert, B. M., Eds.; American Chemical Society: Washington, D.C., 1982; pp 395–497.
6. Crawford, T. C. In *Ascorbic Acid: Chemistry, Metabolism, and Uses*; Seib, P. A.; Tolbert, B. M., Eds.; American Chemical Society: Washington, D.C., 1982; pp 1–36.
7. Seib, P. A. *Int. J. Vit. and Nutr. Res.* **1985**, *27*, 259–306.
8. Reichstein, T.; Grussner, A. *Helv. Chim. Acta.* **1934**, *17*, 311–328.

9. Sonoyama, T.; Tani, H., Kageyama, B.; Kobayashi, K.; Honjo, T.; Yagi, S. *Appl. Environ. Microbio.* **1982**, *43*, 1064.

10. Bielski, B. H. J. In *Ascorbic Acid: Chemistry, Metabolism, and Uses*; Seib, P. A.; Tolbert, B. M., Eds.; American Chemical Society: Washington, D.C., 1982; pp 81–100.

11. Martell, A. E. In *Ascorbic Acid: Chemistry, Metabolism, and Uses*; Seib, P. A.; Tolbert, B. M., Eds.; American Chemical Society: Washington, D.C., 1982; pp 153–178.

12. Blauag, S. M.; Hajratwala, B. *J. Pham. Sci.* **1972**, *61*, 556-562.

13. Hay, G. W.; Lewis, B. A.; Smith, E. In *The Vitamins*; Sebrell, W. H. Jr.; Harris, R. S., Eds.; Academic Press: New York, 1967.

14. Tolbert, B. M.; Ward, J. B. In *Ascorbic Acid: Chemistry, Metabolism, and Uses*; Seib, P. A.; Tolbert, B. M., Eds.; American Chemical Society: Washington, D.C., 1982; pp 101–123.

15. Andrews, G. C.; Crawford, T. In *Ascorbic Acid: Chemistry, Metabolism, and Uses*; Seib, P. A.; Tolbert, B. M., Eds.; American Chemical Society: Washington, D.C., 1982; pp 59–79.

16. Tolbert, B. M., Downing, M.; Carlson, R. W.; Knight, M. K.; Baker, E. M. *Chemistry and Metabolism of Ascorbic Acid and Ascorbate Sulfate*; Univ. of Colorado Press: Colorado, 1974.

17. Crawford, T. C.; Crawford, S. A. *Adv. Carbohyd. Chem.* **1980**, *37*, 79–155.

18. Lee, C. H. *Synthesis and Properties of L-Ascorbate 2-Sulfate*; M.S. Thesis, Kansas State University, 1973.

19. Seib, P. A.; Liang, Y. T.; Lee, C. H.; Hoseney, R. C.; Deyoe, C. W. *J. Chem. Soc., Perkin, Trans.* **1974**, *1*, 1220–1224.

20. Benitez, L. V.; Halver, J. E. *Proc. Natl. Acad. Sci. USA*, **1982**, *79*, 5445.

21. Machlin, L. J.; Garcia, F.; Kuenzig, W.; Richter, C. B.; Spiegel, H. E.; Brin, M. *Am. J. Clin. Nutr.* **1976**, *29*, 825–831.

22. Dabrowski, K.; Kock, G. *Can. J. Fish Aquat. Sci.* **1989**, *46*, 1952–1957.

23. Lee, C. H. *Synthesis and Characterization of L-Ascorbate Phosphates and Their Stabilities in Model Systems*; Ph.D. Dissertation, Kansas State University, 1976.

24. Lee, C. H.; Seib, P. A.; Liang, Y. T.; Hoseney, R. C.; Deyoe, C. W. *Carbohyd. Res.* **1978**, *67*, 127–138.

25. Machlin, L. J.; Garcia, F.; Kuenzig, W.; Brin, M. Am. J. Clin. Nutr. **1979**, *32*, 325–331.

26. Wagner, J. N. *Food Engineering*, **1982**, *6*, 44–45.

27. Stadtman, E. R. *Adv. Food Res.* **1948**, *1*, 325–369.

28. Nursten, H. E. *Food Chem.* **1980**, *6*, 263–277.

29. Underkofler, L. A. *Soc. Chem. Ind. (London), Monograph*, **1961**, *11*, 72-86.

30. Erdman, J. W. Jr.; Klein, B. P. In *Ascorbic Acid: Chemistry, Metabolism, and Uses*; Seib, P. A.; Tolbert, B. M., Eds.; American Chemical Society: Washington, D.C., 1982; pp 499–532.

31. Huelin, G. E. *Food Res.* **1953**, *18*, 633.

RECEIVED July 27, 1993

Chapter 34

Metabolic Pathway for β-Carotene Biosynthesis
Similarities in the Plant and Animal

A. M. Gawienowski

Department of Biochemistry and Molecular Biology, University of Massachusetts, Amherst, MA 01003

Over 75 years ago β-carotene was isolated and assayed in the corpus luteum of the bovine ovary. Its high concentration, the close relationship of steroid biosynthesis to carotenoid biosynthesis, and Porter's isolation of geranylgeranyl diphosphate from pig liver, caused one to question the assumption that only plant tissues can synthesize carotenoids. Subsequently, we reported in 1969 that bovine corpus luteum can synthesize a small amount of β-carotene from acetate. We further reported that the bovine corpus luteum metabolized ^{14}C labeled β-carotene to vitamin A aldehyde. Our continued research on the bovine corpus luteum led to the isolation of phytoene, neurosporene, and β-zeacarotene, which are intermediates in the Porter-Lincoln metabolic pathway to α- and β-carotene in plants. Therefore, the biosynthesis of β-carotene is quite similar in the plant and in the bovine corpus luteum.

Higher plants contain a large number of isoprenoid compounds with a variety of structures and functions. The terpenoid compounds are formed from multiples of the isoprenoid carbon skeleton. They are derived biosynthetically from isopentenyl diphosphate, the compound which was predicted on the basis of chemical structures of a large number of natural products.

In higher plants many of these isoprenoid compounds have important roles in the metabolism and development of the plant. The plant growth regulators gibberellins and abscisic acid are isoprenoid compounds as is β-carotene, the widely present carotenoid.

As Gray (1) stated in his review article, plants have to be able to produce a wide range of isoprenoid compounds in different amounts in different parts of the plant at different stages of growth and development. Since all these compounds are produced by a common biosynthetic pathway, the plant must have excellent control mechanisms to ensure the synthesis of needed compounds at the right place and time (Figure 1).

0097–6156/94/0546–0401$06.00/0

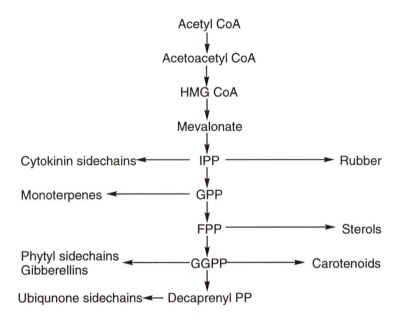

Figure 1. Synthesis of isoprenoids to various terpenoids.

The pathway for the biosynthesis of isoprenoid compounds in plants is based a great deal on the sterol synthesis pathway studied for many years in animals and yeast. In plants, isoprenoid biosynthesis can be seen as a major pathway (Figure 2) from acetyl CoA via mevalonate and isopentenyl diphosphate to long-chain prenyl diphosphates, with a large number of branch points leading to the separate isoprenoid compounds (Figure 3).

Formation of Prenyl Diphosphates

Isopentenyl diphosphate is the key intermediate in the formation of isoprenoid compounds. For most isoprenoid compounds, polymerization of C_5 units (Figure 4) is required to produce longer chain prenyl diphosphates which are the substrates for various enzymes at the branch points leading to the synthesis of the broad spectrum of isoprenoid compounds (Figure 5). According to Gray (1987), labelled isopentenyl diphosphate has been shown to be an excellent precursor of many isoprenoid compounds in extracts of various plants.

Carotenoids were reported in the ovary in 1913 (2) and again in 1932 (3). The bovine corpus luteum contains relatively high concentrations of β-carotene, amounting up to 60 μg/g of tissue weight. Retinal (4) has also been isolated from this tissue. Corpus luteum tissue, when sliced and incubated with β-[15,15′-³H] carotene, yielded radioactive retinal (5). This indicated the ovary possesses the enzymes to synthesize retinal *in situ*, which may have a role in reproductive functions. At ovulation time, specific activity of the carotene cleavage enzymes was two-fold greater in the ovary than in the intestine (6).

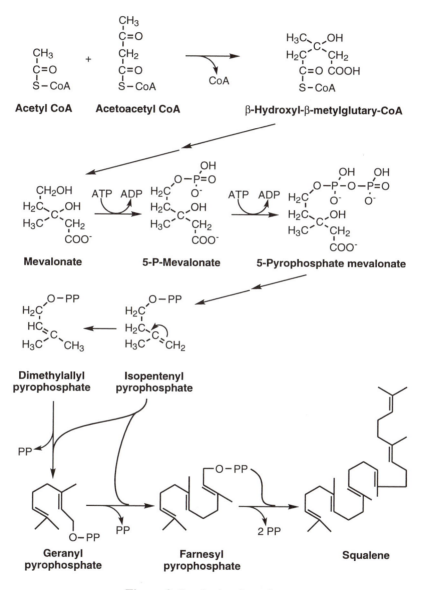

Figure 2. Synthesis of squalene.

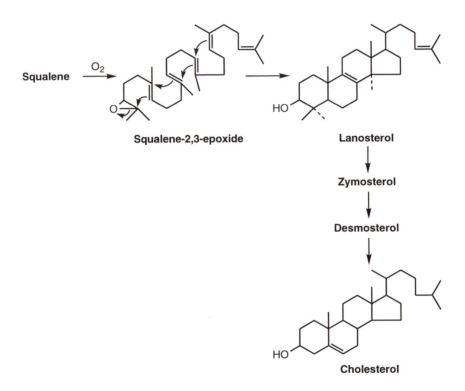

Figure 3. Biosynthesis of cholesterol.

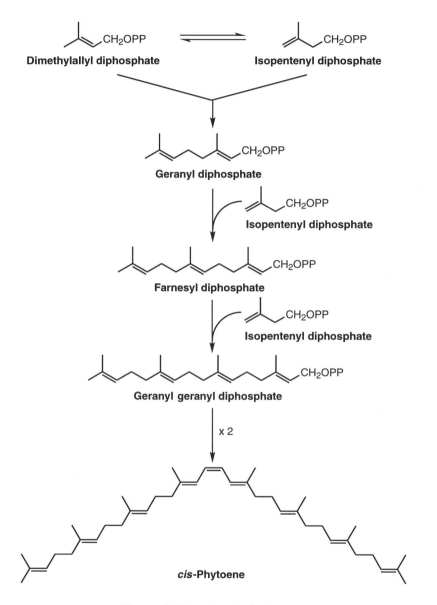

Figure 4. Biosynthesis of phytoene.

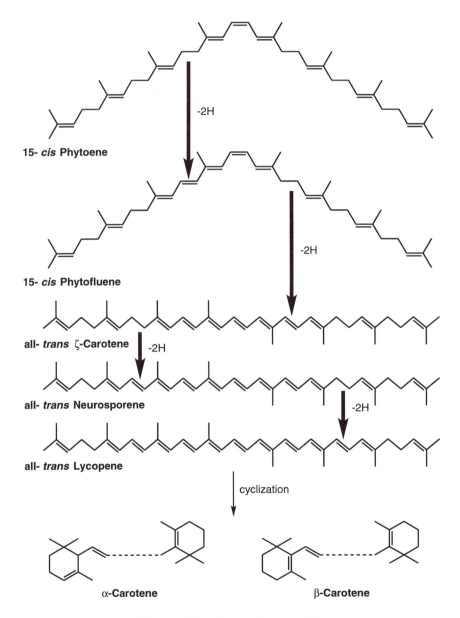

15- *cis* Phytoene

-2H

15- *cis* Phytofluene

-2H

all- *trans* ζ-Carotene   -2H

all- *trans* Neurosporene

-2H

all- *trans* Lycopene

cyclization

α-Carotene

β-Carotene

Figure 5. Synthesis of carotenoids.

It had been suggested that vitamin A may be required for efficient steroid hormone production (7–9) and that it plays a role in reproductive processes, e.g., maintenance of pregnancy (10). It is also of interest that β-carotene-^{14}C was synthesized *in vitro* in the bovine corpus luteum tissue from sodium [1-^{14}C] acetate (12).

The retinoids in vertebrates represent an essential class of nutrients needed for the maintenance of differentiated epithelial structures, vision, reproductive functions and health (13). Retinoic acid, for example, has multiple effects on cell growth and differentiation and appears to be organogenic in embryogenesis. The effects of retinoic acid on gene expression are believed to be mediated by a class of nuclear receptors that function as ligand-dependent transcription factors (14).

Piziak and Gawienowski (9) found that progesterone-4-^{14}C can be metabolized by guinea pig placenta tissue to 20α-hydroxyprogesterone to a significant extent in an atmosphere of 95% oxygen:5% carbon dioxide. Retinol increased this metabolism tenfold.

Vitamin A deficiency is known to have an adverse effect on the size of the ovary and on the integrity of the uterus and would thus have an effect on reproductive ability (15). But a more direct effect on steroid metabolism probably exists. One clue that suggests this fact is the disappearance of vitamin A from the ovary after menopause (16). Ganguly *et al.* discovered that vitamin A deficient rats secreted less than normal amounts of progesterone and 20α-hydroxyprogesterone (17). Juneja *et al.* (8) reported a decrease in conversion of Δ^5-3β-hydroxysteroids into Δ^4-3-ketosteroids in tissues that were only mildly deficient in vitamin A.

Isolation of Phytoene in the Bovine Ovary

In order to understand better the role of the carotenoids in the bovine ovary, we analyzed the corpus luteum for phytoene. Phytoene is a colorless 40-carbon compound that serves as a precursor for many carotenoids in plants. Porter isolated geranylgeranyl diphosphate synthetase, which aids in the biosynthesis of geranylgeranyl diphosphate a precursor of phytoene (18), from pig liver (19).

The formation of phytoene in plants arises from a head to head condensation of two geranylgeranyl diphosphate molecules. Whenever β-carotene has been observed, phytoene has usually been found in substantial amounts (18). Phytoene has also been studied in some mammalian tissues (20,21).

Phytoene was a logical precursor of β-carotene to be investigated in the bovine ovary with and without a corpus luteum. Since phytoene has been well studied in plants, we were able to adapt the plant analytical procedures to ovarian tissue analysis.

Materials and Methods

Standard phytoene was donated by Hoffmann-LaRoche, Inc. and M. Mathews-Roth of Harvard University. Most of the experimental procedures were carried out according to those used by Davies (22). The *N*-bromidesuccinimide method for derivative formation was reported by Zechmeister (23).

Ovaries were obtained fresh from a local abattoir. Ovaries with and without their corpora lutea were homogenized and extracted with a mixture of chloroform-methanol (2:1, v:v). The sample was saponified with a 60% (w/v) potassium hydroxide solution, in the dark under nitrogen. The dried sample was dissolved in petroleum ether and applied to a column (1 x 15 cm) containing 25 g neutral alumina, Brockman activity grade III. The extract was eluted with petroleum ether containing increasing concentrations of diethyl ether.

Visible and ultraviolet absorption spectra of the eluted fractions were determined in a Cary Model 14, Beckman Model 24 or Acta MVI spectrophotometer. Quantitative determinations of phytoene were carried out in known volumes of petroleum ether at 286 nm by their extinction coefficients (Figure 6).

After the quantitative analysis, a N-bromosuccinimide derivative of ζ-carotene was (374, 395, 420 nm) made utilizing the Zechmeister (23) procedure (Figure 6). Phytoene standards were also treated under the same conditions and gave the identical derivative.

Iodine catalyzed photoisomerization of the extract was carried out in hexane in quartz spectrophotometer cuvettes (22). Illumination was performed under two parallel fluorescent lamps (65 W) at a distance of 40 cm.

The elution pattern for phytoene correlates well with the literature. The phytoene eluted from the column in solvent containing between 2 to 5% diethyl ether in petroleum ether, which agrees with Davies (22) and Than et al. (24).

Results

A total of five bovine ovaries were analyzed for phytoene and the results are given in Table I. A large cyst, containing 17.3 ml of fluid, was found in ovary number II which also contained the smallest amount of phytoene per gram of tissue. The corpus luteum of ovary number IV was removed before the analysis. As can be noted, it had the second lowest phytoene concentration.

Table I. Phytoene Concentration in Bovine Ovaries

Ovary:	I	II[a]	III	IV	V
Weight (g)	33.9	50.9	22.2	10.0	21.4
Phytoene conc. (μg/g)	8.20	0.04	6.41	4.22	10.94

[a]Contained a large cyst

Neurosporene

An additional carotenoid was isolated with the Davies (22) procedure. It had an absorption spectrum of 414, 440, 469 and was observed at a concentration of 1.77 μg/g of tissue (Figure 7). On the basis of UV/VIS spectra and elution pattern, its identity was determined to be neurosporene (7,8-dihydro-ψ, ψ-carotene). The cis isomer was converted to the trans form by iodine catalyzed photoisomerization (22). Neurosporene was identified in six bovine corpora lutea by spectra and elution pattern.

Discussion

The higher plants and mammals contain a wide array of isoprenoid compounds which relate to the active isoprene compound, isopentenyl diphosphate. Just as it

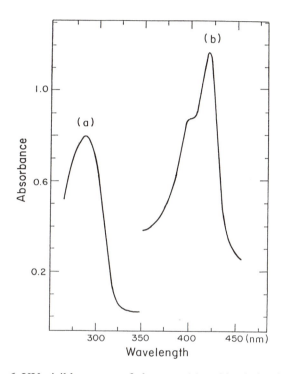

Figure 6. UV-visible spectra of phytoene (a) and its derivative (b).

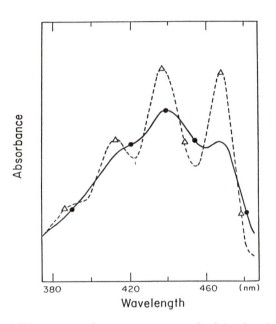

Figure 7. UV-visible spectra of neurosporene standard (– ⟁ –) and sample (—●—).

has been found in plant biosynthesis of β-carotene via the Porter-Lincoln pathway, these tetraterpenoid precursors are apparently similar in the ovary.

With the demonstration of the biosynthesis of acetate-^{14}C to β-carotene-^{14}C in the corpus luteum tissue, one can suspect that plants and mammals have a similar β-carotene synthesis pathway. The isolation of β-carotene, α-carotene, phytoene, β-zeacarotene (*25*), and neurosporene, from the ovary, are good indicators of a Porter-Lincoln type system in the ovary (Figure 8).

The synthesis of retinal-^{14}C from β-carotene-^{14}C in the ovary indicates the enzymes are there for retinoid formation. Retinal has also been isolated from the corpus luteum. This also relates to the known retinoid action on steroid metabolism. Retinoids have an effect on cell growth, differentiation and embryogenesis. Also, the effects of the retinoid on gene expression are thought to be mediated by a family of nuclear receptors that function as ligand-dependent transcription factors (*14*). These functions would only require minute quantities of the retinoids.

Phytoene was isolated and identified in five bovine ovaries. This plus our past analyses of β-zeacarotene (*25*), β-carotene, α-carotene, neurosporene, and retinal in the bovine ovary are good indicators of a Porter-Lincoln type metabolic system in the ovary. The metabolic pathway has been well established in plants, but remains virtually unknown in mammals (*18*).

Glycolysis, the Krebs cycle and other metabolic pathways are well recognized in plants as well as in mammalian systems. Therefore, the similarities of the isoprenoid pathway to β-carotene in plants and mammals should aid in the recognition of another related metabolic pathway.

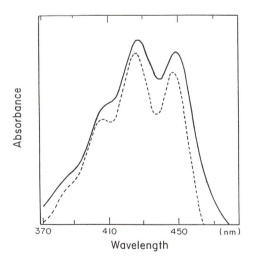

Figure 8. UV-visible spectra of β-zeacarotene standard (- - -) and sample (———).

Acknowledgements

The author thanks N. H. Zawia and J. J. Boniface for technical assistance.

Literature Cited

1. Gray, J. C. *Control of Isoprenoid Biosynthesis in Higher Plants*; Callow, J. A., Ed.; Advances in Botanical Res.; Academic Press: New York, 1987; Vol. 14, pp 25–91.
2. Escher, H. *Hoppe-Seyler's Z. Physiol. Chem.* **1913**, *83*, 198–211.
3. Kuhn, R.; Lederer, E. *Hoppe-Seyler's Z. Physiol. Chem.* **1932**, *206*, 41–64.
4. Austern, B. M.; Gawienowski, A. M. *J. Reprod. Fert.* **1969**, *19*, 203–205.
5. Gawienowski, A. M.; Stacewicz-Sapuncakis, M.; Longley, R. J. *Lipid Res.* **1974**, *15*, 375–379.
6. Sklan, D. *Internat. J. Vit. Nutr. Res.* **1983**, *53*, 23–26.
7. Grangaud, R.; Nicol, M.; Desplanques, D. *Amer. J. Clin. Nutr.* **1969**, *22*, 991–1002.
8. Juneja, H.; Murthy, S.; Ganguly, J. *Biochem. J.* **1966**, *99*, 138–145.
9. Piziak, V. K.; Gawienowski, A. M. *Comp. Biochem. Physiol.* **1972**, *42B*, 201–203.
10. Moore, T. *Vitamin A*; Elsevier: Amsterdam, 1957; p 329.
11. Thompson, J.; Howell, M.; Pitt, G. In *Agents Affecting Fertility*; Austin, C.; Perry, J., Eds; J. & A Churchill Ltd.: London, 1965; pp 32–46.
12. Austern, B. M.; Gawienowski, A. M. *Lipids* **1969**, *4*, 227–229.
13. Goodman, D. S., *N. Engl. J. Med.* **1984**, *310*, 1023–1031.
14. Rajan, N.; Kidd, G.; Talmage, D.; Blaner, W.; Suhara, A.; Goodman, D. S. *J. Lipid Res.* **1991**, *32*, 1195–1204.
15. Truscott, B. L. *Anat. Record.* **1947**, *98*, 111–126.
16. Ragins, A. B., Popper, H. *Arch. Pathology* **1942**, *34*, 647.
17. Ganguly, J.; Pope, S.; Thompson, J.; Toothill, J.; Edwards-Webb, J.; Waynforth, H. *Biochem. J.* **1971**, *122*, 235–239.
18. Britton, G. *The Biochemistry of Natural Pigments*; Cambridge Univ. Press: New York; 1983, pp 46–66.
19. Porter, J. W.; Nandi, D. L. *Arch. Biochem. Biophys.* **1964**, *105*, 7–19.
20. Mathews-Roth, M. M. In *Carotenoids as Colorants and Vitamin A Precursors*; Bauernfeind, J., Ed.; Academic Press: New York, 1981; pp 755–781.
21. Mathews-Roth, M. M.; Crean, C.; Clancy, M. *Nutr. Reports Int.* **1978**, *17*, 581–584.
22. Davies, B. H. in *Chemistry and Biochemistry of Plant Pigments*; Goodwin, T., Ed.; Academic Press: New York, 1976; pp 38–155.
23. Zechmeister, L.; Koe, B. K. *J. Am. Chem. Soc.* **1954**, *76*, 2923–2926.
24. Than, A.; Bramely, P. M.; Davies, B. H.; Rees, A. F. *Phytochem.* **1972**, *11*, 3187–3192.
25. Gawienowski, A. M.; Soderstrom, D. N.; Tan, B. *Biotech. Appl. Biochem.* **1986**, *8*, 190–194.

RECEIVED October 4, 1993

Author Index

413

Affiliation Index

Subject Index

Production: Susan F. Antigone
Indexing: Deborah H. Steiner
Acquisition: Rhonda Bitterli
Cover design: Amy Hayes

Printed and bound by Maple Press, York, PA

Bestsellers from ACS Books

The ACS Style Guide: A Manual for Authors and Editors
Edited by Janet S. Dodd
264 pp; clothbound ISBN 0–8412–0917–0; paperback ISBN 0–8412–0943–X

The Basics of Technical Communicating
By B. Edward Cain
ACS Professional Reference Book; 198 pp;
clothbound ISBN 0–8412–1451–4; paperback ISBN 0–8412–1452–2

Chemical Activities (student and teacher editions)
By Christie L. Borgford and Lee R. Summerlin
330 pp; spiralbound ISBN 0–8412–1417–4; teacher ed. ISBN 0–8412–1416–6

Chemical Demonstrations: A Sourcebook for Teachers,
Volumes 1 and 2, Second Edition
Volume 1 by Lee R. Summerlin and James L. Ealy, Jr.;
Vol. 1, 198 pp; spiralbound ISBN 0–8412–1481–6;
Volume 2 by Lee R. Summerlin, Christie L. Borgford, and Julie B. Ealy
Vol. 2, 234 pp; spiralbound ISBN 0–8412–1535–9

Chemistry and Crime: From Sherlock Holmes to Today's Courtroom
Edited by Samuel M. Gerber
135 pp; clothbound ISBN 0–8412–0784–4; paperback ISBN 0–8412–0785–2

Writing the Laboratory Notebook
By Howard M. Kanare
145 pp; clothbound ISBN 0–8412–0906–5; paperback ISBN 0–8412–0933–2

Developing a Chemical Hygiene Plan
By Jay A. Young, Warren K. Kingsley, and George H. Wahl, Jr.
paperback ISBN 0–8412–1876–5

Introduction to Microwave Sample Preparation: Theory and Practice
Edited by H. M. Kingston and Lois B. Jassie
263 pp; clothbound ISBN 0–8412–1450–6

Principles of Environmental Sampling
Edited by Lawrence H. Keith
ACS Professional Reference Book; 458 pp;
clothbound ISBN 0–8412–1173–6; paperback ISBN 0–8412–1437–9

Biotechnology and Materials Science: Chemistry for the Future
Edited by Mary L. Good (Jacqueline K. Barton, Associate Editor)
135 pp; clothbound ISBN 0–8412–1472–7; paperback ISBN 0–8412–1473–5

For further information and a free catalog of ACS books, contact:
American Chemical Society
Distribution Office, Department 225
1155 16th Street, NW, Washington, DC 20036
Telephone 800–227–5558